STANDARDIZATION

A New Discipline

STANDARDIZATION

A New Discipline

by **LAL C. VERMAN, Ph.D.**

with a foreword by
Francis L. LaQue
President, International Organization for Standardization

ARCHON BOOKS · 1973

Library of Congress Cataloging in Publication Data

Verman, Lal Chand, 1902–
Standardization.

Includes bibliographical references.
1. Standardization. I. Title.
T59.V38 658.5′62 72–8370
ISBN 0-208-01285-0

Published as an Archon Book 1973 by
The Shoe String Press, Inc.
Hamden, Connecticut 06514
Printed in the United States of America

To my wife, Kunti

Contents

Foreword

In his most comprehensive treatment of his subject Dr Verman has very thoroughly documented his concept of standardization as a new discipline. But his book goes much further and more importantly than this. He has recognized and emphasized the extent to which standardization is related to and affects most other disciplines ranging from the arts and the humanities to the most highly specialized engineering practices.

His topics range also from standards as they affect the individual to their relation to the activities of companies, trade associations, local, state and federal governments. They embrace also the concerns of regional groups of nations as well as the more extensive truly international ones.

Dr Verman has devoted most of his professional activities in standardization to the special problems of developing countries based on his leadership role in the Indian Standards Institution. Nevertheless in his book he deals with all the aspects of standardization in catering to a broad spectrum of interest.

His choice of language, the articulate faculty of his exposition and his method of presentation will make the reading of his book an enjoyable as well as instructive experience. This will extend from those concerned with an overall view of what standardization is all about to the standardization specialist seeking information on historical background and specific details of past and current standardization organizations, practices and procedures.

But beyond all this what recommends this book most is the philosophical approach taken by the author. He has provided an eloquent justification for a plea which should be heard by all educators and educational institutions. For far too long they have failed to appreciate the importance of standardization in so many disciplinary fields. Students should leave universities with at least an awareness of the impact that standardization will make on the use they will be making of what they have been taught in their specific areas of interest and in their future activities. This will obviously be the case in the practice of the several branches of engineering. However, it must extend as well to such fields as sociology, political science, economics, law, banking and commerce.

It is to be hoped that universities will recognize the importance of cultivating an awareness of the potential impact of standardization in so many areas of activity. Dr Verman's book has provided an excellent text as an invaluable aid to the cultivation of this awareness.

FRANCIS L. LAQUE
President
International Organization for Standardization
Geneva, Switzerland

April 17th, 1972

Preface

0.1 During recent years many new disciplines of learning have been born as a result of rapid advancement of scientific knowledge about nature and man. Nuclear energy, space travel, electronics, solid state physics, computers and scores of other related and non-related fields present extraordinary challenge and opportunity for deeper study and development. Equally significant are the fields of learning which deal with man's own relationship with modern science and technology. However, it may not be quite unsafe to say that no comparable advancements of spectacular character have been registered during recent years in these latter fields, for example, in economics, social or management sciences, although it has to be admitted that such disciplines do offer an extraordinary challenge for quantification of knowledge tending towards an approach to mathematical precision. All the same, considerable progress has also been made even in these fields during the past few decades, as a result of which terms like cybernetics, econometry, operational research and so on have come to be popularly understood to represent interesting and legitimate challenges to an enquiring mind.

0.2 While an extensive chain reaction in the advancement of knowledge and proliferation of new disciplines has been taking place, the development of the field of standardization as an independent discipline has also been quietly proceeding apace during the past quarter of a century. This development has indeed been so quiet and comparatively so modest that few have dared to think of standardization as a discipline in its own right. Even its exact name would appear difficult to decide. Terms like "industrial standardization," "standards engineering," "standardization engineering" or simply "standardization" have been variously used to express the concept. Then, indeed, there has been quite a lengthy debate on such elementary questions, as to what constitutes a "standard" and what constitutes a "specification." The physical and conceptual standards of measurements and documentary type of standards have often been distinguished as two different entities – the former being the concern of the physicist or more precisely that of the metrologist, and the latter that of the so-called "standards engineer," whose calling requires yet to be given an appropriate and agreed name. Nevertheless, it cannot be denied that both types of standards serve as yardsticks of measurement and thus constitute entities having a common *raison d'être,* so that it would be reasonable for them to be considered as two branches of the same discipline.

0.3 Doubts have often been expressed as to whether it would be wise to classify standardization activity as a branch of engineering, for it may concern itself with fields which could not be considered the legitimate concern of an engineer; such as chemistry, agriculture or management. In certain countries standardization authorities have issued standards dealing even with "good government"; many other standards exist which serve to codify language by standardizing concepts and definitions of terms. On the other hand, it may be argued that the stage is long past when the term "engineering" need apply only to a branch of knowledge dealing with "engines" of some kind or other, for example, consider "human engineering." Similarly, the expression "industrial standardization" hardly encompasses the scope of standardization, for industry is just one of its concerns; agriculture, commerce, transport, communication, management, education, sports, music and even cultural values can hardly be excluded – indeed standardization has been claimed to be able to help regulate any activity of man with which he may be concerned as a member of his social complex. When Shri C. Rajagopalachari,[1] the first Governor General of India after Independence, said in 1949: "Standards are to industry as culture is to society," he was limiting, perhaps intentionally, the regulatory function of standards to just one aspect of man's activity. It has since been shown that properly developed standards can and, in fact, do much more to serve man as a gregarious social animal; so that it might be more accurate to say that "technological standards help regulate society insofar as its technological needs are concerned in the same way as cultural standards serve to regulate society in the sociological sense."

0.4 Though standardization, as an unconscious activity of man, can be shown to have had its origin in prehistoric times, and stray instances exist of a conscious character before the close of the nineteenth century, yet the earliest deliberate effort on an organized scale can be traced back only to the early part of the twentieth century. The First World War provided a good deal of incentive for its development, but it was only after the Second World War that the movement received attention on a world-wide scale to gather momentum enough to stimulate some intensive thinking and extensive application. The process seems just to have begun, as one well-known leader of standardization in the field of electricity, P. Ailleret,[2] President of *Electricité de France,* recently put it: "Whilst the theories of information and management are giving rise to forward looking studies, not enough individual effort has been devoted to outline the theory of standardization, nor to look into its future."

0.5 A great deal has, indeed, been written from time to time about the methodology and techniques of standardization and its organizational aspects. But only a few systematic accounts of the subject as a whole have appeared. John Gaillard[3] was perhaps the first among the leaders of standardization to write a comprehensive and original treatise in 1934 entitled *Industrial Standardization – Its Principles and Application,* in which he treated many theoretical and practical aspects of the subject. For several decades since its publication, this book has pioneered the propagation of the idea of standardization in the world and helped the newcomer to get initiated into the field. Soon after the Second World War, Jessie V. Coles[4] published *Standards and Labels for Consumers' Goods* – a monumental work which deals comprehensively with labels of a special kind, and also treats the current concepts and the practices of standards and standardization, even though largely from a consumer's viewpoint. *Industrial Specifications* by E. H. MacNiece[5] published in 1953 is an excellent presentation of a very specialized and important aspect of standardization which should be read by all those responsible for the writing of standard specifications. Similarly, Benjamin Melnitsky's[6] *Profiting from Industrial Standardization,* written about the same time, is more of a "how-to" book, as the author himself puts it, emphasizing the profit motive, but quite exhaustive. John Perry's[7] *The Story of Standards,* which appeared in 1955, gives an excellent historical account of developments in popular language. The next serious attempt at treating the subject systematically, made a year later, was that of Dickson Reck,[8] who compiled and edited some 33 essays on different aspects of standardization written by leading American authorities in the field and published them under the title *National Standards in a Modern Economy.* By publishing, in 1965, *Standards and Specifications – Information Sources,* E. J. Struglia[9] has rendered yeoman service in collecting together the welter of American sources of standards from what he calls "a jungle enshrouded in light mist." The only non-American attempts at writing comprehensively and systematically are represented by a French treatise *La Normalization* by J. Maily[10] published in 1946 and a British collection of 9 articles edited by C. Douglas Woodward,[11] published in 1956. To these must be added the recent more monumental work in Russian edited by V. V. Tkachenko,[12] entitled *Method and Practice of Standardization,* which gives an exhaustive account of developments from the point of view, and in the background, of the USSR.

0.6 Some concise accounts of the subject as a whole have also appeared in several languages, but mention may particularly be made of *Industrial Standardization* by Leo B. Moore,[13] *Standardization in a Developing Economy* by Lal C. Verman[14] and *La Normalisation* by R. Frontard.[15] There have of course been innumerable articles, brochures, papers and reports treating the subject in parts dealing with specific aspects; but it is clear that only in the United States has some serious attention been paid to an overall codification of the subject.

0.7 In the present work, an attempt has been made to treat the subject of standardization as a unit whole, with a view to bringing into perspective all such activities as may be directly or indirectly related to it, placing particular emphasis on organization, planning and direction, especially from the point of view of national development. The attempt is really in the nature of a search for a method of approach which may help ultimately to make standardization a worthy discipline and an attractive field of work demanding the attention of independent and constructive thinkers. It is hoped that this book will help stimulate active leaders of the profession to undertake further more penetrating studies which may help bring about a deeper understanding of the underlying theories and the guiding principles so essential for the advancement and practical application of any given body of knowledge. With a view to facilitating this process, copious references to published literature have been included, which can by no means be claimed to be exhaustive, but are nevertheless fairly comprehensive. For the same reason, discussion of various aspects has been limited to the basic essentials, leaving details to be studied independently by those interested to pursue certain aspects from the references cited. It has thus been possible to present an integrated treatment of the subject as a whole and the important related matters without dwelling on any of the details.

0.8 The emergence of a large number of new nations since the Second World War has brought about a situation in which standardization could play a significantly important role. Every newly independent country has turned its attention to problems of development on a broad national front which encompasses social, educational, agricultural as well as industrial projects and programmes, aimed at ensuring a better and richer life for the masses of their peoples. In view of the limited resources available to these nations and in order to ensure the greatest possible speed, resort is being made to preplanning all development programmes. While a great deal of assistance has been and is being derived in this effort from more advanced nations, the experience of developed countries cannot, for obvious reasons, be always directly applied to the widely varying and basically different conditions prevailing among the developing countries. Nevertheless, a new branch of economics activity has emerged out of the experience of the past couple of decades in different countries and certain institutes devoted specifically to training in development planning have been created.

0.9 Just as in planning and in executing planned programmes, so in standardization, the developing countries have a great deal to learn from the developed countries, although the conditions prevailing in the two sets of countries do not always lend themselves to a direct transplantation of all the known techniques and the tried-out methods. The developing countries have the advantage of starting from where the latest experience of developed countries leaves off and adapt the lessons thus learnt to their own special needs. For example, one of the significant lessons that the developed countries had to learn at great cost was that in order to gain most from standardization, it was imperative to standardize at an early stage of development of a new technology or a new industry, before a variety of practices became ingrained into the economy of the nation, by which time it became quite costly to bring about order out of the prevailing chaos. The developing countries, starting as they have to, more or less from scratch, are obviously in the most advantageous position to gain a great deal from this lesson.

0.10 Consciously or unconsciously, it was the recognition of this factor which was perhaps responsible during the last quarter of a century for the standardization movement in India to have exhibited considerable vigour and drawn world-wide attention. During this period various contributions have been made through the evolution of new ideas and techniques to facilitate the promotion of standards and standardization in developing economies. For example, the effectiveness of the parallel development of standardization programmes designed to dovetail into the national economic plans has attracted considerable attention at home and abroad. The evolution of the concept of standardization space has helped codify the existing knowledge and push forward its frontiers. These and other contributions from India have naturally influenced the preparation of this book and indeed have, to a large extent, motivated its writing.

0.11 In view of this background, it is but natural that a good deal of the Indian experience has been reflected in these pages. Since this experience has been gained in the context of a developing economy, there is a good chance of its proving useful to other developing countries – particularly to those which are pursuing a deliberate policy of development through national planning. In view of the fact that the author has had direct contact over some years with several other developing countries, it is to be hoped that his appreciation of their individual problems would help to make the book more universally useful to the newly emerging countries in general.

0.12 Basically, problems of standardization are the same whether they arize in the context of developing or developed countries. It is their detailed handling that may differ in the context of the prevailing socio-economic conditions and political circumstances. Many solutions will nevertheless be common, but even the adoption of these has to be judiciously processed, if success is to be assured. Thus, while the book has been written with the world-wide experience in mind, particular attention has been paid to the needs and problems of the newly emerging developing countries and, in doing so, the background experience of one such nation has played a special part. It is the author's hope that it will prove of some service to the various countries interested in furthering the cause of standardization, including those in the advanced guard.

0.13 The author's sincerest thanks are due to many friends all over the world, who have helped make this book a success in many different ways – by helping to collect material, by making this material freely available, by rendering valuable advice and by reviewing and commenting on some of the chapters. Particular mention must, however, be made of the assistance given by Mr S. K. Sen, the present Director General of the Indian Standards Institution, and his devoted band of workers, like Mr R. D. Taneja, Chief Editor; Mr V. P. Vij, Chief Librarian, and others. To Mr

Francis L. LaQue, President of the International Organization for Standardization, author's special thanks are due for his keen interest in the work and for writing its Foreword. Among many of his compatriots and overseas friends, to whom the author feels his grateful indebtedness, are Messrs Olle Sturen, Secretary-General of the International Organization for Standardization; Roy Binney and Gordon Weston of the British Standards Institution; Mr Donald L. Peyton of the American National Standards Institute; Dr P. C. Young of *Comité Européen de Coordination des Normes*; Professor S. Matuura of Hosei University and many others.

REFERENCES TO PREFACE

1. Rajagopalachari, C.: Conference on standardization and quality control, 8–14 February 1948 (Calcutta): Inaugural Address. *ISI Bulletin*, 1949, Vol 1, p. 11.
2. Ailleret, P.: President of the International Electrotechnical Commission (IEC) addressing the Council of International Organization for Standardization (ISO), June 1968.
3. Gaillard, John: *Industrial standardization – its principles and application.* H. W. Wilson Company, New York, 1934, p. 123 (*out of print*).
4. Coles, J. V.: *Standards and labels for consumers' goods.* Ronald Press, New York, 1949, p. 556.
5. MacNiece, E. H.: *Industrial specifications.* John Wiley and Sons, New York, 1953, p. 158.
6. Melnitsky, Benjamin: *Profiting from industrial standardization.* (Conover-Mast Publications Inc, New York, 1953, p. 381.
7. Perry, John: *The story of standards.* Funk and Wagnalls Company, New York, 1955, p. 271.
8. Reck, Dickson: *National standards in a modern economy.* Harper & Brothers, New York, 1956, p. 372.
9. Struglia, Erasmus J.: *Standards and specifications – information sources.* Gale Research Co, Detroit, 1965, p. 187.
10. Maily, J.: *La normalisation* (Standardization). Dunod, Paris, 1946, p. 472 (*in French*) (*out of print*).
11. Woodward, C. Douglas: *Standards for industry.* Heinemann, London, 1965, p. 129.
12. Tkachenko, V. V.: *Method and practice of standardization.* Komitet Standartov, Gosudarstvenniy Soveta Ministrov SSSR, Moscow, 1967, p. 62 (*in Russian*).
13. Moore, Leo B.: Industrial standardization (Section 9 of *Handbood of Industrial Engineering and management*). Englewood Cliffs, NJ, USA, 1955, p. 1203 (*see* pp. 619-651).
14. Verman, Lal C.: *Standardization in a developing economy.* Industrialization and Productivity Bulletin No 7 (United Nations Department of Economic and Social Affairs), New York, 1964, p. 37-51. Also reprinted: *ISI Bulletin*, 1964, Vol 16, pp. 238-245 and 292-298.
15. Frontard, R.: La normalisation. *Notes et Etudes Documentaires*, 1969, No 3593, p. 54 (*in French*).
16. *Fünfzig Jahre Deutscher Normenausschuss (1917–1967)* (Fifty years of German standardization). Deutscher Normenausschuss, Berlin, 1967, p. 117 (*in German*).

17. Jakubiak, T., etc.: *Handbook of standardization.* Polish Standards Organization, Warsaw, 1964 (*in Polish*).
18. Kochtew, A. A.: Technisch-ökonomische prinzipien der standardisierung im maschinenbau (Technical economic principles of mechanical-engineering standardization). Translated from Russian book by A. Kopp and edited by W. Meister. Appearing serially in *Standardisierung.* Feb. 1962. Vol 8, No 2-11 (*in German*).
19. Rago, L. J. von: *Kostensenkung durch Rationalisierung in USA* (Cost cutting via rationalization in the USA). München, Hauser. 1958, p. 193 (*in German*).
20. Sacharov, N. N. and Obraszov, I. G.: *Die technische Arbeitsnormung in Maschinenbau* (Work standards in mechanical engineering). Translated from Russian, Verlag die Wirtschaft, Berlin, 1956, p. 657 (*in German*).
21. Siemens, H.: *Normung-Typing-Gütesicherung; ein Wegweiser für die Praxis* (Standardization, type classification and maintenance of quality; a practical guide). Fachbuch-Verlag, Leipzig, 1953 (*in German*).
22. Small, B. J.: *Streamlined specifications standards.* Reinhold Publishing Corporation, New York:
    V 1 *Architectural,* 1952, p. 980.
    V 2 *Mechanical and electrical,* 1956, p. 494.
23. Starr, C. H.: *Specification and management of materials in industry.* Thames and Hudson, London, 1957, p. 204.
24. *Industrial standardization in developing countries.* United Nations Department of Economic and Social Affairs, New York, 1964, p. 136.

STANDARDIZATION

A New Discipline

Chapter I

Historical Background

1.1 A little reflection will show that standardization in its broader sense has furnished the base on which nature has created the universe. The discreteness of elements and indeed of the fundamental particles constituting the elements, their individual characteristics, their well defined tendency to act and react with each other, amply illustrate what is ordinarily understood to be a well organized standardized pattern of behaviour. The same applies to all other substances occurring naturally or derived from those found in nature. In fact, standardization in nature appears to be of such an immutable character that there exists no possibility of any departure whatever from the laws that control the behaviour of things. The laws may not be understood for the time being, but once they are established, they reveal the complex and yet the utterly simple and logical pattern which nature has adopted to guide its own actions.

1.2 Even among the living, as John Perry[1] puts it, "Natural selection is a process of standardization. Living organisms do not form a continuum, an imperceptible merging of species into species. . . . Each has distinctive characteristics, standard characteristics, passed on from generation to generation." It is, therefore, not unnatural that, without even being conscious of it, all flora and fauna should have a tendency to seek such environments as are most suited to each and to adjust their behaviour accordingly. Birds and animals of all species build their nests and search for their habitats in their own characteristic way; the grunts and growls of animals and the chirps and twitters of birds surely convey some very definite messages. The procedures followed for seeking particular types of foods are characteristic to each species. In short, the study of nature is fascinatingly full of examples indicating how subconscious tendency towards standardization in different species of animals and plants has pervaded the whole of nature, thus making possible their co-existence in harmony.

EARLY EFFORTS OF MAN

1.3 Far from being an exception, man has carried this rule much further. Possessing as he does a better developed brain, he had come to discover early how he could manipulate nature to his own ends and extend the application of the principles of standardization to his own advantage. In the beginning, when he was a hunter and cave-dweller, his pattern of life perhaps was no different from that of other animals around him. However, it could not have taken him long to discover how he could improve his hunting capability with a piece of stone and how he could improve on it further by crudely shaping it to a more lethal form. Thus he began to introduce man-made standards in his daily life. His grunts and growls soon developed into intelligible sounds, perhaps monosyllabic in the beginning, which gradually grew into primitive language as a means of communication enabling him to better convey his thoughts, feelings and messages to his fellowmen and family. Each sound, phrase and syllable carried a given standard meaning. This was the beginning of spoken language which slowly developed through symbols, signs and pictogram stages to the modern written languages which we take for granted today.

PRE-HISTORIC PERIOD

1.4 Perhaps the most striking example of standardization in early prehistory is the form and shape of stone implements, which exhibit an extraordinary degree of similarity whether they have been found in excavations of sites of Europe, Africa or Asia. For the archeologist the differences between the implements may be quite discernible, yet the most striking feature is their general similarity. How these uniform patterns of stone implements came to be established in regions so widely separated will perhaps remain a mystery for a long time to come. One may only hazard a guess that either it was due to the inner nature of man who intuitively thought of carrying out a given task in a certain pre-conceived manner or one may perhaps ascribe the coincidence to long-distance migrations of man spread over long periods of time running into thousands of years. One comes across a similar phenomenon in the universality of bows and arrows all over the world through times immemorial.

HANDAXES CHOPPERS

AFRICA

ASIA

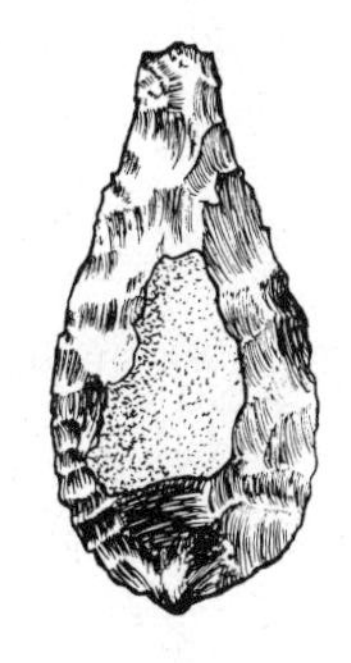

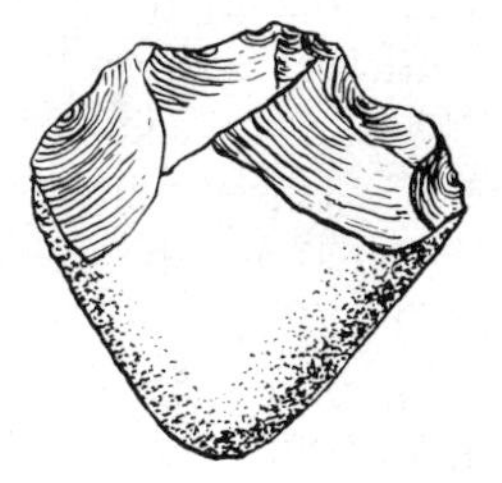

EUROPE

NO CHOPPER

Fig. 1.1 Sketches (approximately one-sixth actual size) of stone implements of pre-historic age from excavations in Africa, Asia and Europe. Note the similarity of design.

1.5 Coming down to later pre-historic times, one finds highly developed civilizations endowed with the faculty of many refined forms of standardization, which would appear not to have been quite so intuitive. Mesopotamia, Sumer, Egypt, Babylon – all offer hundreds of examples in which the pre-christian era civilizations have been found to be using many kinds of standards in their daily life. On the Indian sub-continent, the so-called Indus valley, Mohenjo-daro or Harappa civilization covering an area of about 650,000 square kilometres has left an extraordinary amount of testimony regarding the height to which standardization had been carried about five to six thousand years ago in the fields of town planning, water supply and drainage, house building and even weights and measures.[2, 3]

1.6 The most fundamental standards consciously and deliberately evolved by the ancients were those for weights and measures, which formed the basis of all measurements – an essential prerequisite for any sophisticated form of standardization. Specimens of standardized weights in ratios 1:2:4:8 etc., and a decimally subdivided length scale have been unearthed at several sites of Indus Valley. Their antiquity goes back to about 3500 B.C. Similar finds have been made in most excavations of other ancient civilizations. A striking coincidence is, however, offered by a common feature of the Mohenjo-daro length scale and the one discovered in southwest Babylon.

1.7 The Mohenjo-daro scale, which was found broken at both ends and nine of whose subdivisions have been preserved, has an average distance of 6.7 ± 0.076 mm. between the finely engraved subdivision lines.[2] Forty of these divisions would measure 268 ± 3.04 mm., or let us say, a length varying from 265 to 271 mm. Peculiarly enough, the Babylon scale measures 270 mm. but has only 16 divisions.[4] Considering the accuracy of measurements available in those days, this represents as good an agreement or perhaps better than might be expected. Can this be regarded as a mere coincidence or is it perhaps to be ascribed to inter-communication that might have existed between the two civilizations? If so, it might perhaps be the first example so far discovered of an international agreement on standards of length. Pre-history must be full of the possibility of similar archeological discoveries and it is really up to the standards men of today to sift the available evidence carefully to discover what may be said to be the very roots of their profession. The findings of the archeologist apparently lend some support to the idea that under one guise or another a tendency existed in those ancient times for standardization not only at the national but also perhaps at the international level.

Fig. 1.2 A set of chert weights of the Indus Valley civilization – ratios 1:2:4:8. (*Courtesy*: Archaeological Survey of India).

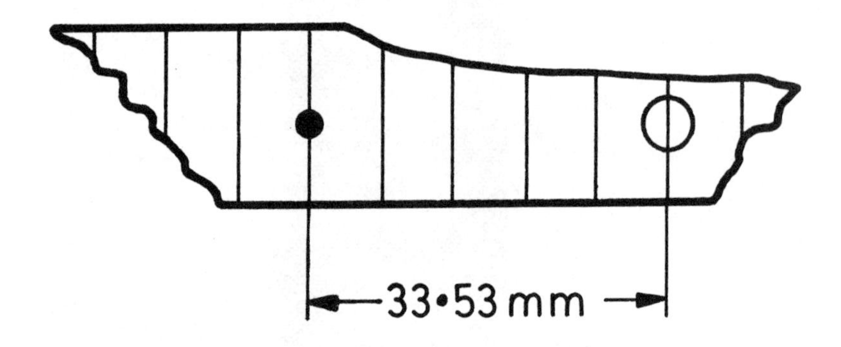

Fig. 1.3 A broken piece of a decimal linear measure of Mohenjo-daro civilization. (*Courtesy*: Archaeological Survey of India).

RELIGIOUS EDICTS

1.8 The importance of standardization apparently became more and more recognized as time went on[1-4] until during the biblical period standards began to find their way into religious edicts:

> The law thou shalt not infringe with the wrong ell, wrong weight, wrong measure. The right balance, right weight, right bushel and right measures shall be with thee.
>
> 3, Moses 19, 35–36

> A false balance is abomination to the Lord, but a just weight is His delight.
>
> Proverbs 11.1

1.9 In the Sanskrit scriptures of about the same period one finds similar references. For example, Manu in his Manusmriti (about 400 B.C.)[5] gives a table of 13 units of weights, and their inter-relationships. He goes on to legislate that "the king should inspect the weights and measures and have them stamped every six months and punish offenders and cheats." In the third century B.C., the Mauryan kings, whose rule extended all over India and who were famous for their efficient central administration, had a special officer of high rank to supervise the administration of weights and measures according to the legal code prescribed by Manu. Kautilya, the Prime Minister of Chandra Gupta Maurya, who is well known for his Artha Shastra (Treatise on Economics)[6] has left an admirable blueprint of the organization he had created for the administration of weights and measures.[7]

THE MIDDLE AGES

1.10 Gradually, as complexities of life increased and clever operators began to learn how to falsify standards, particularly measurement standards, to their own advantage, more extensive legislative measures began to be taken. Laws were promulgated in many lands, basing the standards of length mostly on human body dimensions (width of a finger, the ell, the foot and even the distance between the nose and the tip of one's finger). Same applied to standards of weight and volume. Due to the inaccuracy of definitions and a general lack of adequate inspectorates, confusion continued to prevail until more recent times. Standards for other items were also legalized. For example, in Rome, lead pipes were specified by their dimensions and weight. In England, according to C. F. Innocent,[8] "In the year 1477 the mould for making bricks for building used to be 9 in. long, 4½ in. broad and the bricks when burnt were to be 8½ in. long, 4 in. broad and 2½ in. high. They were further regulated by Acts of Parliament in 1567-68 and in 1625."

1.11 So the story goes through the middle ages in Europe, and no doubt elsewhere, standards being adopted by law, enforced so far as they could be, bypassed by those who stood to gain by so doing. With the invention of the steam engine and the onset of the industrial revolution, however, conditions began to be created for the emergence of standardization in somewhat the same manner as the soil is prepared for cultivation.

But in the purely modern sense, the present-day standards movement may be traced back to the French revolution when the responsibility of standardization passed from the state to the scientists, namely when the French Constituent Assembly formally entrusted the French Academy of Sciences with the task of establishing the metric system of measurement according to a plan proposed by a British engineer – James Watt (*see also* Chapter 13). The advent of mass production, through interchangeability of parts, which took place about the same time across the Atlantic, constituted another important step towards making the development of modern standardization the concern of the engineer and industrialist in his own interest. The experience of Eli Whitney,[9] is often cited as an example:

> In 1798, our [USA] government was in need of more and more arms. Jefferson, then Vice President, signed a contract which bound Eli Whitney to supply ten thousand muskets in two years. At the end of the first year, only five hundred had been delivered, a production of less than two a day. The two years expired and so did Whitney's contract. Necessity became the mother of invention. Urged by the government, Whitney submitted to a board of experts the assembly parts of ten muskets and in their presence assembled from ten identical barrels, ten identical stocks, and ten identical triggers, the first ten standardized rifles. By introducing the principle of interchangeable parts for armament production, he thus became the father of mass production for war purposes.

THE FIRST WORLD WAR

1.12 In this way, men of determination and enterprise extended their operations from war-time to peace-time needs and started to standardize products and processes of industry in general. By the time the First World War broke out, standardization had been well recognized as an industrial process capable of ensuring productivity through interchangeability, not only within a given factory but also from one factory to another, and the importance of creating industry-wide standards and national standards slowly began to be realized. Also, an international standardization movement began to develop early in the twentieth century, when the electrical engineers and physicists concerned with the development of electricity and its uses established the first-ever organization for standardization on a world-wide basis (*see* Chapter 11). The experience of the First World War revealed further potentialities of standardization. Owing to the acute shortage of materials and technical manpower, conservation in every respect became a matter of strategic necessity. In the United States, the

War Industries Board achieved conspicuous results through a process of severe standardization. Restrictions enforced on variety brought about a substantial increase in productive capacity.

1.13 With the cessation of hostilities, industries tended to revert to the traditional freedom of variety but, in the United States, the trend was checked by the publication in 1921 of the Report of the Committee on Elimination of Wastage in Industry appointed by Herbert Hoover[10] as President of the Federated American Engineering Societies. The Committee enquired into the conditions of a large number of typical industries and came to the conclusion that the overall productivity in American industries was not more than 50 percent of the possible maximum. The Committee's report received wide publicity in the United States, particularly in the technical press and in discussions in engineering societies and associations. A nation-wide movement for simplification in industry was started through the agency of the Simplified Practice Division of the United States Department of Commerce. In many cases reduction in variety ranging from 24 to 98 percent was brought about.[11] Thus, standardization – which began merely as an associate of the machine-building processes – developed into a means of ensuring interchangeability and later emerged as a technique of simplification for the conservation of national resources and enhancement of productive capacity.

SECOND WORLD WAR AND POST-WAR PERIOD

1.14 The Second World War brought the urgency of national and international standardization even more pointedly to the forefront. The supply and maintenance facilities of the Allies were severely strained because of differences in standards which prevented interchangeable use of tools and even of common engineering stores like bolts, nuts and screws. Spares for American equipment had frequently to be brought from the United States, which involved considerable loss of war effort at a critical time. Supply management during the war also re-emphasized the importance of standardization and variety reduction of materials and products, and brought about the evolution of many new techniques, including operational research, value analysis, linear programming, statistical quality control and so on.

1.15 In the United Kingdom, a Committee similar to the Hoover Committee of the early twenties was constituted in 1948 under the chairmanship of Sir Ernest Lemon.[12] This was done in spite of the fact that by this time most European countries including UK had well-organized National Standards Bodies actively engaged in the work of preparing standards mostly with, but sometimes without, the help of their governments *see* Chapter 9). Nevertheless, the impact of the Lemon Committee report was significant when it stated that the unnecessary variety of products brings about losses which are not confined to any one stage of manufacture, but extend to the supply of raw materials and components as also to all phases of distribution and to the ultimate user, resulting in higher prices and many other inconveniences and delays.

1.16 About the same time as the Lemon Committee investigated the conditions of British industries, another group of experts was sent to the United States by the Anglo-American Council on Productivity ' 'to secure detailed practical evidence of the benefits which American producers and consumers had derived from a policy of deliberate reduction in variety in manufactured products, whether materials, intermediate components or parts, or end products.[13] Outside the allied camp the war had also brought about similar realization. Along with the hectic post-war reconstruction activity, the preoccupation with strengthening the standards movement and promoting variety reduction became quite universal. Thus France, Germany, Japan, the USSR and most other advanced countries whether on one or the other side of the war, began to take a fresh look at their standards and tried to learn more about how the American industry had been handling the situation.

1.17 Another factor which gave a spurt to the standardization movement at this post-war stage was the creation of the United Nations Standards Coordinating Committee for bringing together the existing national standards bodies into an international forum. Although such a forum had existed earlier under the title of the International Federation of the National Standardizing Associations (ISA), this organization had been rendered inactive during war-time hostilities. The Coordinating Committee of the United Nations met in 1946[14] and brought into being what is today known as the International Organization for Standardization (ISO) (*see* Chapter 11). ISO has since made valuable contributions to the growth of world standards and the development of the standardization movement as a discipline.

1.18 The Second World War had also brought about a transformation of the colonial pattern, as a result of which many countries, one after the other, secured independence during the decade or so that followed. From a meagre beginning of 57 members, the United Nations' membership has now swelled to 132, of which there are nearly 90 in the category of developing countries, most of them newly independent. Anxious as these countries are to develop at a fast enough rate, in the attempt to catch up with the advanced countries, they are also keen to take advantage of every modern technique that can come to their assistance. As a powerful tool, standardization has naturally attracted quite a good deal of their attention. During pre-war times most of them were dependent for their standards needs on the colonial powers controlling them. Today they have realized that their own national interests are not always in line with the latter's and therefore, they must review and re-adapt the old standards and where necessary, produce new standards of their own. This development has greatly assisted in the universalization of thinking on standards and the general advancement of the techniques and procedures involved in preparing and establishing them.

1.19 In the developing countries, therefore, where everything industrial has to be started almost from scratch, the function of standardization as an infra-structure for development could well be appreciated. Most developing countries are relying more and more on a planned development of their economies. It is obvious that in order to realize fully the various advantages which planning has to offer, every phase of planned activity should be so designed as to produce the best possible results with the least possible expenditure of resources, be they financial, material or man-power. This can only be realized if standardization becomes a part and parcel of the planning effort and is pursued simultaneously with all other developmental activity at every level (*see* Chapter 20).

1.20 This idea of standardization being a vital part of national planning is of relatively recent origin as far as most of the developing countries are concerned. But in the socialist countries where planning had become a part of state policy soon after the First World War, standardization had already been playing an important part in advancing their economies. In these states, the use of standards, like any other state laws, was and still is mostly mandatory, and it was quite convenient to ensure this because all means of production and distribution were centrally controlled. In the developing countries, however, though the needs of planning are similar, yet it is not always feasible to make standards mandatory by law. In free or partially controlled so-called mixed economies, standards have by and large to remain voluntary, leaving the state the option to make some of them compulsory and mandatory and to provide adequate administrative machinery for their enforcement (*see* Chapter 14).

THE CONSENSUS PRINCIPLE

1.21 The most remarkable historical development in the realm of standardization has been the evolution of the authority which makes voluntary standards effective instruments for guiding commerce and industry and thus constituting an economic force in national life and also in international trade. This is brought about mainly by following the consensus principle in preparing standards, by which the largest possible agreement is secured among all interests concerned with the use of standards, such as the producer, the user, the trader and the technologist. Once all these interests have agreed and a common ground upon which to base the standard has been found, the standard acquires an authority, possibly much more powerful than a legal instrument might which has secured only a 51 percent majority vote in its favour. While the latter must be enforced by the power of the state, the former would be voluntarily followed by those who had generally agreed to its contents. Standardization through consensus does sometimes mean compromises to be made, but then it is always much more practical to have voluntary standards prevail where constant policing is neither feasible nor necessary (*see* Chapter 14).

1.22 The question has sometimes been asked: if the consensus principle has been so successful in voluntary standardization, why can't it be extended to other spheres of man's activity such as the democratic decisions to be made in his political and social affairs?[15] A time may come when the commonly prevailing 51 percent majority rule may yet be replaced by the consensus principle and man may, as a result begin to live more harmoniously under a new form of democracy in which every voice would carry its own weight, however small, and decisions would be made more on the basis of weightages of the complex needs of society as a whole rather than on what the party holding 51 percent seats might have decided. Exactly how this state of affairs would be brought about is not quite clear, but it may well be the next step in the evolution of democracies that govern the majority of humanity today.

REFERENCES TO CHAPTER 1

1. Perry, John: *Story of standards.* Funk and Wagnalls, New York, 1955, p. 271.
2. Director and Colleagues: Standardization in pre-historic India. *ISO souvenir 1964.* Indian Standards Institution, New Delhi, p. 21-33.
3. Verman, Lal C.: *Standardization in India – ancient and modern.* Shri Ram Institute for Industrial Research (Founder Memorial Lecture, 1965), New Delhi, p. 24.
4. *Fünfzig Jahre Deutscher Normenausschuss (1917–1967)* (Fifty years of German standardization). Deutscher Normenausschuss, Berlin, 1967, p. 117 (*in German*).

5 Acharya, N. R.: *Manusmriti.* Nirmaya Sagar Press, 1946, p. 552.
6. Kangle, R. P.: *The Kautilya Arthashastra (Part I 1960, Part II 1963, Part III 1965).* Bombay University Press, Bombay, p. 283, 606 and 303.
7. Raju, L.: Development of weights in Madras area: Part I and Part II. *Metric Measures,* New Delhi, 1962, Vol 5, No 1, p. 11-16; No 4, p. 3-9.
8. Innocent, C. F.: *The development of English building construction.* 1914.
9. Dickson, Paul W.: *Industrial standardization (company organization, practices and procedures).* National Industrial Conference Board, New York, 1947.
10. Hoover, Herbert: *Memoirs: The cabinet and the presidency. Vol 2. 1920–1933.* Macmillan, New York, 1952, p. 405.
11. Spriegel, W. R. and Lansburgh, R. H.: *Industrial management.* John Wiley and Sons, New York, 1955.
12. *Report of the Committee for Standardization of Engineering Products* (United Kingdom, Ministry of Supply). Her Majesty's Stationery Office, London, 1949, p. 35.
13. *Simplification in industry.* 1949, p. 12. Also *Simplification in British industry,* 1950, p. 13, Anglo-American Council on Productivity, London.
14. *Report of Conference of the United Nations Standards Coordinating Committee together with delegates from certain other national standards bodies.* UN Standards Coordinating Committee, London, 1946, p. 35.
15. Verman, Lal C.: *Standardization – a triple point discipline* (Presidential Address, 57th Indian Science Congress, 1970, Calcutta). Indian Science Congress Association, p. 9. Also printed in adapted form in *ISI Bulletin.* 1970, Vol 22, p. 47-50.

Chapter 2

Semantics and Terminology

2.1 A well-defined specialized terminology is as important to the development of a new discipline as a language itself is to a system of communication. If, therefore, the theory of a discipline is to be seriously outlined and the practices properly codified, it is essential first of all to find a name for the discipline. Would it be appropriate to name it "standardization engineering," "standards engineering," "standards technology," "standardization science" or simply "standardization," or perhaps something else? The question has necessarily to be answered at this stage. There has been no agreed solution so far and it is important to decide the issue for the purpose of elaborating the subject further, even if the decision is regarded as tentative.

2.2 In his Presidential Address to the Indian Science Congress – "Standardization – a Triple-Point Discipline," the author[1] had the occasion to make the following remarks:

> I have chosen to call standardization a triple-point discipline for ready comprehension, but perhaps more accurately it might be described as a multi-point discipline, for it is a discipline in which applied science, technology, industry and economics play extremely important parts. Human psychology, public relations, management and other social sciences are also involved. In very general terms its

object may be described as the regulation of man's relationship to man in respect of the daily exchange of goods and services. Each one of these wide fields of knowledge contributes to standardization and all of them in turn profit by it. Standardization, indeed, furnishes the conditions and the environments under which they can act and re-act on each other to mutual advantage and for the benefit of the community.

2.3 Perhaps most workers in the field would find little to differ with in the above view, but apart from indicating the far-reaching importance of standardization, it does not help to answer the question posed. Before discussing the issue any further it may be pertinent to examine the dictionary meanings of certain relevant terms, such as:

Discipline:

A branch of instruction; a department of knowledge. (Oxford)

A subject that is taught; a branch of learning; field of study. (Webster)

Engineering:

Work done by or the profession of an engineer. (Oxford)

Engineering activities or the function of an engineer. (Webster)

Engineer:

One who contrives, designs or invents; an inventor, a plotter. (Oxford)

(1) A designer or plotter of engines. . . . .

(2) A person who is trained in or follows as a calling or profession a branch of engineering.

(3) A person who carries through an enterprise or brings about a result especially by skilful or artful contrivance.

(4) A person who is trained or skilled in the technicalities of some field (as sociology or insurance) not usually considered to fall within the scope of engineering and who is engaged in using such training or skill in the solution of technical problems.

(5) A person with or without technical training who affects technical knowledge to further his endeavours (as in selling).

(Webster)

Science:

(1) The state or fact of knowing; knowledge or cognizance of something specified or implied; also, knowledge (more or less extensive) as a personal attribute.

(2) Knowledge acquired by study; acquaintance with or mastery of any department of learning.

(3) (a) A particular branch of knowledge or study; a recognized department of learning: often opposite to art;

(b) A craft, trade, or occupation requiring trained skill.

(4) A branch of study which is concerned either with a connected body of demonstrated truths or with observed facts systematically classified and more or less colligated by being brought under general laws, and which includes trustworthy methods for the discovery of new truth within its own domain.

(5) The kind of knowledge or intellectual activity of which the "sciences" are examples. (Oxford)

(1) (a) Possession of knowledge as distinguished from ignorance of misunderstanding: knowledge as a personal attribute.

(b) Knowledge possessed or attained through study or practice.

(2) (a) A branch of department of systematized knowledge that is or can be made a specific object of study.

(b) Something (as a sport or technique) that may be studied or learned like systematized knowledge.

(c) Studies mainly in the works of ancient and modern philosophers formerly taught as a group of field of specialization.

(d) Any of the individual subjects taught at an educational institution in one of the departments of natural science.

(3) (a) Accumulated and accepted knowledge that has been systematized and formulated with reference to the discovery of general truths or the operation of general laws: knowledge classified and made available in work, life, or the search for truth: comprehensive profound or philosophical knowledge, especially knowledge obtained and tested through use of the scientific method.

(b) Such knowledge concerned with the physical world and its phenomena: NATURAL SCIENCE.

(4) A branch of study that is concerned with observation and classification of facts and especially with the establishment or strictly with the quantitative formulation of verifiable general laws chiefly by induction and hypotheses.

(5) A system based or purporting to be based upon scientific principles: a method (as of arrangement, functioning) reconciling practical or utilitarian ends with scientific laws. (Webster)

Technology:
A discourse or treatise on an art or arts; the scientific study of the practical or industrial arts. (Oxford)

(1) The science of application or knowledge to practical purposes: applied science.

(2) Application of scientific knowledge to practical purposes in particular field.

(3) The totality of the means employed by people to provide itself with the objects of material culture.

(Webster)

STANDARDIZATION: SCIENCE, ENGINEERING OR TECHNOLOGY

2.4 It is clear that the two authorities on the English language do not quite agree on the various meanings of certain terms. For example, Webster accepts the generalization of the field of "engineering" to including sociology, insurance, and even salesmanship under certain circumstances. Yet, according to the authors of Oxford Dictionary, this would perhaps be stretching the point. Similarly, Webster is more generous towards the interpretation of the concepts of "technology" by including in it the totality of the means of material culture. But as regards "discipline" and "science" both the authorities would appear to agree in general.

2.5 In the light of these authoritative definitions of the various concepts, it would be justifiable to call standardization almost anything one might choose – engineering, science or technology, and of course a discipline as well. But it would not do for our purpose to leave the matter open like this. In order to arrive at a conclusion we must adopt certain criteria for judgement. For example, the following are suggested:

(1) The decision should not only be acceptable in all English-speaking and English-using countries but also in other countries using other languages – particularly the more important ones such as French, Spanish, German and Russian.

(2) The decision should be consistent with the contents and the scope of standardization as it is understood today and as it may develop in future.

(3) The decision should take full account of the views expressed by standards authorities in the past, though there has so far been no unanimity in the matter among them.

2.6 The expression "Standards engineering" has been used for a very long time particularly in the Americas and more recently even in the UK and on the continent as well. But standardization covers, as it does today, so many other activities, besides those pertaining to engineering, such as terminology, management and government, that the question arises whether it would be justifiable to continue to advocate this usage. A change may be worthwhile, provided a better choice was available. Some authors have strongly suggested the alternative of calling it a science. Seaman,[2] while advocating this move, considers that science is "a field of endeavour wherein comparable results may be obtained by any worker who adheres strictly to the governing precepts or theorems." Thus, he has formulated seven such principles for standardization and shown their validity. But he also admits that "the past experience of many would indicate that the chances of achieving success are never immediate and usually doubtful. These conclusions while generally correct, do not reflect the science-like character aspects of standardization, but indicate instead the lack of scientific approach." The question arises whether the lack of success is always attributable to an incorrect approach or is it perhaps sometimes due to the fact that the approach has of necessity to be subjective in nature, whereas science admits only of the objective approach.

2.7 Meissner[3] has also attempted to reduce standardization to a science and presented another set of seven propositions, which are quite different from the principles of Seaman. If standardization were a science, the chances of the two sets of principles being different would be negligible. Meissner further asserts that "naturally, the scientific character of standardization cannot be identified with the necessity of a scientifically based method of approach." If that is so, then why stretch the point and try to make standardization a science? Let us not forget that the scientific method as defined by one authority demands "the recognition and formulation of a problem, the collection of data through observation and if possible experiment, the formulation of hypotheses, and the testing and confirmation of the hypotheses formulated." It must also be recognized that standardization, to be called a science, should have the possibility of having its theories reduced to mathematical terms and objectively applied in practice. As it happens, a good deal of success in standardization depends, among other things of course, upon a proper subjective approach. Judged from present-day knowledge and the trends of development, it would appear that this would remain so for a long time to come. This view is, by and large, confirmed also by Kvasnitskyi and Levintov's objective analysis of the situation.[17]

2.8 It would perhaps be well not to stretch the point too much and try to get standardization accepted as one of the sciences comparable to natural sciences like physics or chemistry. At best it could aspire to be a science comparable to social sciences like sociology and political science. But is it really worth it, when its comprehensive character enables it to encompass all these sciences plus psychology, economics and management? For similar reasons, calling it a "technology" may be considered most inadequate. It is proposed that for the purpose of discussion in this book, it may simply be called a "discipline." Without being a branch of science, engineering, technology or economics, the term "discipline" is comprehensive enough to include all branches of science and engineering as well as other forms of knowledge. Thus, there would be room for anyone to regard it in a given context as a branch of science, engineering or even technology. In this way the three points of the criterion stated in para 2.5 above will have been fully taken care of.

STANDARDS ENGINEER

2.9 One more related question is in regard to the name to be applied to the workers engaged in the practice of the discipline of standardization. The term "standards engineer" has been in vogue for many years in the USA and now it is being adopted in Europe also. But there has been some hesitancy in its general adoption,[4] because a "standards engineer" is called upon to deal with a vast number of specializations other than engineering, such as chemistry, physics, management and economics. To overcome this objection the term "standardizer" has been suggested by some continental workers. But its acceptability in English-speaking countries is rather doubtful, because it sounds more like a tradesman's designation rather than that of a specialist. Though the final decision will have to be taken by the International Organization for Standardization (ISO), it is significant to note that in the UK considerable thought has already been given to the problem. Standards Associates of the British Standards Institution after due discussion have adopted and defined the term "standards engineer" in the context of his functions within industry as related to the company or in-plant standardization activity. The definition as forwarded to the UK Department of Employment and Productivity for inclusion in the official DEP Occupational Classification register reads as follows:

> *Standards Engineer* develops standardization and implements local, national and international standards within an undertaking. Examines existing processes, procedures and other activities to ensure that, as far as possible, standard practice and optimum procedures throughout the undertaking; consults with research, design and production departments, and formulates specifications in relation to company, statutory and other standards; reviews components and items of equipment to eliminate unnecessary variety

> while retaining suitable alternative choices; reviews standards in the light of improvements in materials, advances in techniques and changes in statutory requirements and in national and international standards. Develops or arranges for the development and implementation of standards concerned with new concepts and original ideas, and compiles new standards as necessary. Maintains an up-to-date library of all relevant standards, keeps abreast of national and international developments in standardization and provides information and advice on all standards matters.
>
> May represent his organization on national or international committees concerned with standardization.

As a result of this effort the term "standards engineer," in spite of its elaborate definition, is being widely adopted in the UK by both private industry and officially. In view of this development, it is quite likely that "standards engineer" would become more and more universally entrenched in English usage and ultimately ISO might perhaps also adopt it in the more general context of the field of standardization as such, as distinct from company or in-plant standardization. Noting Webster's generalized concept of engineering (*see* paras 2.3 and 2.4), such an adoption would perhaps be considered generally acceptable in other languages also. In any case, for the purpose of this book, "standards engineer" will mean a practitioner of the discipline of standardization no matter at which level or in what branch of specialization.

STANDARDIZATION AND STANDARD: DEFINITIONS

2.10 Coming now to standardization activity as such, this term has been variously defined by different authorities, but we shall not concern ourselves with discussing all these versions, nor those given in various dictionaries and encyclopaedias, because it so happens that the international authority on standardization, namely the International Organization for Standardization (ISO) itself has officially defined what standards should mean. According to ISO definition:[6]

> *Standardization* is the process of formulating and applying rules for an orderly approach to a specific activity for the benefit and with the cooperation of all concerned, and in particular for the promotion of optimum overall economy taking due account of functional conditions and safety requirements.
>
> It is based on the consolidated results of science, technique and experience. It determines not only the basis for the present but also for future development, and it should keep pace with progress.

Some particular applications are:

(1) Units of measurement.
(2) Terminology and symbolic representation.
(3) Products and processes (definition and selection of characteristics of products; testing and measuring methods; specification of characteristics of production, for defining their quality, regulation of variety, interchangeability, etc.).
(4) Safety of persons and goods.

2.11 It may be noted that standardization is defined as a process for not only formulating, but also applying certain rules. It is not a cult or faith to be taken for granted, or for that matter a mandate to be imposed by authority. It is simply a process both for making and for implementing certain rules to ensure an orderly approach to any given activity of man, such as manufacturing, selling or constructing. The approach is to be orderly in the sense that in carrying out the activity in question there should be little chance of waste of time, effort or resource. That is why it is expected that the activity carried out under such rules would be beneficial to all concerned, and for this reason such an activity can more readily be pursued without imposition from above and with the cooperation of all concerned. The cooperation, therefore, becomes a basic prerequisite, particularly for formulating the rules, the implementation of which is automatically facilitated, being as it is, to the common advantage of the parties concerned. It will be appreciated how this condition leads to the necessity of securing the largest possible consensus of views among the cooperating parties while framing the rules.

2.12 The definition further goes on to include the concept of the benefit to be derived by all concerned by particularizing the promotion of overall economy. It is always possible that, when considered on a short-term basis or when applied to a limited circle, the benefit may not be all-inclusive. For example, benefit to the seller may not always imply benefit to the buyer as well, though both are concerned in a given transaction. It is, therefore, necessary to emphasize the overall economy aspect of standardization, which not only takes into account the long-term viewpoint but also the fact that the benefit to the community is to be given more weight than benefit to the individual, benefit to the nation more than to the community, and so on. Furthermore, note that the overall economy is to be optimum, which is not the same thing as the maximum economy. Optimum economy only approaches maximum, but may fall somewhat short of it, if some overweighing considerations so require. For example, a case may arise for conserving a strategic material or a site, which may demand departure from maximum economy and adherence to optimum. Lastly, it is enjoined that functional conditions and safety requirements should be satisfied in all standards, for they are the basic elements to ensure serviceability and to safeguard the user and the consumer who in the ultimate analysis offer the prime motive for the production of all goods and provision of all services.

2.13 The second part of the definition of standardization is concerned not so much with defining the concept, as with indicating how the process of standardization is to be carried out by taking full account of all the available information at the existing stage of development of science and technology, and how it is essential to review, from time to time, the rules once laid down, to bring them up-to-date with the advancement of knowledge and experience. In the last paragraph, a few examples are given of the fields of application which must be regarded only as illustrative and by no means exhaustive.

2.14 Having defined and explained standardization, it would be pertinent to take up the definition of a standard, but before discussing the latest ISO definition, it would be interesting to refer to John Gaillard's[7] – definition of 1934 vintage which as a classic has served the profession for many decades:

> A standard is a formulation established verbally, in writing or by any other graphical method, or by means of a model, sample or other physical means of representation, to serve during a certain period of time for defining, designating or specifying certain features of a unit or basis of measurement, a physical object, an action, a process, a method, a practice, a capacity, a function, a duty, a right, a responsibility, a behaviour, an attitude, a concept or a conception.

2.15 It must be admitted that this definition gives a most picturesque panorama of what is covered by the term standard. Nevertheless, feeling that it was somewhat incomplete, the author[8] had an occasion in 1952 of adding the following phrase at the end, namely:

> . . . a concept or a conception, or a combination of any of these, with the object of promoting economy and efficiency in production, disposal, regulation and/or utilization of goods and services, by providing a common ground of understanding among producers, dealers, consumers, users, technologists and other groups concerned.

2.16 It will be noted that most of these ideas were later, in 1956, incorporated neatly in the ISO definition of standardization. But even this definition has been found to be wanting by certain authorities like Sen,[9] who has proposed to define standardization as follows:

> *Standardization* is the process by which systems and values are established in individual, group and social life by natural evolution, custom, authority or common consent which, by remaining (or being kept) invariable over a period of time in a changing environment of unlimited modality, provide the stable basis essential for the growth and attainment of:
> (a) social or group identity and survival,
> (b) communication, understanding and exchange of ideas, goods and services between individuals and groups,
> (c) knowledge and experience for further development, and
> (d) consolidation of social, economic and technological attainments at any point of time so as to release creative energy for the search of higher and better values and systems.

2.17 Sen's definition may be regarded as a definite advancement on ISO, but it has not yet had time enough to be seriously studied by world authorities. Until such time, therefore, as the official ISO definition is modified, for most purposes it will have to be considered authoritative. Thus, a standard was simply defined by ISO in the form of a derived term as follows:[6]

> A standard is the result of a particular standardization effort, approved by a recognized authority. It may take the form of:
> (1) a document containing a set of conditions to be fulfilled (in French "norme")
> (2) a fundamental unit or physical constant – examples: ampere, absolute zero (Kelvin) (in French "étalon")
> (3) an object for physical comparison – examples: metre (in French "étalon").

CONCEPT AND TERM

2.18 This ISO definition of "standard" is quite simple and straightforward and requires little elaboration. But it does bring to the surface an important point of semantics which illustrates the difference between a "concept" and a "term." It will be seen that the term "standard" in the English language comprises three different concepts for which there are two corresponding French terms, namely "norme" and "étalon." In all considerations connected with choice or formulation of terminology, it is extremely important to distinguish between a concept and a term. A concept, according to ISO definition,[10] is "any unit of thought generally expressed by a term, a letter symbol or by any other symbol." Thus a "concept may be the mental representation not only of beings or things (as expressed by nouns), but, in a wider sense, also of qualities (as expressed by adjectives or nouns), of actions (as expressed by verbs or nouns), and even of locations, situations or relations (as expressed by adverbs, prepositions, conjunctions or nouns)." Care, however, must always be taken that concepts are not confused with the individual objects themselves. They are only mental constructs serving to classify the individual objects of the inner or outer world by way of more or less arbitrary abstraction. In no case should concepts be confused with terms which are the linguistic symbols of concepts and which are created by man in a more or less arbitrary manner. On the other hand, a term has been defined (by ISO)[10] – as "any conventional symbol for a concept which consists of articulated sounds or of their written representation (of letters)." Thus a term may be a word, phrase or symbol.

2.19 It will now be clear that the concept "standard" represents the idea of the result of any particular standardization effort, such result being duly approved by a recognized authority. If only the idea is so represented, then it is clearly independent of all languages. The term "standard," on the other hand, in the English language, is the name of this concept and includes all the three kinds of standards that might result from such an effort, namely, a document, a fundamental unit appropriately defined, or a physical object. On the other hand, in the French language, the concept of "standard" has to be represented by one of the two terms, depending upon the particular kind of result. Thus the term "norme" would apply only to the first kind of standard, namely, a document and the term "étalon" would apply to the other two types of standards namely a unit or an object. Such differences between concepts and terms from language to language are not uncommon. Other difficulties also arise; for example, a given word may have several meanings or more than one concept may be represented by one and the same term. Such situations may create misunderstandings between the users of linguistic symbols in the process of communication. It happens often enough when only one language is involved. But when two or more are involved, misunderstandings may assume serious proportions. Thus standardization of terminology has to be given an extremely important and basic position in the whole programme of standardization. Recognizing this need, ISO has created a technical committee, namely ISO/TC 37 Terminology (principles and coordination) to investigate and promote methods of setting up and coordinating national and international standardized terminologies.

2.20 As a result of the labours of this technical committee, ISO has been able to issue a number of recommendations dealing with the basic principles of creating in a coordinated manner vocabularies in various languages to take care of their newly emerging technical needs, particularly in the field of standardization. Following are the titles of some of the recommendations to which reference is invited for further guidance:

(1) Symbols for languages, countries and[11] authorities
(2) Naming principles[12]
(3) International unification of concepts and[13] terms
(4) Guide for the preparation of classified[14] vocabularies
(5) Vocabulary of terminology[10]

DEFINITIONS OF OTHER TERMS

2.21 Another Committee of ISO, the Standing Committee on Principles of Standardization (STACO), has been concerned with defining the terms particularly used in standardization work. The ISO definitions of "standardization" and "standards" given in the above paragraphs are the direct results of deliberations of this Committee. Other definitions recommended by STACO and officially adopted for use in the field of standardization by ISO[6, 15] are the following:

Specification

A specification is a concise statement of a set of requirements to be satisfied by a product, a material or a process indicating, whenever appropriate, the procedure by means of which it may be determined whether the requirements given are satisfied.

Note – (1) A specification may be a standard, a part of a standard, or independent of a standard.

(2) As far as practicable, it is desirable that the requirements are expressed numerically in terms of appropriate units, together with their limits.

Simplification

Simplification (Variety Reduction) is a form of standardization consisting in the reduction of the number of types of products within a definite range to that number which is adequate to meet prevailing needs at a given time.

Unification

Unification is a form of standardization consisting of combining into one, two or more specifications in such a way that the products obtained are interchangeable in use.

Specialization

This term is, properly speaking, not within the sphere of standardization. If it is used it should have the following meaning:

Specialization is a process by which particular production

units concentrate on the manufacture of a limited number of kinds of products.

Functional interchangeability

Functional interchangeability is achieved when those characteristics of the finished product which affect its operation have been standardized to the necessary degree of accuracy. This frequently includes characteristics other than linear dimensions.

Dimensional interchangeability

Dimensional interchangeability is a partial aspect of functional interchangeability and is achieved when the linear dimensions of two finished products are sufficiently close to one another to ensure only interchangeable replacement.

Designation

Designation is a definite and distinguishing name or symbol given to a product of a group of functionally similar products or an abstract matter. It underlines the group similarity but does not bring out the differences between the various members of the group.

Note: The use of two or more names for a single product is confusing and selection of the most appropriate term as the standard should be undertaken at an early stage in the standardization of the product itself (e.g. "capacitor" may be selected as the standard designation of an electrical device for which the term "condenser" has also been used previously).

Identification

Identification is the recognition of the nature of a product or abstract notion so as to be able to apply the proper designation to it. The product may be identified in broad terms only, as one of a certain group, or its characteristics, size, etc., may be identified, thus distinguishing it from other members in the group.

To facilitate identification, marking by means of a number, colour or other such indication, may be used.

Code

Code is a particular form of identification marking or reference which serves the dual purpose of establishing in a systematic manner the complete identity of an individual product, bringing out at the same time its similarity with other products. It may consist of a brief and systematic combination of letters, numerals and/or symbols. A systematic code, used in conjunction with a key for decoding, can assist identification of a product and enable its characteristics to be known.

Coding

Coding is the action of devising such a code.

Marking

Marking is the action and the result of stamping, inscribing, printing, labeling, etc., marks, symbols, numerals, letters, etc., upon a product itself or on its package, for the purpose of identifying the product and its origin and giving its basic characteristics, its intended use, etc.

Note: In the manufacturing of products, the term "marking" is also used in another meaning when it regards: marking of outlets, terminal markings.

2.22 Apart from the work of ISO/TC 37 and STACO, every other technical committee of ISO is entrusted with the task of creating its own vocabulary and defining its own terminology as may be required for its own specific field of work. Thus, the work of ISO in this field has been quite considerable and is proceeding apace. There is however, a limitation from which it has suffered. Most of the participants in ISO technical committees, particularly TC 37, are from the European countries and the USA. This is so because countries in other parts of the world far removed from Europe have hardly advanced far enough to contribute significantly to the basic work of this character on terminology in better known languages. On the other hand, it is this latter group of countries, including a large number of newly developing countries, which are most in need of new vocabularies to express technical thought in their own respective languages in all fields of learning, including science and technology and of course standardization. In this connection, while ISO has been understandably able to do little to help these countries, the efforts of the United Nations Educational, Scientific and Cultural Organization (UNESCO) are worthy of note. In 1968, UNESCO called a Conference[16] in Kuala Lampur to discuss the question of modernization of Asian languages. This Conference recommended the creation of a centre or an institute under the joint auspices of UNESCO and the national governments of Asian countries for, among other things, dealing with the question of terminology and developing multilingual glossaries.

TERMINOLOGIES FOR DEVELOPING COUNTRIES

2.23 Creation of technical terminologies in the language of any developing country is a tremendous undertaking, because firstly, a start has to be made almost from scratch and secondly, the whole vast field of technology which is closely linked with all sciences is required to be covered. For the purpose of standardization, however, the task may be considered somewhat limited, yet it is not quite so, for it must be borne in mind that every branch of technology is a concern of standardization and has to be catered to. Thus, development of terminology for standardization in any local language would have to be carried out in close liaison with the authorities concerned with the development of technical terminology as such, which in some countries may be government ministries or departments, in others, universities or other learned bodies. Whatever the form of a national set-up, it is imperative that the national standardization authorities should be assigned a leading role to play. They hold a key position. On the one hand, they are in a position immediately to implement the decisions on the national level and introduce the newly adopted technical terms into the mainstream of national life, and on the other, to coordinate the national work in the field with the international developments.

2.24 Some developing countries have always used one or the other European language for standards work, such as the Latin American countries; others prefer to continue to use the language of their erstwhile rulers even after independence, such as the Philippines and Singapore. But a large number of the new nations have chosen to develop their own languages for scientific and technological purposes. A few of these latter cases of countries are known to have adopted a set of guiding principles for developing their terminology and have achieved a measure of success, as for example India, Iran and Thailand. There is a tendency among some of these countries to attempt to coin new words for concepts for which internationally used terms may be available. This is a tendency which must be curbed, for mankind today is moving more rapidly than ever before towards one-world concept. Anything which would hamper this movement should be discouraged, especially in the field of standardization, where all effort is to be focussed on international unification in the interest of maximum economic advantage.

2.25 It would be worthwhile, therefore, for standards authorities in developing countries to adopt a set of principles or rules for guiding the selection and coining of terms for use in standardization work. In view of the wide variety of languages prevailing in the different countries, it is naturally most difficult to frame any rigid rules which could be generally applicable. Each country would have to frame its own guiding principles in line with the genius of its own language. Some general guidelines which may be helpful in framing these rules may, however, be found useful. Based on the experience of a few developing countries, the author would venture to suggest that the following points could usefully be borne in mind while framing the guiding principles:

(1) A close study may be made of all the ISO recommendations dealing with the principles of formulating standardization terminologies. This would be helpful in coordinating the national work with the international effort.

(2) As a basic principle it must be recognized that whether a given word is adapted or borrowed, or is freshly coined, the genius of local language and its character should be maintained throughout.

(3) Words that have found extensive usage in the popular language of the country for expressing technical concepts should always be given the highest priority for adoption in the standards terminology. Such words may have been borrowed from other languages or may have been coined indigenously by popular selection.

(4) The next preference may be given to the adoption of words which are in common international usage in several foreign languages. Such words would have to be transliterated into the local scripts but the transliteration should not necessitate the introduction of new signs or symbols or modification of the local script in any way. However, they may be otherwise adapted to conform to the genius of the local language.

(5) Failing a satisfactory solution under these two choices, attempt may be made to coin new words, but in doing so, care should be taken to construct them in such a way that their meanings may be obvious and self-explanatory to the largest possible proportion of users. Many languages offer such possibilities in which commonly used roots and derivatives can be employed to coin new words of this character.

(6) As a last resort, attempt may be made to derive new words from roots of the classical parent language from which the modern language may have descended. This is recommended as the last choice because in following this procedure it often happens that the resulting words sound more strange than even the foreign words. There is always a possibility of new words being interpreted differently by different readers. Thus the specific meaning given to each such word has to be firmly and formally established. In this connection, it may be well to remember that purism of language is not a worthy goal, if it entails sacrifice of intelligibility and moves a country unnecessarily away from internationalism.

2.26 Apart from adopting a set of guiding principles as suggested above, it would be useful if the authority concerned with national standardization appoints a high-level learned body of men representing various disciplines to advise on the framing of local language terminology for use in standards. It should be this body's first responsibility to deliberate on the guiding principles to be adopted. It will also be its concern to cooperate and collaborate with other similar bodies which may be concerned with terminologies in other spheres of national life.

2.27 In order to avoid overlaps, repetition and other forms of confusion in the official vocabulary during the evolution period, it would be extremely useful to maintain a card index of all the terms adopted. Each term together with its definition and other contextual material may be allotted an independent card and the whole collection of cards arranged in one alphabetical series. Each time a new proposal for adopting a term comes under consideration, reference to the card index would be found most useful. Similarly, for those concerned with writing standards the index would furnish the authentic source for officially adopted terms. Such an index would grow with time and become a valuable permanent asset to any standards organization.

REFERENCES TO CHAPTER 2

1. Verman, Lal C.: *Standardization – a triple point discipline* (Presidential Address, 57th Indian Science Congress, 1970. Calcutta). Indian Science Congress Association, p. 9. Also printed in adapted form in *ISI Bulletin.* 1970, vol 22, p. 47-50.
2. Seaman, Ellsworth F.: Is standardization a science? *Journal of the American Society of Naval Engineers.* 1953, vol 65, p. 37-47.
3. Meissner, Erwin: Ist Standardisierung eine Wissenschaft? (Is standardization a science?) *Standardisierung.* 1968, vol 14, p. 257-259 (*in German*).
4. Verman, Lal C.: What is standardization? *ISI Bulletin.* 1970, vol 22, p. 291-293.
5. Verman, Lal C.: Standardization in a developing economy. *Industrialization and Productivity Bulletin No 7* (UN Department of Economic and Social Affairs), New York, 1964, p. 37-51. Also reprinted: *ISI Bulletin.* 1964, vol 16, p. 238-245 and 292-298.
6. *Definitions 1 – Standardization vocabulary: basic terms and definitions.* International Organization for Standardization, Geneva, 1971, p. 5.
7. Gaillard, John: *Industrial standardization: its principles and application.* H. W. Wilson Company, New York, 1934, p. 123 (*out of print*).
8. Verman, Lal C.: Standardization – its principles and development in the world and the ECAFE region. *ISI Bulletin,* 1952, vol 4, p. 3-9.
9. Sen, S. K.: Defining standardization. *ISI Bulletin.* 1971, vol. 23, p. 389-390.
10. *ISO/R 1087–1969 Vocabulary of terminology.* International Organization for Standardization, Geneva, p. 23.
11. *ISO/R 639–1967 Symbols for languages, countries and authorities.* International Organization for Standardization, Geneva, p. 15.
12. *ISO/R 704–1968 Naming principles.* International Organization for Standardization, Geneva, p. 15.

13. *ISO/R 860–1968 International unification of concepts and terms.* International Organization for Standardization, Geneva, p. 16.
14. *ISO/R 919–1969 Guide for the preparation of classified vocabularies* (example of method). International Organization for Standardization, Geneva, p. 15.
15. ISO Council Annual Meeting, 1962 and 1963. *ISI Bulletin.* vol 14, p. 335-337 and vol 15, p. 339-344.
16. Modernization of Asian languages. *ISI Bulletin,* 1968, vol 20, p. 184.
17. Kvasnitskyi, V. N. and Levintov, A. G.: Trends in the growth of scientific basis of standardization (for discussion). *Standarty i Kochestvo.* 1970, No 5, p. 39-41 (*in Russian*).

Chapter 3
Aims and Functions

3.0 In order to get a correct perspective of the aims and functions of standardization, it is important first of all to examine the inter-relationship of the various attributes of standards. This could be done conveniently by considering what may be called "standardization space," with "subject," "aspect" and "level" constituting the three axes of reference.[1, 2] *(See* Fig. 3.1.)

STANDARDIZATION SPACE

3.1 *Subject* Standards cover almost all aspects of all economic activity of man engineering, industry, construction, agriculture, commerce, forestry, and so on; and some aspects of cultural activity as well, such as sport and music. Each of these fields deals with a large number of items; for instance, under engineering we have electricity, steel, machines, fasteners, etc. Each item may be further subdivided into subjects suitable for being covered by a standard; for example, under fasteners, we may have a standard on screw threads, bolts and nuts, washers, rivets, etc. Corresponding to the subject matter of a standard, or rather the product or process covered by it, we may assign a point on the "X-axis" of what we have called the standardization space.

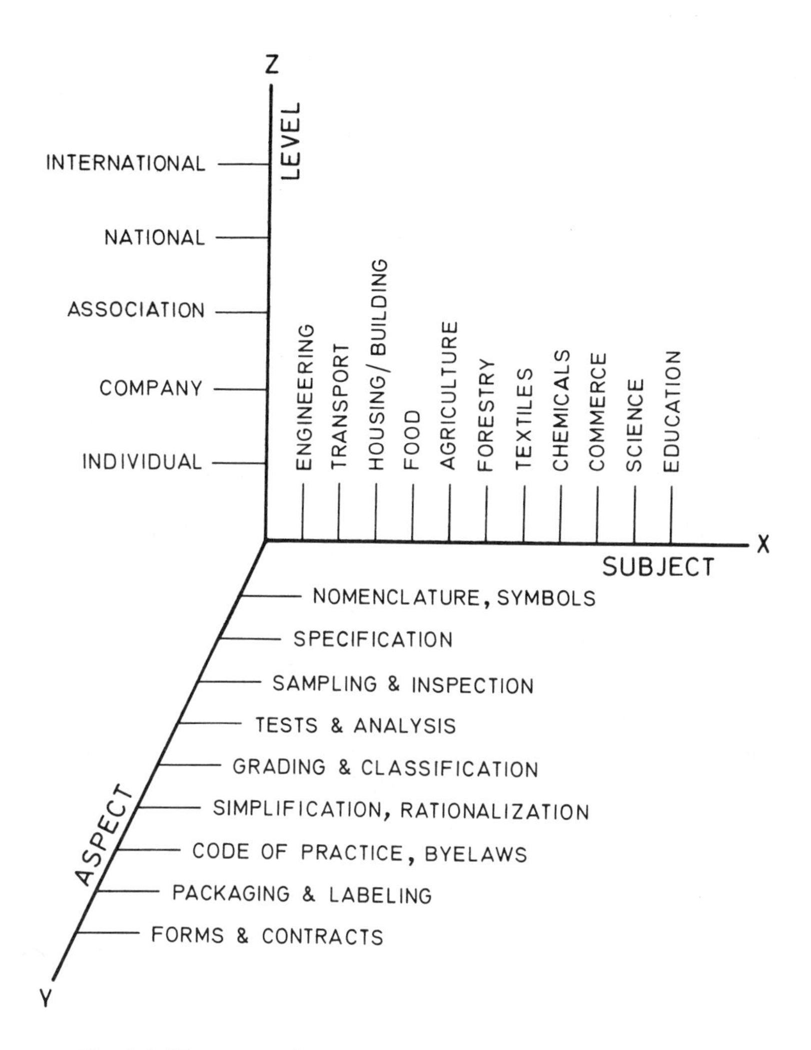

Fig. 3.1 Diagrammatic representation of standardization space.

3.2 *Aspect* Standards differ in form and type depending on the particular aspect of a subject that may be covered. The aspect may be:

(1) a set of nomenclature or definitions of terms for a given field of industry;

(2) a specification for the quality, composition, or performance of a material, an instrument, a machine or a structure;

(3) a method of sampling or inspection to determine conformity to a specified requirement of a large batch of material by inspection of a smaller sample;

(4) a method of test or analysis to evaluate specified characteristics of a material or chemical;

(5) a method of grading and grade definitions for natural products, such as timber, minerals, etc.;

(6) a scheme of simplification or rationalization, that is limitation of variety of sizes, shapes or grades designed to meet most economically the needs of the consumer. This also includes dimensional freezing of component designs to ensure interchangeability;

(7) a set of requirements for packaging and/or labeling;

(8) a set of conditions to be satisfied for the supply and delivery of goods or the rendering of a service;

(9) a code of practice dealing with design, construction, operation, safety, maintenance of a building, an installation, or a machine, conservation or transport of material or goods, model byelaws, etc.;

(10) a model for routine use or a model contract or model agreement; and so on.

There may be other aspects besides those enumerated above, but it will be appreciated that any or all aspects may be applicable to any or all of the "subjects" which we have arranged along the X-axis of the standardization space (*see* Fig. 3.1). Thus let us assign a point on the Y-axis to each of these aspects that may be of interest. We now have an X–Y space in which all discrete points corresponding to each "subject" – "aspect" combination assume significance. For each such point there may be a standard.

3.3 *Level* The third or the Z-dimension is concerned with the operational levels of a standard or, in other words, it defines the domain to which a standard may be applicable. The level is determined by the group of interests creating and using the standard in its day-to-day operations. Thus, the standard may be:

(1) an individual standard, specially laid down by an individual user, builder, a government department or a corporate body to suit his or its specific needs, such as a specification for a piece of furniture, a design for building a house, a dam, or for constructing a bridge, or erecting a factory;

(2) a company or in-plant standard, prepared by common agreement between various departments of a concern for guiding its design, purchase, manufacture and other operations;

(3) an industry or trade standard prepared by an organized group of related interests in a given industry or within a given trade or profession

(sometimes this level is called association level because the activity originates with associations of trade or industry or other professional bodies);

(4) a national standard established after consulting all interests concerned within a country, through a national standards organization which may be a government department, a non-government unit or a quasi-government body;

(5) an international standard or an international recommendation for standardization, resulting from an international agreement between independent sovereign nations having common interests.

On the Z-axis of the "standardization space" points may be allotted to each one of these levels.

3.4 Parenthetically, it may be explained that a government of a country may be responsible for any or all types of standards so far as their level or domain is concerned. For example: (a) most governments have need for individual standards for particular projects or special jobs which are not likely to be repeated; (b) a government-owned production unit or a commercial undertaking will require company standards for its internal needs; (c) purchase specifications issued by a government and used by all or a number of its departments and agencies for procurement purposes would correspond to what are here termed industry, trade or association standards; (d) government may enact laws and promulgate regulations incorporating standards for public safety and health, which, being applicable to the whole nation, would be considered as national standards; (e) some governments have been known to seek the acceptance of their standards by other nations for the purpose of promoting mutual trade; the standards thus framed would fall under the category of international standards.

3.5 In addition to the standardization activity described above at the five different levels, namely the international, national, association, company and individual, another level has evolved during the past two decades, namely the regional, sub-regional, or multi-national group level. This activity takes many different forms and has various objectives in different groups (*see* Chapter 12). Though the activity at this level can and often does serve a useful purpose, yet there is a potential danger that it may lead to an unnecessary vested interest which may impede the international movement. With proper safeguards, however, such a danger can be averted and the regional movement made actually to assist in the promotion of international standardization. Since the reasons for its existence are diversely motivated, it is considered advisable not to include the regional level in the regular scheme of standardization space, but to discuss it separately (*see* Chapter 12).

3.6 This three-dimensional characteristic of standards is diagrammatically illustrated in Fig. 3.1, with X, Y and Z respectively covering the three dimensions described above. In this diagram any standard can be fixed by a discrete point, provided the diagram were made sufficiently extensive and detailed in the X-direction. Certain standards may have to be represented by more than one point if they happen to deal with more than one subject or aspect, or if they happen to be adopted at more than one level.

3.7 A fourth dimension would be necessary, if it were desired to plot all the standards in existence in the world on this diagram as unique points. So far as international standards are concerned, however, the fourth dimension is not required, for, in the strictest sense of the word, there can be only one international standard for a particular subject (a point on the X-axis) and a particular aspect of that subject (a point on the Y-axis). But there can be as many national standards for a given point in the X–Y plane as there are nations, and this would require a fourth dimension to take this multiplicity into account. Similarly, the multiplicity of associations, companies, and individuals will require the fourth dimension, in which each point will particularize an association, company or individual. Riebensahm[3] has made an attempt to depict this pyramidal multiplicity of points in standardization space in a diagrammatic form (*see* Fig. 3.2); but clearly this is merely diagrammatic and special care must be taken in reading it.

3.8 Yet another and fifth attribute of a standard, which need hardly be considered as a fifth and distinct dimension of standardization space, is the character of a standard in respect of its applicability with reference to time. It must be recognized that for several reasons no standard can be regarded as final or absolute. It is often considered advisable that a new standard dealing with a fresh subject may be issued as a tentative standard for a prescribed period of time, for the purpose of being tested in actual practice. During this period the experience of the industry and trade concerned is accumulated and pooled, and in the light of the data collected, the tentative standard is amended and altered to what may be termed, for the sake of convenience, a "finalized" standard. In some organizations, national and others, it has become a normal part of their procedure to issue all new standards as tentative in the first instance. Still, in other cases, it is found that certain organizations would rather delay the issue of their standard or postpone its implementation, until all pre-

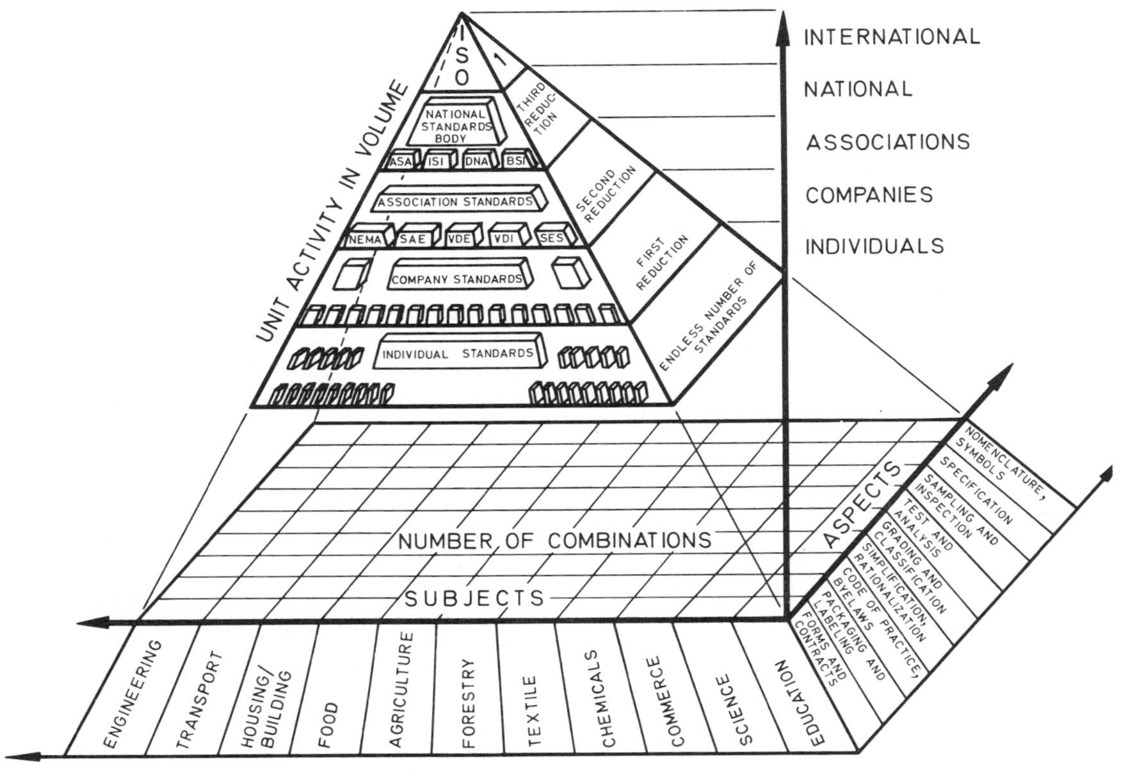

Fig. 3.2 Pyramidal multiplicity of points in standardization space.

liminary stages of eliciting opinions, experimental investigation, trial in practice, etc., have been passed. In any case, whatever the details of procedure may be, the finalized standard is not only expected to reflect the general consensus as to its acceptability, but also its proved practicability in actual use. Thus, the time element need not be introduced here to further complicate the concept of standardization space.

3.9 A word of caution at this stage is called for. The finalized standards, after having duly been declared tentative for a period or after having been otherwise tried out, are by no means to be considered as final, for changes in industrial practice and in prevailing trade conditions with passage of time would require that standards be continuously brought up-to-date. This circumstance imposes the necessity for periodic issue of amendments to the finalized standards and for their major revisions from time to time. Such amendments or revisions may be necessitated by any one or more of the following circumstances:

(1) progress in the art or technique of manufacture or production, leading to improved or altered quality of the product;

(2) demand by consumers for an improved or somewhat different type of product, necessitated by a change in their needs or tastes;

(3) advancement in knowledge or technique, making possible a more precise specification or improved methods of evaluating quality characteristics;

(4) changed conditions in respect of availability of raw materials; for example,

- (a) war-time shortages, making certain materials scarce or unavailable;
- (b) discovery of new deposits or sources of natural raw materials;
- (c) discovery of new types of man-made or synthetic materials;
- (d) reduction in cost of erstwhile more costly, but more desirable materials;

(5) changes in economic conditions, market demands, consumers' tastes, pattern of trade and other factors.

3.10 Thus it will be seen that a standard is a live organism and, in common with other forms of life, it is subject to continual change and adaptation so as to remain in harmony with its environments and maintain its utility to the community it serves.

3.11 From the above enunciation of standardization space, it will be clear that the field of standardization today has become much more extensive than it was, let us say, about half a century ago, when it first began to develop as a part of engineering activity aimed chiefly at simplification and limitation of variety. Multiplicity of variables along the aspect-axis of standardization space shows the growth of this complexity during the past few decades, and there is no indication that this growth has reached a saturation point. On the contrary, judging from the rate of progress of present-day science and technology, there is every reason to believe that a great many new variables will continue to be added to the subject as well as the aspect axis of standardization space, although we may take it that the level-axis may remain more or less as now conceived.

3.12 It is obvious that standardization space as described above cannot be taken as a mathematical space of either continuous or discrete variables; it is to be regarded merely as a convenient model to illustrate the three important attributes of standards.

AIMS

3.13 It will now be evident that there can be no hard and fast division line between the aims of standardization at, say, the national level as distinct from the aims at, say, the association level. Ultimate aims of standardization may be enumerated as being applicable to all levels, yet a distinct functional character may perhaps be ascribed to standards at each specific level. It is also important to consider standardization as an activity, and to distinguish clearly between the aims of this activity on the one hand and on the other the methods and means employed for achieving the aims, and the resulting effects obtained by these means.

3.14 Aims of standardization in general, applicable to all levels, individually and collectively, may thus be listed as follows:

(1) To achieve maximum overall economy in terms of:
- (a) cost,
- (b) human effort and
- (c) conservation of essential materials as opposed to more readily available materials.

This involves judicious choice of raw materials and the adoption of production and handling practices known or expected to be most economical.

(2) To ensure maximum convenience in use.

It is this objective of standardization which leads to simplification, rationalization, interchangeability of parts and freezing of dimensions of components. Increased productivity, elimination of unnecessary waste, and reduction of inventories are the consequential benefits.

(3) To adopt the best possible solutions to recurring problems consistent with (1) and (2) above and taking into account all the available scientific knowledge and up-to-date technological developments.

This objective of standardization is aimed at facilitating design procedures and guiding the formulation of research and development programmes. It involves standardization of basic terminologies, codes of practice, model forms of contract and so on.

(4) To define requisite levels of quality in such a manner that practical evaluation of quality and its attainment are consistent with (1) and (2) above.

This aim leads to the standardization of sampling procedures, test methods, grading schemes and quality specifications in general.

Aims under (2), (3) and (4) above have the effect of providing a common medium of communication between contracting parties, resulting in the elimination of disputes, and if they arise, in their speedy settlement, thus contributing to the achievement of the objective under (1).

3.15 Coming now to specific aims of standardization at different levels, aims at the international level may now be defined simply in terms of the ISO objectives, namely, "to facilitate international exchange of goods and services, and to develop mutual cooperation in the spheres of intellectual, scientific, technological and economic activity."

3.16 As a means to this end, effort is to be concentrated on the coordination of national viewpoints represented in national standards or otherwise, with a view to evolving ISO or other international recommendations and standards. To make such an effort successful within a reasonable period of time, due consideration has to be given to the following points.

(1) Every attempt should be made to discover a solution which represents the closest approach to the ideal solution calculated to achieve the ultimate aims of standardization, divergences being permitted only to the extent dictated by practical consideration and not by the established national practices.

(2) Failing achievement of coordination on the basis mentioned above, such solutions may be adopted as would represent the least overall deviation from existing national practices, when considered on the basis of

world-wide economy, consistent with the capacity of developing countries and smaller nations to make necessary adjustment.

(3) Proper emphasis should be placed on technological and economic needs, and not so much on the means used for describing the needs, such as language, terminology, system of units of measurement, and so on.

All participating countries should voluntarily agree to the points mentioned above as the policy of ISO and of other international bodies when dealing with standards. It will be useful if the countries concerned would explicitly instruct their delegates to be guided by this policy during their deliberations on technical and other matters concerned with the coordination effort.

3.17 Aims at the national level may be considered to be the same as those at the international level, if for the words "international" and "world-wide," we read "national" and "nation-wide;" and for "national" we read "association" or "industry" or "company," as may be applicable. One further feature, however, enters at this stage, namely the requirement that national standards should, as far as possible, be in line with the international recommendations and standards. Where relevant international recommendations or standards are not available, their probable future trends and existing international practice or usage should be borne in mind in drawing up national standards. Such considerations are necessary to ensure the achievement of the aims of international standardization in respect of existing standards and also to facilitate future coordination effort aimed at harmonizing new national standards at the international level.

3.18 Aims at the association level are again similar to those at the national level, except that the national interests are now replaced by association interests and international interests by national interests. In addition to giving due consideration to national standards, it is important that the international standards and recommendations be kept in view, wherever they are available.

3.19 Aims at the company level (and the individual level) are similar to the above except that they are one step removed from the corresponding preceding levels. In either case, it is important to give due consideration to the international, national and association standards (and in the case of individual standards, to company level standards, if relevant, keeping in view the needs for ultimate coordination at the next higher level immediately above).

3.20 For convenient reference and comparative study, a brief statement, summarizing the aims and means discussed above are recapitulated in Table 3.1.

TABLE 3.1

BRIEF STATEMENT OF AIMS OF STANDARDIZATION

(*see* 3.20)

Note: For detailed statements please refer to text paragraphs of corresponding numbers.

Ultimate aims (*see* 3.14)		(1) Economy of cost, human effort and essential materials;	(2) Convenience of use;	(3) Solutions to recurring problems; and	(4) Quality definitions and evaluation.
Specific level aims					
International (*see* 3.15 *and* 3.16	Aim	To facilitate international exchange of goods and services and to develop mutual cooperation in the spheres of intellectual, scientific, technological and economic activity.			
	Means	(1) Prefer ideal solution to national practices;	(2) Failing (1), coordinate national standards on basis of worldwide economy; and	(3) Emphasize technological and economic needs	
National (*see* 3.17)	Aim	To facilitate national exchange of goods and services, to promote international standardization and to develop mutual cooperation in the spheres of intellectual, scientific, technological and economic activity.			
	Means	(1) Prefer ideal solution to industry practices;	(2) Failing (1), coordinate association standards on basis of nationwide economy;	(3) Emphasize technological and economic needs; and	(4) Align with international standards.

Level					
Association (*see* 3.18)	Aim	To facilitate industry-wide exchange of goods and services, to promote national standardization and to develop mutual cooperation in the spheres of technological and economic activity.			
	Means	(1) Prefer ideal solution to company practices;	(2) Failing (1), co-ordinate company standards on basis of industry-wide economy;	(3) Emphasize technological and economic needs; and	(4) Align with national and international standards.
Company (*see* 3.19)	Aim	To facilitate company-wide interchangeability of materials and to coordinate all operations, to promote industry-wide and national standardization and to develop mutual cooperation in the spheres of technological and economic activity.			
	Means	(1) Prefer ideal solution to individual practices;	(2) Failing (1), co-ordinate individual standards on basis of company-wide economy;	(3) Emphasize technological and economic needs; and	(4) Align with association, national and international standards.
Individual (*see* 3.19)	Aim	To facilitate meeting the needs of the project in hand and promote the use of existing standards of various levels as may be appropriate.			
	Means	(1) Prefer ideal solution to existing practices;	(2) Failing (1), adopt most economic solution;	(3) Emphasize technological and economic needs; and	(4) Align with company, association, national and international standards.

COMMUNITY OF INTERESTS

3.21 Having discussed the aims of standardization and the means to be employed at various levels to achieve them, it may now be worthwhile to examine the detailed manner in which standards operate to help achieve these goals and assist those concerned with using them. While all sections of society are concerned with standards in one way or the other, the community of interests, for convenience of discussion, may be divided into the following four groups:

(1) *Producers,* including large scale, small scale and cottage industry producers; as also processors, assemblers, constructors, erectors, servicing agencies, contractors and the like.

(2) *Traders,* including wholesale and retail merchants as also exporters, importers, commission agents and so on.

(3) *Consumers,* including users, whether they are industrial consumers using goods and services for further production, or the ultimate consumers and users of goods and services like the housewife.

(4) *Technologists,* including scientists and engineers concerned with research and investigation connected with standardization, as also technologists and others engaged in testing and certification of goods and services.

At first sight, these groups would appear to be quite distinct, but it must be recognized that the interests of a given party might overlap with others. For example, a producer would more often than not be a consumer of the products he buys for his own use; almost everyone in any group would be a consumer of articles of everyday use; and so on. Nevertheless, for operation of standards, the functional interests of these groups operate as distinct from one another and are served through standardization in a characteristic manner.

3.22 Among all the groups, the interest of the producer group perhaps is the most highly developed. In every department of his activity, standards play an important role. Briefly stated, for the producer, standards:

(1) provide short cuts to design procedures by furnishing ready-made and generally accepted solutions to recurring problems;

(2) make possible longer production runs with fewer changes in production line and reduce tooling and set-up time;

(3) streamline inspection and testing and enable quality control procedures to reduce rejections and re-working;

(4) enable the procurement of raw materials, interchangeable parts and components from ready stocks without loss of time;

(5) reduce stocks and inventory of materials, parts and end-products;

(6) simplify servicing and maintenance;

(7) facilitate training of staff and operators;

(8) reduce overheads on clerical and administrative work;

(9) facilitate marketing and winning of consumer confidence; and

(10) as a consequence of all these factors, lead to higher productivity in every department, which means reduced costs, lower prices, higher sales and greater profitability.

3.23 So far as the trader interest is concerned, some of the benefits derived by the producers are equally shared by him; for example:

(3) streamlining inspection and testing;
(4) easy procurement;
(5) reduced investment in inventory;
(6) simplification of servicing;
(7) facility in training;
(8) low overheads;
(9) facility in market expansion; and
(10) greater turn-over and higher profit on investment.

In addition, the trader enjoys certain other benefits; for example:

(11) standards provide authoritative documents on the basis of which procurement of stocks and their sales can be readily arranged to meet clearly defined customer needs; and

(12) standards enable all parties concerned to avoid, or reduce to a minimum, the possibility of misunderstanding leading to unnecessary trade disputes and arbitration proceedings – a great asset, particularly in international trade.

3.24 To the consumer and the user, standards also operate to bring several benefits, some of which are common to the producer and trader groups. In this respect, the industrial consumer is particularly interested in the following items from among those enumerated above:

(3) streamlining inspection and testing;
(4) easy procurement;
(5) reduced investment in inventory;
(6) simplification of servicing;
(7) facility in training;

(8) low overheads;
(10) reduced prices;
(11) basis of transactions; and
(12) reduction of disputes.

A great benefit to the consumer is that of the low price resulting from the economical operations of the producer and the middleman, and of course the quality and reliability of the goods and services he purchases.

3.25 The ultimate or everyday consumer – the common man – being a small-scale buyer of his daily needs has been somewhat limited in his ability to profit fully from the benefits conferred by standards. Perhaps it would be more accurate to say that he was so limited, until he began to organize himself into consumer cooperatives and consumer testing organizations and until certification marks schemes became more generally in vogue. But we shall revert to this subject more in detail later (*see* Chapters 15-18). Suffice it to say here that though he may be somewhat limited in deriving the full scope of the benefits as compared to his counterpart, the industrial consumer, he nevertheless shares with the latter the possibility of deriving the same or similar benefits.

3.26 As regards the technologist interest, it must be recognized that this is more of a service group. Though he does not derive any personal benefits from standards, yet they do assist him professionally in a positive manner to serve the community more effectively. For example, to the technologist, standards:

(13) furnish an authoritative basis of judgement to facilitate comparable and reproducible results to be obtained in evaluating goods and services;

(14) assist in meeting difficult performance specifications and other unusual requirements of special items;

(15) provide accurately defined tools, apparatus and equipment and generally agreed procedures to be used and followed in technical evaluations;

(16) furnish generally agreed and acceptable solution of recurring problems and enable him to concentrate more effectively on vital and fundamental issues of original character concerning design, development and research; and

(17) furnish starting-points for research and development for further improvement of goods and services.

3.27 It will thus be noted that though there are some specific benefits to be derived by individual groups of interests, a good many benefits are common to several groups. As indicated earlier, the group interests are not so very distinct either – they do overlap a great deal. It is in this background that the whole community of interests represented by all sections of society becomes primarily involved in the standardization movement.

REFERENCES TO CHAPTER 3

1. Verman, Lal C.: Standardization – its principles and development in the world and the ECAFE region. *ISI Bulletin.* 1952, vol 4, p. 3-9.
2. Verman, Lal C.: Standardization in a developing economy. *Industrialization and Productivity Bulletin No 7* (United Nations Department of Economic and Social Affairs), New York, 1964, p. 37-51.
3. Riebensahm, Hans E. (UNTAO expert): *Report on initiation and promotion of company standardization among Indian industries.* Indian Standards Institution, New Delhi, 1966, p. 31.
4. See reference 2 of Chapters 7 and 8, pp. 92 and 103.

Chapter 4
Subjects

4.0 Of all the axes of standardization space discussed in the previous chapter, the subject or X-axis is the most populous. Almost any material, process or action having an economic value or even cultural value for which a standard can be usefully written has a legitimate claim to be assigned a point on the X-axis. It is for this reason that the large multiplicity of points belonging to the different branches of human activity make it a formidable task to devise a scheme for subdividing this axis in a reasonably sound and yet comprehensive manner. Besides all the branches of engineering and technology, we have to deal with agriculture, chemistry, medicine, biology, public health, documentation, linguistics, management and even governmental byelaws having technical implications. However, since standardization space is merely a scheme for graphic representation, it is hardly necessary to divide the X-axis any more formally than simply to recognize that any subject on which a standard has been issued or has been decided to be prepared may be assigned a point on it. Nevertheless, for restricted fields of endeavour, the X-axis could bc rationally demarcated for groups of subjects, as for example, electric cables, pesticides, household appliances and so on.

4.1 But while a formal division or demarcation of the subject axis may not always be considered necessary or feasible, it is important in actual practice to subdivide the work of standardization within a given organization according to some system or other. This need arises especially in national standards institutions and may also arise in an international body like the International Organization for Standardization (ISO), all of which are concerned with subjects covering the full spectrum of national and international economies. In the case of associations and company standards organizations, the subdivision of the subjects and their allocation to specialist staff is a much simpler matter, for the range of subjects is bound to be limited to the relevant industry or the profession concerned in the case of an association, and to the product or products dealt with by a company. In the case of individual level standards, the question of subdivision of the X-axis hardly arises, since in this case it is always a matter of dealing with very few items at a time or, let us rather say, one-subject-one-aspect combination at any given time.

GROUPING OF SUBJECTS

4.2 The task of subdividing the subjects and allocating them to teams of specialists in a national or international standards body corresponds somewhat to that of organizing a large university where almost all disciplines of human knowledge have to be catered to, and where a separate department may have to be created for each discipline. But the resources available to standards bodies are insignificant compared to those of the universities; nor is the volume of work involved in each specialized subject comparably as large, though the degree of specialization may be equally great. This means that, depending upon the size of the standards organization and the pattern of economy of the country, several subjects could be lumped together in a given division or department, staffed with a limited number of specialists. This becomes quite convenient and feasible because more highly specialized knowledge and experience for standardization work is normally contributed by members of technical committees, who represent the different economic group interests (*see* para 3.21).

4.3 The staff members of the standards organization, though specialists in one field of technology or another, serve mainly as committee secretaries and furnish the background material and services for decisions which are formally taken mainly by the committees. In this way, a mechanical engineer, for example, may be able to take care of committees on building construction and electrotechnology in addition to those in his own field of work. But as the work load grows and additional manpower becomes necessary, new staff may have to be appointed to look after additional lines of work. When the quantum of work increases even further, it may become justified to create independent divisions or departments to take charge of different disciplines. But at no time can it be envisaged that the staff of a standards organization will be complete enough in the sense that for each speciality dealt with there would be at least one staff member who is a specialist. This is so because the number of subjects dealt with is limitless and dependence on committee members for specialized knowledge and experience is the only economical, verily sensible, approach.

4.4 With a view to facilitating internal distribution of subjects and their proper classification, certain National Standards Bodies have adopted the practice of dividing all subjects of interest into domains and denoting each domain by a letter of the alphabet. These bodies include the Standards Association of Australia,[1] the Canadian Standards Association,[2] the Japanese Industry Standards Committee[3] and the American National Standards Institute.[4] In the absence of an international agreement, the practice that has come to prevail in this regard is not quite uniform among the various countries. Certain letters have come to denote uniformly the same subjects in all the four countries, for example A, B, C and D denote civil, mechanical, electrical and automotive engineering respectively, and G and H ferrous and non-ferrous metals and metallurgy. On the other hand, other designations differ considerably, for example the letter S in Australia means domestic economy, in Canada structures, in Japan domestic wares, and in the USA accoustics, vibration, mechanical shock and sound recording. Such letter designations are also made as part of the number designations of corresponding standards (*see also* Chapter 22, para 22.5). Most other standards bodies do not find it necessary or convenient to designate the subject domains in this manner. In view of the new developments taking place daily and new domains of subjects being constantly created, it would be evident that any system of classification would require frequent revision. Under the circumstances it is for consideration whether the system is to be regarded as indispensable, particularly for new standards bodies.

4.5 In any country, particularly if it is small or developing, a standards organization would generally start in a small way with a limited staff and a limited number of subjects to be dealt with. The subjects would naturally be chosen with an eye on the economy of the country, depending upon what sectors need be given high priority, in which standardization could be most useful and where it was seriously lacking. Thus one or perhaps two technical divisions, depending on the estimated work load, may be created in the first instance. In creating these divisions, the allotment of the subjects may be made purely arbitrarily, depending upon the specialities and interests of the available staff. As the demand for new standards increases and diversification of subjects begins to take place, new divisions may be created with the help of such existing staff as could be spared, together with additional staff specializing in the new lines of work.

4.6 In this context it would be of interest to consider the evolution of the divisional structure of one organization in a developing country. The organization in question, the Indian Standards Institution (ISI), was started in 1947, when it initiated work in the Textile and Engineering Divisions and allocated work on all the subjects on hand to one or the other of the two divisions. Thus the Textile Division dealt with chemical subjects as well, and the Engineering Division with building construction, metals and so on, in addition of course to all other branches of engineering. A year later, a Chemical Division was created out of the Textile Division and five years later the Building Division was bifurcated from the Engineering Division. The process went on in this fashion and continues even today. During the course of this process, as specialization increased, the names of certain divisions were also changed. Thus Engineering Division became Mechanical Engineering Division and Building Division was renamed Civil Engineering Division. In Fig. 4.1 is illustrated the growth of the existing nine major divisions of ISI that have grown out of the original two divisions through such a progressive bifurcation process. Even now there are fairly large sections handling important specialized work within some of the existing divisions, which may one day be decided to be separated into independent divisions. The multi-purpose project section in the Civil Engineering Division and the surgical instruments section of the Consumer Products Division are but two such examples.

ALLOCATION OF SUBJECTS

4.7 Even when a number of fairly clear-cut discipline-wise divisions has been created within a standards organization, questions arise as to which division could a given subject be allotted. For example, should household refrigerators go to the mechanical engineering division or to the electrical, or perhaps to the consumer products? Do paints belong to the chemical division from the production point of view or does the building division, representing a large user interest, have a claim on the subject? Such problems arise more frequently than may appear at first sight and decisions have to be made in a rational manner, the objective always being the expeditious disposal of the work on hand. Some of the principles that

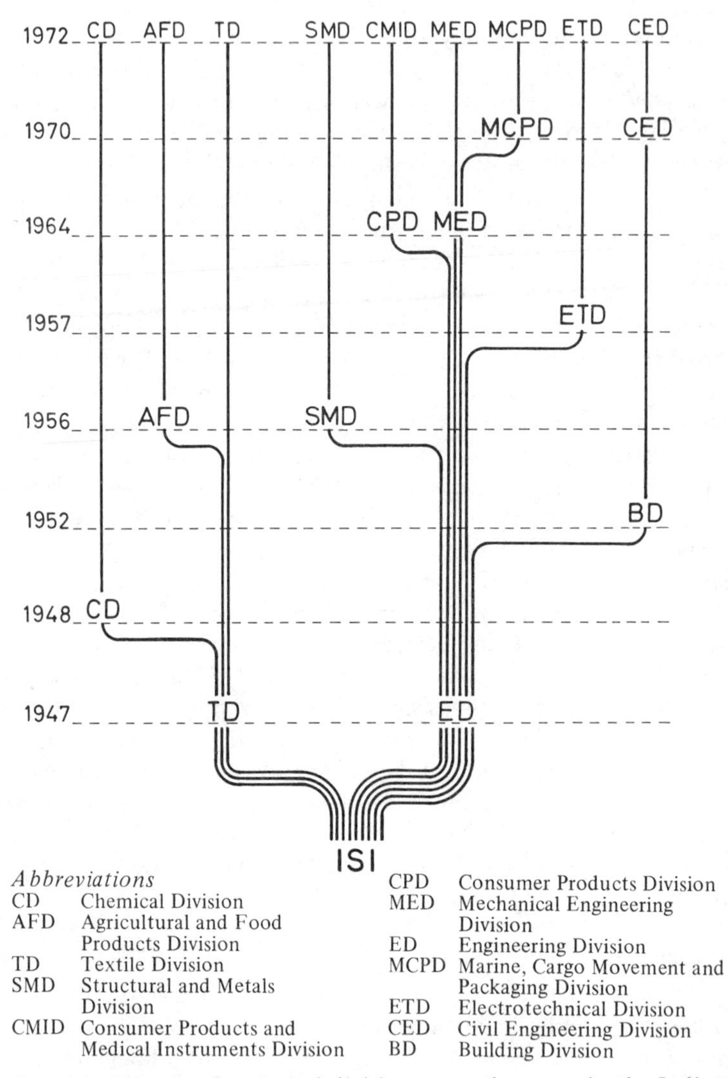

Abbreviations

CD	Chemical Division
AFD	Agricultural and Food Products Division
TD	Textile Division
SMD	Structural and Metals Division
CMID	Consumer Products and Medical Instruments Division
CPD	Consumer Products Division
MED	Mechanical Engineering Division
ED	Engineering Division
MCPD	Marine, Cargo Movement and Packaging Division
ETD	Electrotechnical Division
CED	Civil Engineering Division
BD	Building Division

Fig. 4.1 Growth of technical divisions over the years in the Indian Standards Institution. (*Courtesy*: ISI)

are recommended for guidance in taking such decisions may be enunciated as follows:

(1) Standards for major raw materials or products required more or less exclusively for one industry, which are not normally required by any other

industrial group, are best handled by the division or section dealing with the major consumer industry, in collaboration, if need be, with the division concerned with the producer industry.

Example: Raw cotton, wool, silk and other natural fibrous materials would thus be assigned to the division dealing with textiles and not the one dealing with agricultural products. Similarly, in the case of man-made fibres, there may be some question whether to allot them to the textile division or to the one dealing with plastics and chemicals. In the same way, fertilizers, pesticides and farm implements would perhaps better be dealt with by the division concerned with agriculture rather than the one with chemicals and mechanical engineering respectively.

(2) Raw materials and processed goods not used exclusively by one single industrial group but by several diverse industries should be dealt with by the division concerned with the producer industry.

Example: Electric motors would better be dealt with by the division dealing with electrical products rather than any other engineering division which represents a consumer group. Similarly, bolts and nuts used by all sections of industry had better be assigned to the division dealing with mechanical engineering.

(3) It is necessary to make a distinction between raw materials on the one hand and goods processed or manufactured from these raw materials on the other. The dividing line may be drawn on the basis of whether a given raw material during its conversion undergoes a major industrial operation or whether the operation involved is one of minor character. In the former case, the allotment of the item should be made to the major industrial group concerned with processing or manufacture, but in the latter case, it may be assigned to the division dealing with the producer group of enterprises.

Example: Leather and hides and skins could thus be assigned to the division dealing with agricultural and animal products, while shoes made from leather should go to consumer products division.

(4) In the case of items which are clearly unrelated to any of the major industries covered by the different divisions within an organization, it will be necessary to make an *ad hoc* decision, because no set of guiding principles could be enunciated to take care of each and every case. In such cases, choice may sometimes be made on the basis of availability of a staff member in one division or another who may be more knowledgeable on the item than others, or who possesses a particular enthusiasm for handling it.

Example: Gymnastic equipment, banking and trading documents and other such odd items of this character may come under this category.

(5) In making decisions on the above basis, several borderline cases will come up which could offer two equally justifiable solutions. In such cases, the decision may be based on attempting to maintain a balance between the work load among various divisions of the organization.

Example: A given item may justifiably be assigned either to the division dealing with chemicals or to one dealing with agriculture. If the accumulated work load of the chemical division is less than that of the agricultural division, the former may be allotted the item.

(6) In no case should divided responsibility be tolerated and the subject assigned partly to one division and partly to another.

Example: There is sometimes a tendency to divide responsibility either aspect-wise or level-wise. For example, while specifications for a material may be looked after by one division, methods of tests may be assigned to another; or else it might happen that the national standards may be prepared by one division and the international work may be assigned to another. This sort of division of labour can lead to a great deal of unnecessary duplication of efforts and waste of time, and should always be avoided. Aspect-wise splitting of responsibility within committee structure may be found useful under certain exceptional circumstances, but level-wise splitting in committee structure is not at all advisable.

(7) In case a subject does fall within the scope of two or more divisions and it is not considered advisable to allot it exclusively to any one of these divisions, the possibility of creating a joint-responsibility technical committee should be explored. In this case the committee would be jointly appointed by and be responsible to the two or more divisions concerned, and its secretariat could be assigned to the directorate staff of one of these divisions.

Example: Household wiring code may have to be made the joint responsibility of both the building division and electrotechnical division.

4.8 There may be considerable divergence of opinion on the validity of the principles enunciated above. But according to experience in certain developing countries like India and Iran, they have been furnishing very helpful guidance. They are, however, flexible enough to suit many actual situations. Nevertheless, each standards body would do well to consider them critically and evolve a set of similar principles for its own individual guidance.

4.9 However efficiently the division of work and allotment of subjects may be carried out, there will always remain the need for every division within an organization to work in close collaboration with all the other divisions. There are so many cross-linkages and inter-connections between one subject and another that an effective organization would perhaps require the drafts of every division to be vetted by every other division, before they are circulated widely to outside interests. But this is not always feasible. The best that can be done to avoid contradictions and divergences between different standards of the same organization is perhaps the reliance on mutual consultation at the initial stage of drafting and circulation of drafts among the staff at the time of wide circulation (*see also* Chapter 10).

4.10 Situations arise in developing countries, when subjects have to be taken up for standardization which have not been thought of being covered by standards in more advanced countries of the world. Thus, no guidance or assistance can be derived from previous work in the field and the main reliance has to be placed upon indigenous knowledge and experience. Thus Iran had to develop its own specifications on raisins, apricots and other dried fruit – an important class of export item. India had to prepare original standards to regulate its multi-purpose river valley projects, which comprehensively cover the whole field, from the initial survey stage through construction up to operation and maintenance. Philippines standards on manila fibres, Thai standards on fish sauce, and scores of other examples could be cited in this category. All such subjects have to be found an appropriate place in the otherwise conventionally organized divisional structure.

4.11 Another class of subjects for which the developing countries have to break new ground are those for which standards in industrially advanced countries may have existed for many years and experience of decades may have accumulated, but certain conditions of manufacture or use in developing countries may be so different as to require substantial variations. Such was the case in regard to steel and cement composition specifications in India, where the raw materials available for their production contain certain unavoidable impurities which cause overseas standards to become inapplicable without appropriate amendments. This subject of how developing countries may adopt a systematic procedure to take advantage of the existing international and national overseas standards in pursuing their work will be dealt with in more detail later (*see* Chapter 19).

PRIORITIES

4.12 It may also be pertinent to consider here the question of priorities which may be assigned to various groups of subjects. Obviously, the first and foremost attention should be paid to the system of weights and measures, which forms the basis of all standards. Prevailing conditions differ in different countries in this regard, but by and large it may be taken for granted that, if in a country standardization movement is just being initiated, then it would be its weights and measures situation which would need immediate attention. Most newly independent countries would have borrowed the system of weights and measures from their colonial rulers. Thus, in some countries, the foot-pound system might have been introduced, while in others, the metric system. Yet more often than not, side by side with either of these systems, there would continue to exist the use of indigenous units of weights and measures. Except perhaps in a few countries, therefore, there is always the need for introducing some sort of regulatory measures so that the desirable uniformity in weights and measures is attained at an early date. Without it, the programme of standardization will find it very difficult to become effective (*see also* Chapter 13).[5]

4.13 Even in countries where the metric system might have been introduced, it is often found that its implementation is not so thorough as might be desired. One of the problems to be faced may be that of a lack of strong administration for effectively enforcing the legislation for regularly inspecting and marking, at specified intervals, the weights and measures actually used in commerce and industry. Sometimes, this situation is ascribable to lack of funds and administrative backing at high levels, but on other occasions it may be due to non-availability of trained personnel for deployment. In either of these cases it should be the concern of the standards organization of the country to move their governments to strengthen the administration adequately. For it is only thus that the standards movement itself can be strengthened.

4.14 In other countries where the metric system has not been legally adopted, it will be extremely advisable, first of all, to move for legalization of the metric system on an exclusive basis and then to organize its administration. This and other connected problems of introduction of the metric system and creation of administrative and training apparatus for its implementation in one of the developing countries has been exhaustively dealt with in a recent book entitled *Metric Change in India* published by the Indian Standards Institution in 1970[5] (*see also* Chapter 13).

4.15 Next, in order of priority, would follow the basic standards (*see*

Chapter 5, paras 5.2, 5.6) and then the subjects considered to be important in the light of the economic structure of the country and its immediate plans of development. For example, in a predominantly agricultural country, the agricultural produce and agricultural inputs should be given the highest priority, next to weights and measures; whereas in a country aspiring to become industrially developed as well as agriculturally self-sufficient, priorities will perhaps be somewhat modified. In this latter case, minerals and other raw materials required for industrial development will have also to be considered for high priority assignment.

4.16 In practice it has been found that the assignment of priorities to various subjects in standardization work can only be partially helpful. This is so because the priorities can only guide the central standards organization to take up work in a given order, but its successful completion depends on so many other factors beyond the direct control of a standards organization, for example, consideration of drafts by various outside interests concerned and the need for arriving at a consensus of viewpoint among all the participants concerned. This part of the process somehow cannot be reduced to priority handling. It has to take its own course. But the priorities originally assigned at the central office level do help to push the work forward to some extent and exert their own influence, though such influence is somewhat limited.

4.17 In order to keep a record of the subjects, their priorities and progress, it would be well to maintain a card index of all subjects arranged in one alphabetical series, with different coloured cards being used for different priorities. For example, green cards may be used for the highest priority, red ones for medium and yellow for low priority. In maintaining these cards, cross-reference cards should be liberally used if the index is to be really useful. For example, when the subject of "Machine Screws" is taken up, a master card marked "Screws, Machine" may be made out, but simultaneously several other cards should also be introduced, such as for "Machine Screws," "Fasteners," "Threaded Fasteners," "Bolts," etc. On each of these supplementary cards, which may all be in one colour, say white, reference to the master card should be given by some sort of indication, say, "*See* Screws, Machine." All relevant data and progress may, however, be recorded on the master card only.

REFERENCES TO CHAPTER 4

1. *Annual list of SAA publications.* Standards Association of Australia, Sydney, 1971, p. 287.
2. *List of publications.* Canadian Standards Association, Rexdale (Ontario), 1971, p. 38.
3. *JIS Yearbook 1971.* Japanese Standards Association, Tokyo, p. 181.
4. *Catalog 1970.* American National Standards Institute, New York, p. 128.
5. Verman, Lal C. and Kaul, Jainath: *Metric change in India.* Indian Standards Institution, New Delhi, 1970, pp. 529, xxvii.

Chapter 5

Aspects

5.0 The concept of "aspect of a standard" was introduced in Chapter 3 when discussing standardization space. Aspect was described as one of the several facets of a subject matter which may be dealt with in a standard. Thus, an aspect may be:

(1) nomenclature or a set of symbols,
(2) specification,
(3) method of sampling or inspection,
(4) method of test or analysis,
(5) method of grading or classification,
(6) scheme of simplification or rationalization,
(7) packaging and labeling requirements,
(8) supply and delivery conditions,
(9) code of practice including model byelaws, or
(10) forms including contract forms.

As mentioned earlier, this is not a complete list of all the possible aspects which a standard may cover. With the progress of technology and continued ingress of standardization into fresh fields of endeavour, other aspects of standards are bound to be added to the list. Some such possibilities are indicated towards the end of this chapter.

5.1 It is always possible and sometimes even desirable to deal with more than one aspect of a subject in a single standard, particularly when it is a specification of a material or a product. In such cases, methods of sampling and testing may be included as an integral part of a specification, which may also cover the grading and classification scheme and in some cases dimensional simplification as well. Such broad-based standards provide a most convenient method of prescribing all the essential requirements of a given material or product and are considered most useful from the point of view of the user of standards. But quite often it is found advisable to deal, in a given standard, with only one aspect of a subject in depth, such as in the case of a set of nomenclature, a code of practice, model byelaws and so on.

BASIC STANDARDS

5.2 Nomenclature type of standards cover glossaries and definitions of terms and serve to introduce uniformity and cohesion in all other types of standards. Sometimes, short glossaries of terms with their definitions are included in a specification itself. But whenever a separate glossary on a given subject exists in a standard form, a mere reference to it in the specification is considered adequate. As regards the basic principles to be followed in writing terminology standards, some information has already been given in Chapter 2.

5.3 Terminological standards are included as an item in the aspect list given above merely as an important example of a larger class of standards which deal with the basic issues affecting all other standards. Examples of other basic standards in this class are:

(1) system of units,
(2) symbols – letter symbols and graphic symbols,
(3) preferred numbers, and
(4) style manual for drafting standards.

5.4 Among these examples, the system of weights and measures is perhaps the most important and a separate chapter has been devoted to it (Chapter 13). The question has also been referred to in Chapter 4 and discussed as a "subject" for receiving the highest priority (*see* paras 4.12–4.14). It is mentioned here again merely as an item which is to be considered as basic to most standards like terminology; though in the strictest sense, unlike terminology, it cannot be treated as an "aspect" of a subject. It truly represents a whole domain of subjects, which may be covered by the science and practice of "metrology." Letter symbols and graphic symbols have been dealt with in many standards – national, international and others. They are of tremendous assistance in abbreviating many a technical narrative which would otherwise become most unwieldy. For example, how much simpler it is to write:

$$a{:}b{::}c{:}d$$

than to say at length that:

> the quantity *a* bears the same relation
> to quantity *b* as quantity *c* does to quantity *d*

Similarly, with the help of graphic symbols, how simple it is to indicate the location of an electric switch on a drawing by simply using a graphic symbol. Such symbols (*see* Fig. 5.1) have become common in many fields of technology today, but it is absolutely necessary to unify them on a global basis so that they serve as a truly universal language without words.

5.5 Preferred numbers which constitute guidelines when standardizing a series of products varying in size, capacity or some other parameter are also discussed elsewhere (*see* Chapter 23). Style manuals or guides for drafting standards are a type of basic standards which are indispensable to any standards organization at any level. Some of the more important ones are included in the bibliography at the end of this chapter[1–6], and the subject as such is discussed in Chapter 22. There may be other kinds of documents which could also serve as basic standards such as those dealing with documentation matters generally.

5.6 It is important that a standards organization gives the highest of priority to working out its basic standards at the earliest stages of its formation, if possible, even before subjects of economic importance are taken up, or at least simultaneously with such subjects. Once the basic standards become available and the staff and committee members become familiar with their use, it will be found that much time would be saved in making many routine decisions during the actual drafting of standards and processing them through the committees.

SPECIFICATIONS

5.7 The specification type of standards are the most common for they cover the requirements of a material or a product in a comprehensive manner and provide the user full guidance in producing, processing, selling as well as buying and using. The requirements covered may be of three types – obligatory, optional or recommendatory, and informative. Obligatory requirements help the producer in incorporating into the product

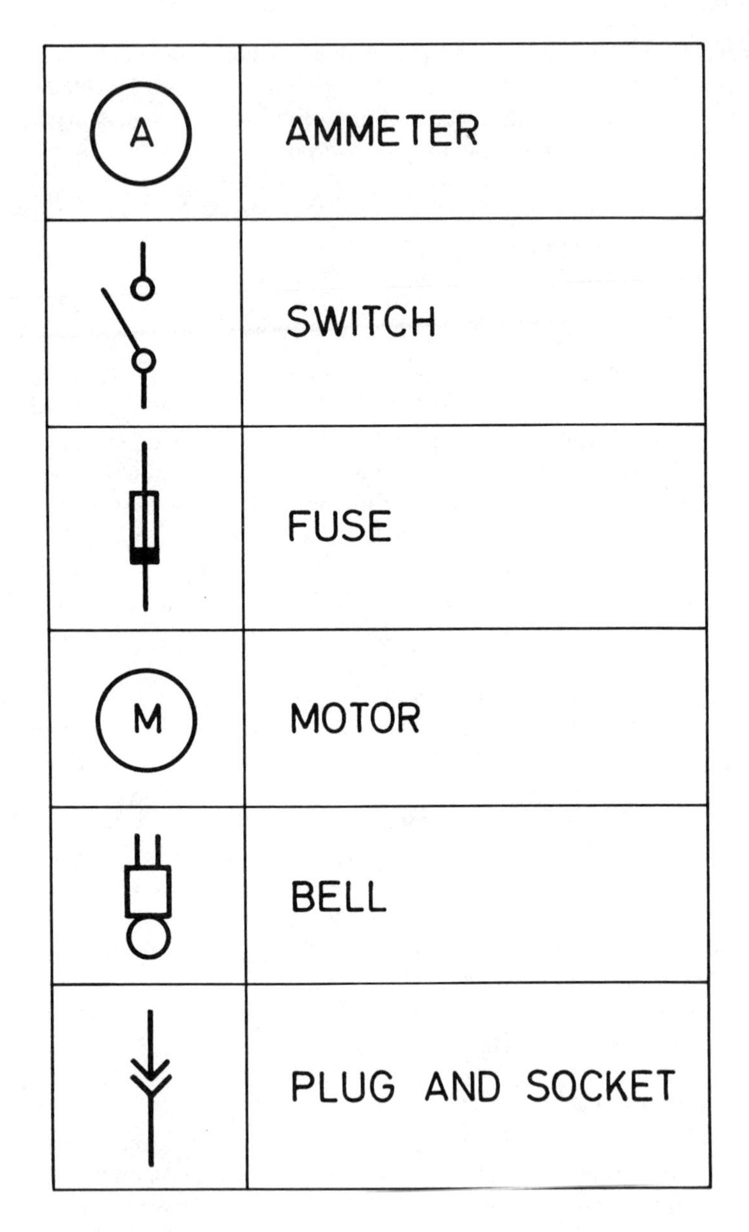

Fig. 5.1 An illustrative set of IEC graphic symbols for electro-technology.

the basic essential qualities which will ensure its usefulness and satisfy the customers' needs. Optional or recommendatory requirements help improve further the serviceability of a product and sometimes help meet the peculiar needs of a particular type of customer having special demands to meet. Informative requirements are more in the nature of additional information which may help to connect up the information given in a specification with other sources of information, thus extending the usefulness of the standard.

5.8 It is the specification type of standards that are used as the basis of contracts in commercial dealings, because they are broadly based and most often include methods of test and other essential elements. They furnish all the technical conditions and stipulate in well-defined terms the various requirements that the product must satisfy in order to meet the needs of a buyer. In another sense, they assume the character of a legal document and are often cited in court disputes. It is important, therefore, that they be properly worded in unambiguous terms, capable of no more than one interpretation. A loosely or badly worded standard, instead of being a help, can often become a hindrance to the legal authorities and bring disrepute to the standards organization concerned with its preparation.

5.9 Besides serving legal needs, specification standards serve a wide variety of users. To the designer, they furnish guidelines for designing the product; to the purchase department, they help outline the requirements of materials to be produced for manufacturing the product; to the salesman, they serve as a source material for his sales promotion; to the procuring agency, they constitute ready-made documents incorporating all the technical requirements that are necessary to be stipulated in purchase contracts,[7] and so on down the line.

SAMPLING AND INSPECTION

5.10 Methods of sampling and inspection are quite often incorporated in the specification type of standards, but sometimes they have to be spelt out separately. Until recently these methods have been devised on an *ad hoc* basis without any valid mathematical or other sound justification. More recently, however, statistical theories have been brought to bear on the subject, the object being to ensure that a given sample, appropriately drawn and subsequently tested or inspected, will yield results which will reasonably represent the quality of the whole of the consignment from which the sample was originally taken. These matters are discussed in further detail in Chapter 23.

TEST AND ANALYSIS

5.11 Methods of test and analysis are also quite often incorporated in specification standards. But many occasions arise where extensive treatment becomes necessary, in which case they have to be published separately as independent standards. In either case, the purpose of delineating in precise terms the details of testing and analytical methods is to ensure reproducibility of results when a given sample is tested in different laboratories by different workers, so that the results, no matter where obtained, remain consistent and comparable. It is also important to ensure repeatability when the same material is tested or analysed at different times within the same laboratory and with the same equipment. Both reproducibility and repeatability are important characteristics of any method of test and are further discussed in Chapter 23.

5.12 In order to ensure a fairly high degree of reproducibility and repeatability, it must first of all be recognized that a large majority of testing is based on empirical measurements. It is, therefore, extremely important to ensure consistency of the conditions prevailing during any determination. These conditions include very many variables. Firstly, the apparatus and equipment used must be accurately described, constructed and calibrated. It is more often than not the case that variations in apparatus and equipment used introduce variations in results. The same applies to chemical reagents – particularly in regard to their chemical purity and physical condition. Thirdly, the procedure employed must be precisely outlined step by step and scrupulously followed. Fourthly, whenever necessary, the atmospheric conditions prevailing during a determination should be controlled for quite often they influence the results of measurements. Since most of the results are empirical in character and the laws of their variability with atmospheric conditions are not always known, they cannot be readily converted from one set of conditions under which they are obtained to a standard set of conditions. It is important, therefore, to specify in certain cases that the surrounding atmosphere be properly controlled within prescribed standard limits. Sometimes it becomes necessary to subject the test specimens or samples to an exposure in a special or a standard atmosphere so as to bring them to a normal stable condition suitable for being tested. ISO and IEC have adopted standard atmospheric conditions for test,[8-10] which cover the requirements of all countries in the temperate, tropical and sub-tropical zones.

5.13 Test methods have thus to be drafted keeping many factors in view as briefly indicated above. Another important element of a test method which is more often ignored than observed is the criterion for judgement of results and the manner of their statement. This is also a statistical question and is discussed in greater detail in Chapter 23.

GRADING AND CLASSIFICATION

5.14 Methods of grading and classification are sometimes dealt with in the body of specifications for the material or product itself. But for many classes of materials, particularly those handled in bulk, separate methods of grading and classification become essential, such as for coal, metallic ores, etc. In such cases, sampling and statistical techniques play a very important and significant part in defining the grades and classes of the materials in question. But grades and classes may also apply to manufactured products, for example, classes of insulation of electrical equipment, grading of processed mica, etc.

5.15 In assigning designation to grades or classes, the usual practice is to make use of numerals or letters of the alphabet; for example, Grade A, Grade B, etc., or Class 1, Class 2, etc. This device serves the purpose neatly when the gradation list is intended to express superiority of one grade or class over the next. But often the objective of grading and classification is not to indicate qualitative differences, but to bring out other parameters such as the origin or end use, dimensions or sizes, capacities or outputs. In such cases if A, B, C type of designations are not acceptable, some other designations based on the relevant parameter may be used, for example, industrial alcohol may be graded by percentage of ethanol, bolts and nuts by dimensions, electric motors by horse power rating and so on. An interesting example is that of classifying electrical insulation where class A has the lowest performance in terms of temperature rise, which improves with Class B, C and onward. In making such decisions, it would be advisable, however, to keep consumer's interest always in view, for it is he who is the least informed and most likely to be misled (*see also* Chapter 18).

SIMPLIFICATION AND RATIONALIZATION

5.16 Schemes of simplification and rationalization pertain mainly to the limitation of variety of a product or grades and classes of a material. Such schemes are often made a part of specifications but may be treated separately if circumstances so require. Limitation of variety may be affected in respect of and guided by any of the characteristic variables involved, for example, dimensions and sizes, capacities and outputs, qualities and grades, or other attributes. In working out these schemes, some basic guidelines are available in the preferred number series (*see* Chapter 23), but in using these numbers a good deal of discretion has to be exercised, keeping in view the requirements of the consumer and the convenience of the producer. Too much restriction of variety can be as uneconomical from the user's point of view as absence of restriction could be from the producer's angle. Overall economy demands a proper analysis of each situation before decisions are taken. If such an analysis can be quantitatively made, it would be most useful. But, unfortunately, most situations do not lend themselves to quantitative handling and qualitative judgement remains the only alternative. Even so, if objectivity is maintained, good solutions are possible. Considerable guidance can be often obtained from demand-frequency data collected from past experience, as exemplified in the plot of Fig. 5.2 pertaining to the Indian Standard I-beams. A mathematical approach to variety reduction and its effect on economy has been ably attempted by Matuura,[25] which points to a new avenue of approach to the further development of the subject (*see also* Chapter 21).

PACKAGING AND LABELING

5.17 Packaging and labeling requirements are more often than not made a part of specification standards. Nevertheless, during recent years, packaging has become an important subject in its own right[26] and many standards are now being issued to serve the newly arising needs. Labeling is also assuming many forms – for example, informative labeling which has become, during the past couple of decades, a new branch of activity closely related to standardization. This matter is further dealt with in Chapter 16.

SUPPLY AND DELIVERY

5.18 Supply and delivery conditions are also quite often made a part of the specification type of standards. Though they constitute more of a non-technical set of requirements, they have necessarily to be linked with technical requirements, such as with those concerning sampling, inspection, packaging and labeling. Whenever they are separately dealt with in an independent standard, they constitute mainly the contractual obligations.

CODES OF PRACTICE

5.19 Codes of practice including model byelaws constitute a very important sector of standardization, quite distinct from the specification sector; so much so, that in some standards institutions, codes of practice

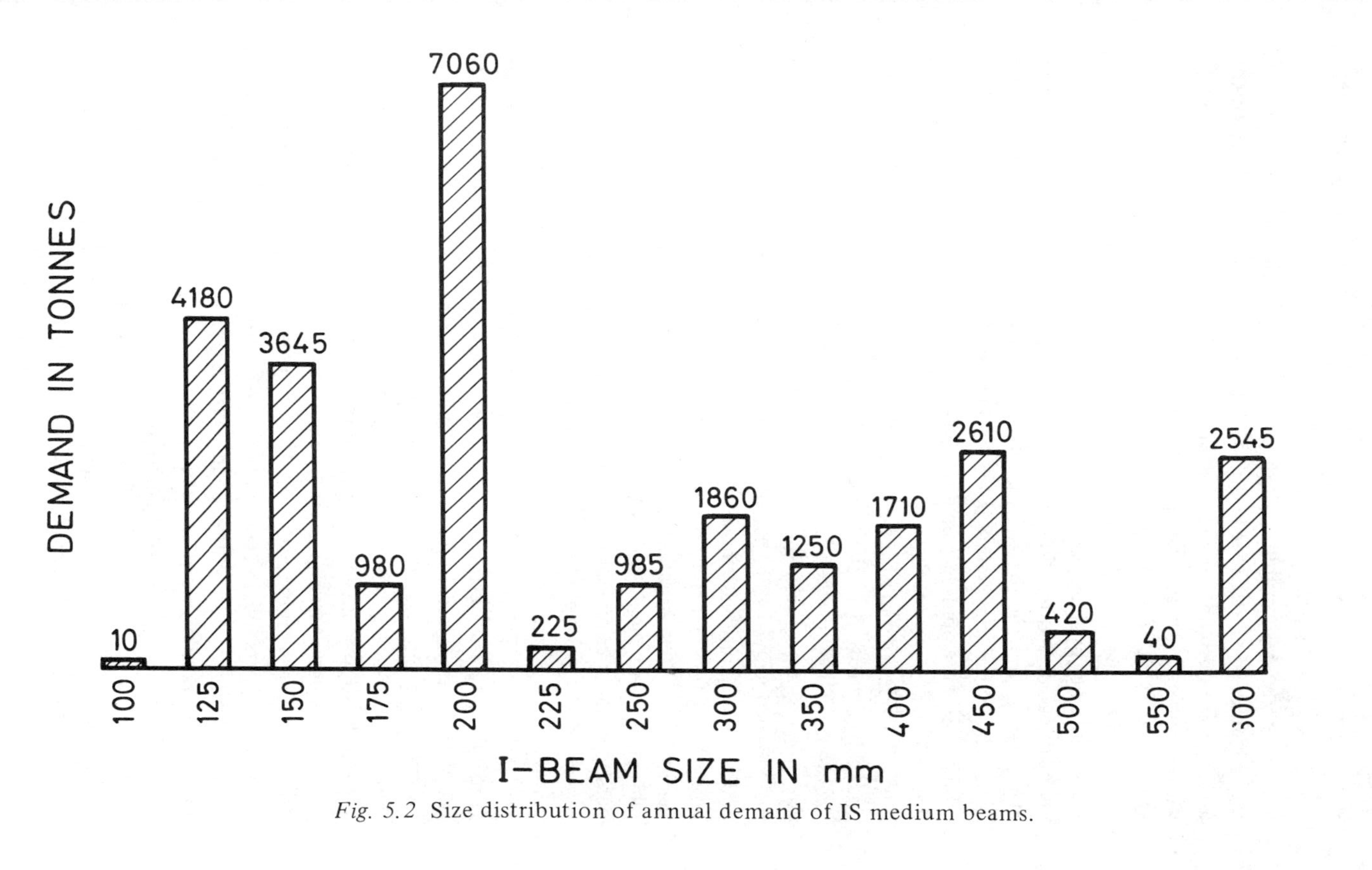

Fig. 5.2 Size distribution of annual demand of IS medium beams.

are made the responsibility of a separate major division with its own independent committee structure. But this is perhaps not always justifiable, because a code of practice may deal with almost any one of the various subjects on the X-axis and should, therefore, be more appropriately made the responsibility of the division connected with the specialized subject in question. For example, a code of practice dealing with building byelaws should be the concern of the building or civil engineering division; while the one dealing with the installation of electrical equipment that of the electrical division. A code may even be concerned with the maintenance of factories or a class of factories, in which case it would legitimately belong to the mechanical engineering division.

5.20 Codes of practice deal with a wide front of technological activity and help regulate it in the interest of safety and efficiency, and expedite decisions. They provide proper safeguards for health and safety of personnel as also for the optimum utilization of the equipment and machinery involved, consistent, of course, with overall economy. Design codes assist in rationalizing designs of equipment, buildings and structures, and provide for the degree of safety that is acceptable both technically and legally. Codes for installation, maintenance and operation of machinery and equipment help expedite the related work and ensure safety of personnel and equipment. Codes of practice for other operations may include those for packaging, labeling, transport, handling and storage of goods and products. Codes of practice for safety help regulate industrial practices in the interest of safety of personnel and equipment. Codes for controlling factory effluents, disposal of sewage, treatment of potable water and others are important for maintaining public health. Codes for store-keeping and classification of stores[28] help regulate inventories of a complex stock of goods and maintain them in an ever-ready condition to be able to supply any one of the hundreds of items at a moment's notice. Codes for organizing and conducting large conferences[11] are perhaps amongst some of the novel ones of more recent origin. But those for the training of personnel, commonly known as curricula and courses, are perhaps amongst the oldest codes. Office manuals represent codes for effective disposal of routine office work.

5.21 Whatever form of operation or process a code of practice may deal with, it has to stipulate the use of many kinds of materials and processes, and a variety of products and equipment. It is here that references to the various existing standards, to which these materials, products, etc., may be required to conform, help the code writer to line up his work, with that of the specification writer, thus making a unit whole of the process of standardization.

BYELAWS

5.22 Model byelaws for building construction in urban areas are gradually becoming the responsibility of standards organizations, largely because of the progressive realization of the need for ensuring uniformity in byelaws within a given country in which the byelaw-making power is distributed over smaller units such as the states, districts, counties or corporations. It will be appreciated that in a country where national standards are available for general use, wide variations in building byelaws would have the effect of nullifying the possible advantages that could accrue from national standards. Unification of byelaws over a given country also contributes to the overall economy of the nation by assisting the unification of practices concerned with design, erection and maintenance, and for training of personnel. Besides, it also ensures the most economical utilization of the material, manpower and financial resources of the nation.

5.23 Even after model byelaws have been produced through a general consensus of opinion of all the central and local authorities concerned, it is not always certain that they will be adopted all over the nation uniformly. Local authorities may be guided largely by them and introduce only minor changes considered necessary for local purposes. Some of these variations in a large country may even be dictated by differences in climatic conditions or other geographical factors. But more often than not these variations emanating from such factors are provided for in the model byelaws, or at least should be so provided for. On administrative and jurisdictional grounds, however, the local authority would always exercise its prerogative, but the less such prerogative is exercised the better it is for the nation.

5.24 Some countries have issued national building codes covering all phases of building and construction activity, presenting a comprehensive collection of material which could form the basis of all types of local byelaws for construction. The codes attempt to present what may be considered good practice while byelaws lay down minimum standards to be adhered to. In ultimate analysis, however, they both serve the same purpose. Outstanding examples of comprehensive codes are from Canada,[12] India,[13] New Zealand[14] and South Africa.[15]

FORMS

5.25 Forms including contract forms have also recently become the concern of standardization activity, largely because of the chaos that has resulted from the lack of uniformity in several fields of economic endeavour. The most significant demand for standardizing forms is perhaps for the unification of the international banking documents[16] and the international shipping documents, both of which require world-wide attention. The International Organization for Standardization Technical Committee (ISO/TC 68) Standardization in the sphere of banking, as also the International Chamber of Commerce are actively engaged on standardizing banking documents. Contract forms, however, have not attracted international attention but some national standards bodies have issued standards relating to contract forms, such as in France, Germany, New Zealand and elsewhere. Quite often there is a difference of opinion whether a contract form could or should be standardized on the national basis. Though there may be procedural and administrative difficulties involved, yet technically speaking, there is nothing against standardizing contract forms, provided of course the subject is dealt with in a rational and objective manner, like most other subjects that are submitted to the standardization process. In contrast to the comparative indifference at the national and international levels, the work on standardization of forms at the company or in-plant level may be considered quite extensive as well as intensive.

NEW MANAGEMENT TECHNIQUES

5.26 During the Second World War, and as a result of a general drive for greater efficiency and higher productivity in industry and government, several new management techniques were developed and widely employed to assist management in reducing decision-making to an objective process based, whenever possible, on mathematical deductions. Though these developments have been largely independent of the formal standardization movement, yet, in the overall picture, they strongly supplement standardization. The various techniques evolved may even be considered novel aspects of standardization space. Among those worthy of interest are the techniques of (1) Value analysis, (2) Materials management, (3) Programme evaluation and review technique (PERT) and (4) Linear programming. A great deal of information on these techniques is available in literature, but perhaps a brief description of each will be of interest here.

5.27 Value analysis[17, 18] involves a systematic approach to the determination of the intrinsic value of any thing or any process or a constituent part thereof, and a search for ways and means to enhance this value in terms of the cost involved. Quite often, in industry as also in other phases of life, things are used or operations are gone through as a matter of routine, without enquiring whether they are essential or could be improved upon or replaced with advantage. Value analysis technique helps the executive and the engineer to get the most value out of the effort and resources expended, by subjecting every situation to a systematic analysis. The importance of value analysis to the process of standardization would be readily appreciated, as a tool for proving the adequacy or otherwise of any requirement or test method or any other constituent part of a standard. Thus, after being thoroughly analysed by this process, many standards may be found wanting. They may have to be discarded, modified or replaced with new standards. This can happen at any one of the levels and with any aspect of any subject that may be dealt with in a standard.

5.28 The technique of materials management[20, 27, 29, 30] has attracted a great deal of attention in recent years because of the spectacular successes it has scored in the control of materials in respect of their quality, quantity, timing and cost. The materials component of the cost of a product, which may mount up to 70 to 80 percent in some industries, is particularly susceptible to economic measures through materials management. Any economies affected in this area can lead to major reductions in the cost of production of the finished product. Inventory control[19] is just one part of materials management, while specifications, coding, classification, variety reduction, quality control and several other management techniques all contribute to its effectiveness.

5.29 Programme evaluation and review technique,[20, 21] briefly known as PERT, constitutes a methodology for executing efficiently the extensive and complex projects involving a number of individual operations, all aimed at achieving an ultimate goal. It consists of overall planning, sequencing and scheduling events and the accompanying activities and timing them appropriately in relation to one another along well-defined paths including the crucial critical path. Its great merit lies in the facility it affords the management to keep the progress under constant review with the possibility of rescheduling the programme from time to time with changing conditions. It makes possible the most effective allocation and utilization of resources and ensures the most economical results. The technique itself constitutes a standard method of approach to the execution of any project, large or small, and, in turn, can serve standardization projects themselves to be handled in an efficient and expeditious manner.

5.30 Linear programming[22-24] is a mathematical procedure developed in recent years particularly for solving difficult types of problems involving multiplicity of variables, which are commonly met with in industry and commerce. These problems are characterized by processes involving a number of operations constituting a programme, each operation of which could be expressed by a linear equality or inequality. One such problem of interest to standardization may be that of finding the most economical fertilizer mix having the requisite ratios of nutrient elements, nitrogen to phosphorus to potassium (N:P:K), made up by compounding together several of the available chemical fertilizers each containing a known percentage of one or more of the nutrients and costing a known amount. It would be clear how useful it would be if linear programming was made an integral part of the standardization process, whenever problems of this character involving maximization of benefits or minimization of costs or time may arise.

5.31 There are many other management techniques which have proved their usefulness such as the queueing theory, simulation, learning curves, ABC analysis, review analysis of multiple projects and so on. They are all discussed in the literature cited above and have some direct or indirect bearing on standardization. To discuss them all here in any detail would be an uncalled for digression.

REFERENCES TO CHAPTER 5

1. *Guide for the presentation of ISO recommendations.* 2nd ed. International Organization for Standardization, Geneva, 1971, p. 15.
2. *Guide to the drafting of IEC documents.* International Electrochemical Commission, Geneva, 1961, p. 25.
3. *PD 6112–1967 Guide to the preparation of specifications.* British Standards Institution, London, p. 17.
4. *DIN 820–1966 Standard procedure, sheet conditions for test purposes.* International Electrotechnical Commission, Geneva, (DNA), Berlin.
5. *IS:12–1964 Guide for drafting Indian standards (2nd revision).* Indian Standards Institution, New Delhi, p. 57.
6. *SAAMP 15–1971 SAA style manual (revised ed.).* Standards Association of Australia, Sydney, p. 46.
7. Verman, Lal C.: Standards guide purchasing: purchasers guide standardization. *Eastern Purchasing Journal,* New Delhi, 1960, vol 1, pp. 33-37.
8. *ISO/R 554–1967 Standard atmospheres for conditioning and/or testing: Standard reference atmosphere: Specifications.* International Organization for Standardization, Geneva, p. 4.
9. *ISO/R 558–1967 Conditioning atmosphere: Test atmosphere: Reference atmosphere: Definitions.* International Organization for Standardization, Geneva, p. 4.

10. *IEC-PUB 160–1963 Standard atmospheric conditions for test purposes.* International Electrotechnical Commission, Geneva, p. 11.
11. *Guide for organization of ISI conventions* (internal document). Indian Standards Institution, New Delhi, 1964, p. 34.
12. *National building code of Canada* (one document in 9 parts). National Research Council, Ottawa, 1965, p. 513.
13. *SP:7–1970 National building code of India* 1970 (in 10 separate parts). Indian Standards Institution, New Delhi, p. 525.
14. *NZSS 1900 Model building byelaw* (in 11 chapters). Standards Association of New Zealand, Wellington, 1963–70.
15. *Comprehensive model building regulations* (in 11 separate parts). South African Bureau of Standards, Pretoria, 1951–62, p. 750.
16. International Organization for Standardization Technical Committee ISO/TC 68 Standardization in the sphere of banking, as also the International Chamber of Commerce, are currently engaged in this task.
17. Miles, Lawrence D.: *Techniques of value analysis and engineering.* McGraw-Hill, New York, 1961, p. 275.
18. Miles, Lawrence D.: "How to analyse value." pp. 11-1 to 11-60 in Alijian, George W.: *Purchasing handbook.* McGraw-Hill, New York, 1958.
19. Buchan, Joseph and Koenigsberg, Ernest: *Scientific inventory management.* Prentice Hall, Englewood Cliffs, 1962, pp. 462-505.
20. Menon, P. G.: Materials management and operational research in India. *Material Management Journal Publications,* New Delhi, 1968, p. 341.
21. *PERT: a dynamic project planning and control method.* International Business Machine Co, New York, p. 28.
22. Chadda, R. S.: *Inventory management in India.* Allied Publishers, New Delhi, 1964, pp. 38-42.
23. Hadley, G.: *Linear programming.* Addison-Wesley, Reading (USA), 1962, p. 520.
24. Naylor and Byrne: *Linear programming.* Wadsworth, Belmont, 1963, p. 183.
25. Matuura, S.: Measurement of the effect of standardization. *ISI Bulletin,* 1968, vol 20, pp. 305-309.
26. Coles, J. V.: *Standards and labels for consumers' goods.* Ronald Press, New York, 1949, p. 556.
27. Palit, A. R.: *Outline of materials management.* Academic Publishers, Bombay, 1970, p. 200.
28. Evans, B. Agard: Rationalized store-keeping. *ISI Bulletin,* 1959, vol 11, pp. 205-208.
29. Palit, A. R.: The scope, purpose and techniques of materials management. *The Eastern Purchasing Journal,* New Delhi, 1962, vol 3, pp. 245-248.
30. Venkataraman, R.: Materials management in Bhilai Steel Plant. *ISI Bulletin,* 1965, vol 17, pp. 157-160.

Chapter 6
Individual Level

6.0 In Chapter 3, individual level standards were described as those specially laid down by an individual consumer or user, a builder, or constructor, a government department or corporate body, in fact, any one to suit his own specific needs. Such a standard may be a specification for making a piece of furniture; a design for building a house, or constructing a dam; a project for erecting a bridge, or a factory. The documents that actually correspond to the individual level of standardization space such as project plans or plant layouts are not commonly recognized as standards in the traditional sense of the word, even though they serve in their own way the purpose of standards. In fact, there exists an amount of disagreement whether the individual level should at all be included in the scheme of standardization space. The main argument for this viewpoint is that, standardization today having become an organized activity and individual level standards being the concern of an individual, they cannot reasonably form a part of the organized activity. It cannot, however, be denied that the object of standardization is to serve the community which consists of individuals, and the welfare of the community and its individual members has necessarily to be considered indivisible.

6.1 Several other more cogent arguments could be cited in favour of maintaining the individual level as the base level of standardization on the Z-axis of the space diagram. For example:

(1) In order that an individual level standard may adequately serve the purpose of the individual concerned, it is necessary that it be fully in line with standards at all other levels, including company, association, national and international.

(2) If standards at various levels are to serve their due purpose, it is necessary that, having been appropriately coordinated at all levels (*see* Chapter 3, Table 3.1), they should be extensively and intensively implemented; and individual level happens to be the one level at which such implementation can be pushed a very long way.

(3) The structure of standardization space would remain incomplete, if the importance of individual level was ignored, because in any theoretical treatment, the main object is to build a logical structure which would cover almost every case that might arise without exception. And

(4) By recognizing individual level as an integral part of standardization space, the activity at this level could be more readily absorbed into and become a significant part of the discipline of standardization.

6.2 Individual standards are needed by almost anyone concerned with economic activity, be it a private or public unit or enterprise or even an individual. If the originator of an individual standard happens to be a company or a government department, naturally a very high priority would be given to reliance being placed on company or departmental standards, which, in turn, must have already been aligned with standards at higher levels. But for an independent individual originator it will be well to place preferential reliance on national standards, which will assist him in getting competitive bids from contractors and suppliers. He will have to choose from a wide spectrum of materials and equipment covered by national standards. Even in making use of codes of practice for design and construction, he will have to exercise an amount of discretion wherever alternative choices in good practice are allowed in the national codes. In the absence of coverage in a national standard of a particularly special item, he will have to depend upon association or international level standards, since he is not likely to have access to any company standards. Failing all these resources, he will have to draft his own requirements in the light of his personal experience, knowledge and needs.

6.3 Thus, it is clear that individual level standards are distinct from standards at other levels in one important aspect. Most other standards deal with the solutions of recurring problems, evolved for the convenience of a community of interests, while individual level standards are characterized by their being designed for non-repetitive situations intended for the use by the originator of the standard for his own purpose. The originator himself or even another individual may, of course, re-use the standard after its first use and after due revision in the light of experience, but such a contingency would seldom arise. By the time a given project is finished, technology would have advanced and the requirement considerably changed. A new project would seldom be the same as the previous one. That is not to say that no two bridges or two factories could be built identically alike, but in that case the standard in question would no longer remain an individual level standard. It could readily be graduated to company or in-plant level and made a part of the regular accepted practice of the originator concerned. In this form it becomes readily recognized as a standard in the conventional sense of the word, offering a solution to a repetitive situation. This clearly demonstrates how thin the line really is that divides individual level standards from company or in-plant level standards.

6.4 Whether an individual level standard is a plan or a specification for a building, a structure or a factory, or whether it is a comprehensive design for the execution of a large river valley project, the basic requirements to be fulfilled are the same, and may be outlined step-by-step as follows:

(1) Definition of the objectives to be achieved,

(2) Determination of requirements to be fulfilled,

(3) Making a feasibility study and preliminary assessment of the resources required,

(4) Carrying out of the surveys and execution of designs,

(5) Drawing up specifications and estimation of quantities and costs,

(6) Arrangement for procurement of materials and equipment and proceeding with construction,

(7) Marshalling of the machinery and setting the project in operation, and

(8) Post-construction surveys and evaluation of the results.

6.5 So far as objectives (item 1) and requirements (item 2) are concerned, there appears to be little in which standards can give any guidance. They are mainly the concern of the originator or the owner. Take, for instance, the case of a multi-storey apartment house to be built. Its requirements have to be first fixed in terms of the number of apartments, their size, and facilities to be provided. The order of magnitude of the financial resources required (item 3) can then be roughly estimated by the use of trade norms or trade standards of the locality. Having estimated the resources and planned to provide for them, the surveys and design (item 4) can proceed. Here is where formal standards begin to play a major role.

6.6 In countries where national standards exist for surveying instruments, the designer of the project can derive considerable assistance in obtaining accurate measurement of the site (item 4). Surveys become particularly important in certain cases, such as a dam project, where standard methods become invaluable in collecting the precipitation data and making other hydrological measurements, extending over several years, to provide adequate enough historical records to be able to forecast the future trends. In certain countries these operations are covered by national standards, such as in India; in others, by in-plant standards of the national authority concerned with natural resources, such as in the USA. After the surveys and investigation, follow the actual design for which codes of practice are generally available for practically all types of jobs – building, structures, dams, factories, electrical installations and so on. Wherever national codes are available they should be used. Otherwise codes of professional societies would be considered authoritative for the purpose.

6.7 In the next step for drawing up specifications and estimation of quantities and costs (item 5), which usually go together, material standards as well as the codes play a dominant role. Specifications for each material used, such as bricks, cement, glass, steel and cables, and every type of equipment needed, such as motors, cranes, machine-tools and concrete mixers, are generally covered by national standards and often certified with national standards marks. In both cases, this coverage facilitates the estimation of costs because competitive prices can readily be ascertained and material and equipment procured on their basis, without risk of obsolescence and with the assurance of achieving best possible economy.

6.8 During actual construction and installation of equipment (item 6), PERT methods and other management techniques (*see* Chapter 5, para 5.29) discussed earlier could be of invaluable assistance. For dealing with components of construction, such as air conditioning or heating systems, codes of practice again would come in quite handy. Among the whole range of codes, the concrete codes, wiring codes, and welding codes are particularly basic to all construction works. Installation and maintenance codes are also often available to help put the project in operation (item 7). For the final step of evaluation (item 8), no standard may be available as a rule, but having based so many foregoing steps on standards, considerable further assistance may be expected to be derived from them in framing the plan for evaluation of results.

6.9 Individual standards find a large number of applications in every field of economic life of a nation; but perhaps more so in the building and civil engineering fields. In fact, a specialized profession of specification writers has recently emerged in the USA which fomulates individual level standards for building and civil engineering projects based on designs prepared by the architects or engineers. These individual standards, known commonly as specifications of work, form the basis of contracts for construction and generally include the necessary legal clauses. In certain European countries such as Sweden[1] and UK[2], a move is afoot to provide special assistance to architects and contractors by providing a central or national building specification capable of being used in this manner and with computers.

6.10 Thus, individual standards may range from the most comprehensive river valley project plans to transport networks, structures, residential houses and so on down the line to the most elementary forms, such as a drawing or simply a sketch for a piece of furniture, the style of a dress or a design of a piece of jewellery. Even in these relatively simpler cases, national or other standards may be found useful; for example, the test methods for determining robustness of furniture, the specification for a piece of cloth or the size of garments for a dress and the purity of gold for the jewellery.

6.11 The individual level standards may even be wholly unwritten and unsketched, particularly when they express preferences for a particular taste in style or fashion and other personal preferences. However informal or personal they may be, they are nevertheless standards of a sort in the intrinsic sense of the word. In this area, the link with any form of formal standard disappears entirely. In fact, formal standards are almost impossible and sometimes inadvisable to lay down for some of these things, at least at the present state of development of technology. For example, the nature of fragrance of a perfume, or the organoleptic quality of foods, cannot yet be quantitatively measured or objectively classified, though some standard methods exist to assist subjective judgement.[3 4] Individual level standards concerned with personal preferences would continue to serve the purpose without assistance from formal standards until such time as some radical advance has been made in technology for an objective determination of certain qualities which so far have lent themselves to subjective methods only. But certain things like fashions, colour preferences, tastes in food and choice decoration styles will continue to remain within the realm of the individual.

REFERENCES TO CHAPTER 6

1. Nielsen, F.: Central specification. *Proceedings of the third CIB congress "Towards industrialized building" (Copenhagen).* Elsevier Publishing Co, Rotterdam, 1966, p. 153.
2. National building specification. *Royal Institute of British Architects Journal.* 1969, vol 76, pp. 314-315.
3. *IS:2284–1963 Method for olfactory assessment of natural and synthetic perfumery materials.* Indian Standards Institution, New Delhi, p. 10.
4. Johnson, F. C.: Combined gas-liquid chromatography for flavour formulation and assessment. *Food Manufacture,* 1970, vol 45, pp. 45-52.

Chapter 7

Company Standardization

7.0 The two terms "company standardization" and "in-plant standardization" have so far been used interchangeably. But for convenience of further discussion it would perhaps be desirable to choose one, and "company standardization" appears to command preference. This term has been in use for many decades, particularly in the USA, and in recent times in Europe as well. The term "in-plant standardization" appears to be somewhat restricted, because the activity as such would cover fields other than plants and factories, such as offices, work-sites, merchandising houses and public utility services. It is, therefore, necessary that the term "company standardization" should be understood in its wider sense to cover all sorts of enterprises, including government departments, project authorities, institutions, municipalities, city corporations and so on. In other words, "company standardization" may be translated more accurately to mean "enterprise standardization." It would be applicable in all cases whatever the size or character of the enterprise may be – large, small or medium; private, official or semi-official.

7.1 It should also be emphasized that company standardization should be understood to cover every aspect of the working of an enterprise whatever may be the nature of its business. Thus, it may deal with engineering standards, production standards, administrative and financial norms, codes of practice for manufacturing and maintenance, and even codes for conducting market surveys and estimating costs – in fact, standards relating to any one of the various activities of the enterprise whatever. Of course in different types of enterprises, emphasis will have to be placed on certain specific groups of standards, for example, in an engineering firm, design data, drawing practice and manufacturing process standards would be considered of basic importance, while in a government procurement agency purchase specifications will receive predominant attention.

7.2 In the complex realm of standardization, company standardization holds a particularly important position. Though it is the next higher level above the bottom-most individual level, it is the first level at which the corporate interest begins to function and organized effort becomes predominant. Company level functions both as source and sink of standards. Besides originating company standards for its individual needs, company standardization effort helps collect a good deal of practical data and experience necessary for the next higher levels of standards, namely, the industry and national levels. In addition, as a sink, company level constitutes the most prolific organized user of higher level standards – industry, national and international. No standards programme of any enterprise can really be successful without its having to rely heavily on other standards – national standards in particular.

7.3 As indicated earlier, company level standardization has specially been well developed in the USA. However, during the last two decades or so, it has also become quite widespread in the UK and the continental countries of Europe. Among other parts of the world, Australia, India and Japan may be mentioned as having also established a viable movement towards all-out company standardization. A great deal of development has taken place and a large number of papers has been written on the subject. Apart from hundreds of articles that have been published in the technical press and scores of papers delivered and discussed in conferences and seminars, some excellent books[1-3] have been specially devoted to company standardization. The movement has become so widespread in many countries that a large number of professional people have become engaged as standards engineers at the company level. Together with those active at other levels of standardization, company standards engineers have organized themselves into several professional societies[4] in many countries. Two of these societies issue regular periodicals.[5] These societies, through their meetings and conferences and their journals, are consistently helping to advance the theory and practice of company standardization and to promote the cause of standardization in general. Some of these societies are regularly organizing seminars and courses of training for company standardization engineers. In some countries like India and Japan, the national standards bodies are also taking initiative in the matter of training company standards engineers (*see* Chapter 25). The work of these societies has become so widely appreciated and supported that serious consideration is presently being given to the formation of a world-wide federation of all the active national standards engineers societies, with a view to developing a worldwide movement.

SCOPE

7.4 It would be appreciated that the need for standards in individual enterprises would differ and depend largely on the type of business they are engaged in. For example, one company may be manufacturing producer goods while another may be engaged in marketing a whole variety of consumer goods. A municipal corporation's need for maintaining public utility services would differ considerably from those of a river valley authority engaged in developing multi-purpose projects. Thus, while the subjects of interest would differ from enterprise to enterprise, their interest in various aspects of standardization would nevertheless be universal. This is illustrated in Table 7.1, in which the interest of a representative sample of the different kinds of enterprises in different aspects of standards is indicated. The very few blank spaces found in this table demonstrate the universality of company standards interest in almost all aspects of standardization.

7.5 On the other hand, the programme of company standardization and the organization best suited for developing the programme would be determined largely by the particular needs of the enterprise and its own internal organizational structure. It has, however, been found advantageous by experience that, irrespective of the nature of the enterprise, wherever the executive head of company standards activity is made responsible directly to the top management, the results obtained in terms of economy and efficiency have been considerable and the difficulties and problems in executing the programme reduced to the least. This can readily be explained by the fact that company standardization programme, in order to be effective, has to be all-pervading, and being dependent on voluntary cooperation by all concerned, top-level backing of the basic policy of standardization is essential for its success.

7.6 Whatever the nature or the size of an enterprise, the head of its company standards department has necessarily to collaborate closely with

TABLE 7.1

INTEREST OF VARIOUS TYPES OF ENTERPRISES IN DIFFERENT ASPECTS OF STANDARDIZATION

Type of Enterprise	ASPECTS										
	Nomenclature, Symbols, etc.	Specifications	Sampling and Inspection	Tests and Analyses	Grading and Classification	Simplification, Rationalization	Codes of practice and Bye-laws	Safety	Packaging and Labeling	Supply and Delivery	Forms and Contracts
Manufacturer	+	+	+	+	+	+	+	+	+	+	+
Agricultural Producer	+	+	+	+	+		+	+	+	+	+
Processor	+	+	+	+	+	+	+	+	+	+	+
Trader	+	+	+	+	+	+			+	+	+
Government (Civil)	+	+	+	+	+	+	+	+	+	+	+
Government (Defence)	+	+	+	+	+	+	+	+	+	+	+
Municipality	+	+	+	+	+	+	+	+	+	+	+
Project Authority	+	+	+	+	+	+	+	+	+	+	+
Public Utility	+	+	+	+	+	+	+	+	+	+	+
Institution	+					+	+			+	+
Administrative Office	+	+	+			+	+			+	+
Technical Consultant	+	+	+		+	+	+	+	+	+	+

the heads of every other department or division, because there is always one or more aspects of standardization work which touches upon, or is concerned with, the work of every department or division. Take, for instance, a manufacturing unit of a fairly large size. Apart from the various technical departments which are naturally concerned with all kinds of standards on technological subjects, even its administration and personnel departments would have a need for many other types of standards on non-technical matters, such as those for forms, office procedures, nomenclature and so on. Table 7.2, which has been adopted after Hussey[6] and Milek,[7] illustrates how the various aspects of standardization would be of direct interest to different departments. It is important, therefore, that

TABLE 7.2

CONCERN OF THE VARIOUS DEPARTMENTS OF A MANUFACTURING ENTERPRISE WITH THE DIFFERENT ASPECTS OF WORK OF ITS STANDARDS DEPARTMENT

Departments	ASPECTS										
	Forms and Contracts	Supply and Delivery	Packaging and Labeling	Safety	Codes of practice and Bye-laws	Simplification, Rationalization	Grading and Classification	Tests and Analyses	Sampling and Inspection	Specifications	Nomenclature, Symbols, etc.
Accounting	+						+			+	+
Administration	+						+			+	+
Advertising and Sales		+							+	+	+
Design and Engineering	+	+	+	+	+	+	+	+	+		+
Maintenance and Construction		+		+			+	+			+
Manufacturing and Production	+	+	+	+	+	+	+	+	+	+	+
Materials Management	+	+	+	+	+	+			+	+	+
Quality and Reliability Control		+	+	+	+			+	+		+
Research and Development	+	+	+	+	+	+	+	+	+		
Safety and Health				+			+	+			
Shipping							+	+	+	+	+
Standardization	+	+	+	+	+	+	+	+	+	+	+

each department for its part should maintain collaboration with the standards department during the course of development of standards of interest to them. At this stage it may be emphasized that if collaboration is established at the earliest stages of consideration for the setting up of a standards project and continued through the drafting and finalization stages, then widespread acceptance and automatic implementation of standards could readily be assured.

7.7 The case illustrated in Table 7.2 is that of a manufacturing unit; but with minor modification, similar tables could be drawn up for any other kind of enterprise. For example, in a city corporation or a river valley project, instead of advertising the departments concerned may be public relations; surplus disposal, instead of sales; and so on. The basic considerations, however, apply in every case, namely every department of an enterprise is bound to be concerned with one or more aspects of standardization, in developing which it must cooperate with the standards department, in the interest of efficient and economic working of the enterprise.

NEED

7.8 In spite of the considerable growth that the company standardization movement has registered during the post World War decades, all over the world, many top-level executives and administrators are not quite convinced about the need for establishing a separate department to take charge of this activity. The chief argument is that national and industry or association standards are available to all departments for use from which they can always choose their requirements. In a case where a special standard for the company's use is required, the department concerned, being the most knowledgeable and competent party, could always draw it up. There are many flaws in this *laissez-faire* approach, as illustrated by the following examples, which it may be worthwhile to consider:

(1) National and association standards are written to take care of a large clientele of users and cover all their needs. A given company will have need for a limited number of items; for example, in the range of bolts and nuts, in the variety of non-ferrous or ferrous alloys. If each department of an enterprise were to make its own selection without coordinating its needs with those of other departments, there is bound to be unnecessary overlap and greater variety of choices than if the needs of all departments were centrally pooled and assessed.

(2) Each department making its own selection of standards for its

individual use more often than not would leave the decisions either unrecorded, or implicitly or invisibly covered by other decisions outlined in documents relevant to the subject in question, such as a drawing dealing with the design of a particular piece of equipment. Such practice would make it difficult to follow up one department's decisions on standards by the other departments. Nor would such a situation facilitate future reference to the decisions taken earlier within the same department.

(3) Independence of each department in making its own decisions on standards would lead to multiplicity of practices within the same company. Cases are known where serious contradictions have arisen, leading to wasteful operation.

(4) Lack of systematically and centrally recorded standards or decisions on standards would make it most difficult to train newly recruited staff in accepted company practices.

(5) In most companies working on this basis, it has been invariably found that their inventories of stocks are so unnecessarily large as to be highly uneconomic.

BENEFITS

7.9 A well-organized company standards department led by a capable chief under the patronage of top level management can render invaluable service in bringing order in the day-to-day working of the enterprise and smoothen the way for its further advancement. This is brought about in many ways. For example:

(1) The standards department acts as the central repository and clearing house of all standards and makes them available to everyone in the enterprise interested in consulting them.

(2) It records every decision of the enterprise on standards matters which is arrived at after due consultation with other departments concerned.

(3) It organizes collective thinking in collaboration with representatives of other departments on all problems which lend themselves to standardization.

(4) It assists in finding solutions to all repetitive problems and incorporates these solutions in company standards for the purpose of future reference by all concerned.

(5) In regard to non-repetitive problems, it assists the parties concerned with their solutions and fulfils their needs for standards.

(6) It organizes the collective study of the available national and

association standards of direct interest to the company's business, and in collaboration with other departments, culls from them the types and varieties of products and practices considered desirable for adoption by the company as a whole as its own standards.

(7) Wherever suitable national, association or other standards are lacking, it organizes the preparation of original company standards for its use.

(8) It organizes the participation and representation of the enterprise in the collective thinking of the interested associations and the national standards body of the country; and, if called upon, of the international organizations.

(9) It promotes research and investigation and the collection of practical data of direct interest to the formulation of standards at the company, association and national levels and sometimes even at the international level: by so doing, it enhances the prestige of the enterprise.

(10) It guides and promotes, in conformity with company standards, the production in ancillary industries which supply the needs of the enterprise.

(11) It helps to organize the training of new personnel of the enterprise as also that of the ancillary industries so as to make them familiar with the standards practices of the enterprise.

In short, as Melnitsky[1] puts it, "in any and all events, the standards department can be viewed as a missionary to spread the gospel of internal standardization throughout the company," and it may safely be added, outside as well, whenever an opportunity arises.

GAINS

7.10 The effects of the activities of the company standards department on the working of the enterprise are very far-reaching. Though all the benefits reaped cannot always be assessed quantitatively, their qualitative impact has been widely acknowledged.[8-10, 39-54] Consider, for example, the order company standards bring in all phases of working of the enterprise. The designer knows what materials and parts to utilize in his designs, which will automatically be available in the stocks of the company and for which the purchasing department would already have worked out its routine to keep the inventories continually replenished. The workers would be familiar with such parts and their uses. Communication in terms of standard terminology, symbols and drawings would be so facilitated that that there would be no need for apprehension or misunderstanding arising between one department and another. Inventories to be maintained would be minimal and their sources of supply and lead-time would have been accurately ascertained. All purchases could be based on written specifications and all buying could be made highly competitive. The most important feature would be that the overall know-how of the enterprise as a unit would be reduced to writing. There would be no danger of its being lost or frittered away with the change of personnel. It would be available to assist any newcomer who had to be initiated into the firm's operations.

7.11 It is natural that all this order would reflect strongly in an increase in productivity as also in production, and lead to considerable economies and high profitability. It is obviously not always possible to assess the various gains quantitatively, though some attempts[11-14] have been made from time to time in this direction, motivated chiefly with a view to carrying conviction with the top management, which sometimes becomes necessary. Frontard,[12] dealing with a simple example of variety reduction concludes that 10–25 percent saving in production cost can be achieved after deducting all the costs involved in introducing standardization. Sometimes, the calculations are made on the basis of the ratio of cost of standardization incurred to the total ascertainable gains or the savings effected. This latter estimate is not always concerned with the actual magnitude of the cost of production itself, but is quite useful to management for making decisions, in regard to the continuation or expansion of standardization programmes. Gupta[11] records the American and Japanese experience, where the cost-to-gains ratios have reached as high as 1 to 20, but a 1 to 10 ratio is found to be common. This can mean millions of dollars saved in operational costs of large size enterprises and substantial savings for small and medium size units, which may spell the difference between affluence and penury – between profit and loss.

ORGANIZATION AND PROCEDURE

7.12 Depending on the size and character of the enterprises, company standards organizations with regard to their framework would differ considerably from one another.[15-19] In smaller firms the responsibility for approved standards programmes may be entrusted to a divisional chief in charge of some other functional group, for example design or engineering. But in a larger firm, the pattern often assumes the form of an independent standards department at the same level as other functional departments. In either case, the official in charge of standardization should, as mentioned earlier, be made responsible and report to the top management. Some firms give this activity such an importance that one of the vice-presidents of the company is made responsible for it. In such a case, the company standards programme has the best chance of success and makes a worthwhile contribution to the working of the enterprise.

7.13 The procedures followed for preparing company standards also differ a great deal from firm to firm, but one feature, which is essential to ensure the effectiveness of standards, is common to all; that is, that all standards are prepared by a consensus of opinion of all those who are concerned with their use in their day-to-day work. This basic requirement has to be reflected in any one of the operational patterns that may be adopted. Two major types of patterns are available, namely, the committee pattern and the consultation approach. In the first case, the standards department nominates committees for each line of work of interest. Each committee would consist of representatives from those departments of the company which may be concerned with the work allotted to it. These representatives are expected to reflect the views of their departments, which implies that some prior consultation among the individual departmental groups would be called for. Obviously, the committee structure is a slow device and may not always be justified, but in large firms and in important cases affecting several departments, it may be the only approach.

7.14 The more direct and simpler method, which is sometimes followed in smaller firms, and in simpler cases in big firms as well, is the consultative method. In this case, a draft standard may be prepared by the standards department or someone designated by it. The draft is then circulated for views and comments to key personnel of other departments concerned, before it is issued as a company standard. It may sometimes happen that contradictions among the comments received as a result of the consultation cannot be resolved simply by the standards department. In such a case, a brief meeting of all the interested parties may have to be called to work out a common consensus on specific points of divergence. Thus we have a blend of committee-cum-consultative methods. This is perhaps the most desirable approach in most complicated cases and perhaps quite expeditious. The main object should be to adopt that methodology which best suits the needs of each situation as it arises.

7.15 For the final approval of the standard before it is issued for company use, it is not considered necessary to secure the seal of approval of top-level management. This is a matter of detail which could be left to those directly concerned, working together under the aegis of the standards department. Top level management should, however, support the programme officially. Perhaps the best way to indicate this support would be to endorse a general policy directive and also perhaps issue a mandate concerning the working procedures of the standards department. For approval of individual standards it has been found most useful to create what may be called a standards policy committee or a standards executive committee, consisting of all the heads of the divisions or departments of the enterprise concerned with standards; that really means all the divisional or departmental heads within the enterprise. Such a policy committee, whose secretariat should reside in the standards department, would be responsible to top-level management in all matters and may also render advice on the details of standards policy and its working. The policy or executive committee should be entrusted with the task of approving new projects for standardization and also deciding when old standards may be revised. In fact, it may undertake the review of every operation and item of interest to determine the subjects on which standards may usefully be prepared. It is claimed by Mohr[17] that in respect of a manufacturing enterprise "somewhere between 20 and 40 percent of all that is being done is standardizable."

PROCESS

7.16 When a new project for standardization is approved, the first concern would be to collect the data relating to the company's existing practices and its needs, present and future. The next steps will be to study carefully the relevant national standards to determine the extent to which these needs could be met on the basis of the decisions already incorporated in them, and to determine in which manner the guidance given by them could be followed to rationalize company practices. If, however, national standards are found not to cover the subject, or they prove to be otherwise inadequate for the purpose, then search may be made among association or industry level standards for guidance. Sometimes it does happen that friendly companies offer assistance by making their own standards available to one another. These would come in handy on occasions. But cases might arise where no guidance is available and where new ground has to be broken. This is quite often the case with product standards and special items. In such cases all the know-how and expertise available would have to be marshalled from within the company and from outside consultants. Much valuable original work has been done in this area at company level, and this has led the way to formulating rather complex standards at higher levels – even the international level.

7.17 Formats of company standards documents are usually presented in an extremely simplified[18-20] form, so that they indicate essential

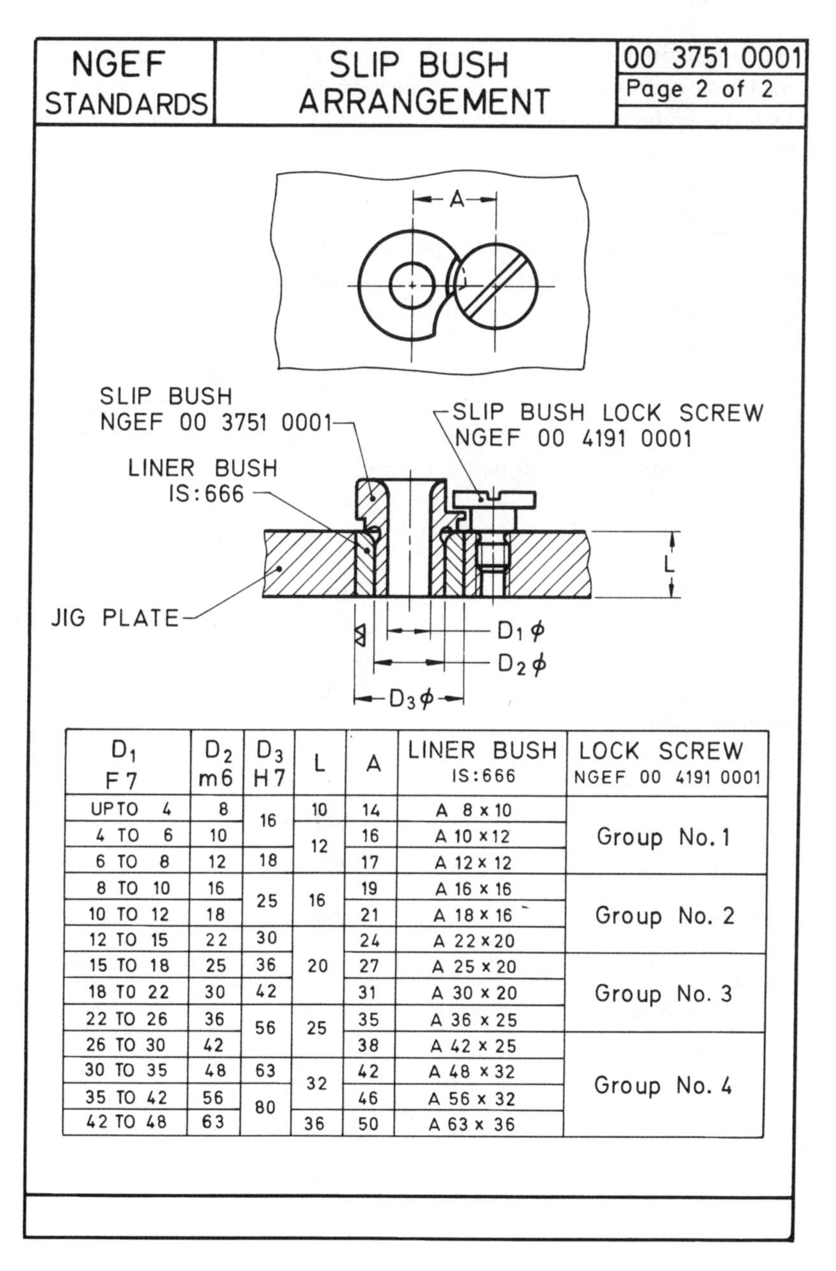

NGEF STANDARDS	SLIP BUSH ARRANGEMENT	00 3751 0001 Page 2 of 2

D_1 F7	D_2 m6	D_3 H7	L	A	LINER BUSH IS:666	LOCK SCREW NGEF 00 4191 0001
UPTO 4	8	16	10	14	A 8 x 10	Group No. 1
4 TO 6	10		12	16	A 10 x 12	
6 TO 8	12	18		17	A 12 x 12	
8 TO 10	16	25	16	19	A 16 x 16	Group No. 2
10 TO 12	18			21	A 18 x 16	
12 TO 15	22	30	20	24	A 22 x 20	
15 TO 18	25	36		27	A 25 x 20	Group No. 3
18 TO 22	30	42		31	A 30 x 20	
22 TO 26	36	56	25	35	A 36 x 25	
26 TO 30	42			38	A 42 x 25	Group No. 4
30 TO 35	48	63	32	42	A 48 x 32	
35 TO 42	56	80		46	A 56 x 32	
42 TO 48	63		36	50	A 63 x 36	

Fig. 7.1 A typical company standard sheet.

requirements in a concise form to facilitate their use in office as well as in workshop or laboratory. Most often they are printed in a loose-leaf form which can be bound together readily and added to or replaced from time to time. A typical company standard sheet is illustrated in Fig. 7.1.

7.18 Constituting as it does the basic organized level of standardization, the company standards movement plays an important part in directly shaping the national standards and indirectly the international standards as well. Most leading firms of importance, sometimes including even those having no formal standards department of their own, participate directly in the technical work of the National Standards Bodies and assist the latter in setting up national standards. By being subscribing and committee members of the national bodies, these firms act as pillars which support the national standards effort. As such they are called upon by the National Standards Bodies to assist in their country's participation in the international standards work. In certain countries where industry level standardization is well developed, such as in the USA, company standards departments often work through the medium of associations of trade, industry and professions. But in the newly developing countries such cases seldom arise. Thus, from the individual level right up to the international level, standardization presents a continuous two-way traffic of support through participation upward and through supply of authoritative technical documents downwards.

REFERENCES TO CHAPTER 7

1. Melnitsky, Benjamin: *Profiting from industrial standardization.* Conover-Mast Publications, New York, 1953, p. 381.
2. *Bedrijf en norm* (Company standardization). Nederlands Normalisatie-Instituut, The Hague, 1962, p. 233 (*in Dutch*).
3. *La normalisation dans l'entreprise* (Company standardization). Association Française de Normalisation, Paris, 1967, p. 295. (*in French*).
4. List of known professional societies and organizations of standards engineers in different countries:
   (a) Belgium — Comité de la Normalisation d'Entreprise (CNE) c/o Institut Belge de Normalisation
   Avenue de la Brabançonne 29,
   B–1040, Bruxelles-4
   (b) Czechoslovakia — Central Commission for Technical Standardization of the Central Council of the Czechoslovakian Scientific and Technical Society,
   Prague
   (c) France — Association des Cardes de Normalisation (ACANOR),
   108 Rue Montmartre,
   Paris-2e

(d) Germany	Normenpraxis, c/o DNA, 4-7 Burggrafenstrasse, 1 Berlin 30
(e) Hungary	Commission of Hungarian Standards Engineers (Federation of Technical and Scientific Societies), Hungarian Office for Ṡtandardization, Magyar Szabvanyugyi Hivital, Ulloi-ut 25, Budapest 11
(f) India	Institute of Standards Engineers, 9 Bahadur Shah Zafar Marg, New Delhi-1
(g) Japan	Japanese Standards Association, 1–24, Akasaka 4, Minato-ku, Tokyo
(h) Netherlands	Commissie Bedrijfsnormalisatie (COBENO) c/o Nederlands Normalisatie Instituut Polakweg 5, Rijswijk (ZH)-2106
(j) Sweden	Association of Swedish Standards Engineers, Allmanna Svenska Elektriska Aktiebolaget, Fack, Vasterås 1,
(k) UK	Standards Associates Section, British Standards Institution, 2 Park Street, London W1
(m) USA	Standards Engineers Society, Post Box 7507, Philadelphia, PA 19101
(n) USSR	All Union Council of Scientific and Technical Societies (VSNTO) Committee for Standardization, Moscow

5. The two journals issued by the standards engineers societies of the USA and India are respectively:
   (a) *Standards Engineering* (two-monthly)
   (b) *Standards Engineer* (quarterly).

6. Hussey, George F.: Definitions of the areas of standardization. *Proceedings of the Fifth National Conference on Standards.* American Standards Association (now American National Standards Institute), New York, 1954, pp. 48-50.
7. Milek, John T.: Role of management in company standardization. *Magazine of Standards,* New York, 1962, vol 33, pp. 107-114.
8. Sen, S. K.: Development of company standards. *Indian Finance, ISI Supplement,* New Delhi, 1963, pp. 13-14.
9. Gokhale, M. S.: How standards can help. *RCA Value Engineering Seminar,* 1958.
10. Singh, Brij B.: Necessity, object and savings by company standardization. *United Nations Inter-Regional Seminar on Promotion of Industrial Standardization in Developing Countries,* Denmark, 1965, vol 1, pp. 315-322.
11. Gupta, A. K.: Company standardization in USA and Japan. *ISI Bulletin,* 1964, vol 16, pp. 287-291.
12. Frontard, R.: Economic aspects of standardization. *ISI Bulletin,* 1966, vol 18, pp. 441-446.

13. Riebensahm, H. E.: Cost savings through in-plant standards in lubrication and lubricants – a case study. *ISI Bulletin,* 1965, vol 17, pp. 445-448.
14. Goswami, H.: Programme evaluation and review technique help cost reduction through standardization. *ISI Bulletin,* 1966, vol 18, pp. 316-319.
15. Sedgwick, Harry K.: Engineering standardization – friend or foe. *Magazine of Standards,* New York, 1966, vol 37, pp. 225-229.
16. Robb, H. *et al.*: Factors involved in organizing a company standards department. *Proceedings of American Standards Association* (now ANSI), New York, 1953, pp. 41-51.
17. Mohr, Ernest E.: Starting a standards programme. *Magazine of Standards,* New York, 1960, vol 31, pp. 238-239.
18. *PD 3542-1970 Operation of company standards department.* British Standards Institution, London, p. 22.
19. White, Arnold B: Development of a dynamic standards programme. *Standards Engineering,* 1957, vol 9, no 3, pp.1 and 7-10.
20. Willard, Charles O.: Preparation of a standards document. *Standards Engineering,* 1968, vol 20, no 3, p. 15-18.
21. Rosenwald, Arnold M.: Let's talk sense about company standards. *Standards Engineering,* 1962, vol 14, no 6, pp. 3-7.
22. Verman, Lal C.: Standardization in a developing economy (in two parts). *ISI Bulletin,* 1964, vol 16, pp. 238-245 and 292-298.
23. Ghosh, A. N.: Economic effects of standardization – national and international. *Tenth Indian Standards Convention (Ernakulam) Souvenir.* Indian Standards Institution, New Delhi, 1966, pp. 23-28.
24. Gokhale, M. S.: Techniques of standardization. *Proceedings of the Annual Meeting of the Standards Engineers Society,* Chicago, 1957.
25. Gokhale, M. S.: Design and standards. *Proceedings of the Annual Meeting of the Standards Engineers Society,* Chicago, 1957.
26. Boulton, B. C.: Co-ordinating standards work with design and production. *Product Engineering,* USA, Sept 1943.
27. Goeltz, Phillip H.: The company's standards manual. *Magazine of Standards,* New York, 1960, vol 31, pp. 268-270.
28. Palit, A. R.: The scope, purpose and techniques of materials management. *Eastern Purchasing Journal,* New Delhi, 1962, vol 3, p. 245-248.
29. Venkataraman, R.: Materials management in Bhilai Steel Plant. *ISI Bulletin,* 1965, vol 17, pp. 157-160 and 184.
30. Callan, Philip J.: An identification numbering system for purchased materials. *Proceedings of the Fourth National Standardization Conference and Thirty-fifth Annual Meeting of the American Standards Association* (now ANSI), New York, 1953, pp. 52-56.
31. Nagaraja, Y. R.: Materials classification and coding as a means to reduce inventory. *Eleventh Indian Standards Convention (Chandigarh). Doc S-4/6.* Indian Standards Institution, New Delhi, 1967, p. 9.
32. Gaillard, John: The company standards engineer. *Standards Engineering,* 1958–59, vol 10, no 6, p. 6.
33. Woerter, Everett: Job description for standards engineer. *Standards Engineering,* 1961, vol 13, no 4, pp. 3-8.
34. Gardner, Sherwin: A standards section – function and responsibility. *Standards Engineering,* 1964, vol 16, no 3, pp. 3-5.
35. Trowbridge, Roy P.: The General Motors standards programme. *Magazine of Standards,* New York, 1963, vol 34, pp. 67-72.
36. Hall, John C.: The Ford approach to manufacturing standards. *Standards Engineering,* 1962, vol 14, no 4, pp. 6-9.

37. Bhattacharya, G. C. and Bhatia, S. K.: Company standards unit as a centre for co-ordination of the various service functions within an organization. *Tenth Indian Standards Convention (Ernakulam), Doc S-3/1.* Indian Standards Institution, New Delhi, 1966, pp. 1-7.
38. Rangashai, M.: Role of company standards in mechanical engineering industry. *ISI Bulletin,* 1968, vol 20, pp. 209-212.
39. Bisbee, R. F.: Methods of evaluating savings from standardization. *Second National Standardization Conference of the American Standards Association* (now ANSI), New York, Oct 1951, pp. 44-49.
40. *Kummer, W. et al.*: Dollar savings through standards. *Proceedings of the Seventh National Conference on Standards of the American Standards Association* (now ANSI), New York, Oct 1956, pp. 34-42.
41. Griffin, Thomas F. *et al.*: Here's how standards make money for my company. *Proceedings of the Ninth National Conference on Standards of the American Standards Association* (now ANSI), New York, Nov 1958, pp. 51-66.
42. Berg, Winfred M.: Standardization reduces cost. *Proceedings of the Tenth Annual Meeting of the Standards Engineers Society,* Chicago, 1961, pp. 5-13.
43. Berry, Richard B.: When does standardization cost money? *Proceedings of the Tenth Annual Meeting of the Standards Engineers Society,* Chicago, 1961, pp. 71-73.
44. McAleer, James A.: Improved profits through better buying. *Proceedings of the Thirteenth National Conference on Standards of the American Standards Association* (now ANSI), New York, Feb 1963, pp. 61-62.
45. *Cost reduction through standardization – 19 papers for Session S-2 of the Ninth Indian Standards Convention (Bangalore).* Indian Standards Institution, New Delhi, 1965:
    - S-2/1.1 Riebensahm, Hans E.: Cost savings through in-plant standards on lubrication and lubricants – a case study.
    - S-2/1.2 Goswami, H.: PERT helps in cost reduction through standardization.
    - S-2/1.3 Sathyanarayana, K. and Tolpadi, S. G.: How many holes a second – a case study in standardization.
    - S-2/1.4 Mehta, V. N.: The establishment and implementation of a company standards programme.
    - S-2/1.5 Trivedi, M. P.: Cost reduction through standardization.
    - S-2/1.6 Sundara Raju, S. M. and Banerjee, B.: Standardization via cost reduction.
    - S-2/2.1 Lodh, M. R.: How standardization helps cost reduction.
    - S-2/2.2 Bandyopadhyay, M. M.: Standardization: an aid to the way of better and economic performance.
    - S-2/3.1 Tandon, L. N.: Cost reduction in design and manufacture of jigs and tools through standardization.
    - S-2/3.2 Malik, J. N.: Cost reduction through standardization.
    - S-2/3.3 Meswani, N. V.: Cost reduction through standardization in automobiles.
    - S-2/4.1 Chatterjee, D. P. and Mukherjee, G.: Cost reduction through standardization in alloy steels plant.
    - S-2/4.2 Allabaksh, M. and Thirumalarao, S. D.: Standards for and test of evaluating expellers.
    - S-2/5.1 Iyengar, K. N. S.: Cost reduction through standardization in building industry.
    - S-2/5.2 Hafeez, M. A. and Mathur, G. C.: Cost reduction through standardization in mass housing programmes.

S-2/6.1 Padmanabhan, S.: Cost reduction through standardization with special reference to company standardization in fertilizer industry in India.

S-2/7.1 Bhimasena Rao, M. and Anantakrishnan, C. P.: Cost reduction through standardization.

S-2/8.1 Rao, P. H.: Cost reduction through standardization in leather manufacture.

S-2/9.1 Sreedhara Rao, K.: Scope for cost reduction through standardization in thermal power stations.

46. Joynt, Gerald C.: Standards pay dividends. *Standards Engineering,* 1966, vol 18, no 3, pp. 1 and 11-14.
47. *Economic effect of the introduction of standards: Study of standardization patents and marketing. Industrial development in Asia and the Far East,* vol 1, 1966. Economic Commission for Asia and the Far East, Bangkok. pp. 261-263.
48. Economic aspects of standardization. *La Normalisation dans l'Entreprise.* Association Francaise de Normalisation, Paris, 1967, pp. 23-27 (*in French*).
49. Piper, Paul A.: Observations on standardization cost and returns. *Proceedings of the 17th Annual Meeting of the Standards Engineers Society,* New York, Sept 1968, pp. 108-119.
50. Willets, Walter E.: Standardization as a major factor in cost reduction in procurement. *Proceedings of the 17th Annual Meeting of the Standards Engineers Society,* New York, Sept 1968, pp. 120-127.
51. Standardization as effective factor in materials economy. *Standardisierung,* 1969, vol 15, pp. 251-254 (*in German*).
52. *Dollar savings through standards. Report of survey to obtain data on savings derived from the use of standards by American industry.* American Standards Association (now ANSI), New York, 1955, p. 40.
53. *Company Standardization – organization, costs, savings.* American Standards Association (now ANSI), New York, 1959, p. 26.
54. Arnold, W. F.: The tangible benefits of standardization. *Standards Engineering,* Jan 1965, vol 17, no 2, pp. 4-6.

Chapter 8

Industry Level Standardization

8.0 On the next rung of the ladder after company standardization lies "industry level" or "association level" standardization, the two terms being used interchangeably. This represents the collective activity of groups of companies and other enterprises concerned with a given industry or trade. It also includes the standardization activity of certain professional bodies of engineers and the societies concerned with the advancement of science and technology. In some countries, the terms "association level" is preferred to "industry level," because the actual standardization activity is carried out mostly by associations of industrial enterprises or professional institutes. Associations of trades as well as those of dealers and wholesalers are also sometimes concerned with standardization. For this reason such standards may sometimes be referred to as trade standards. For the purpose of this book, however, the terms "industry standards" and "industry level standardization" are preferred because of their general acceptability in most countries of the world.

8.1 Apart from associations of industrialists and traders, professional bodies of engineers and architects, and sometimes even learned societies of scientists, find it necessary to prepare standards applicable to their own respective fields of work. Such occasions arise particularly when rapid advances are being registered in a given field of knowledge or in its application to industry. When new technologies are developing, new concepts have to be created for which new terminology becomes necessary. The same thing happens when new processes are introduced or new products begin to be manufactured. Such was, for example, the case when television first came into being or when atomic power and computers were first developed. So, whenever frontiers of existing knowledge are pushed forward or the application of new knowledge is advanced, there arises the need for new and original standards, and in the first instance such standards are often evolved by scientists and engineers who are actually involved in the new developments. This happens mostly at the industry or professional level, but ultimately these standards, like most others, get absorbed into the national and international structure.

8.2 Industry level standardization serves to integrate the company level standards and unify them in the interest of the industry as a whole. On the other hand, industry standards serve as the basis for an overall integration at the national level. It is clear that when an industry association is engaged in a standardization project of its own, its main concern is with the interests of the particular industry and its members. How the project would affect other industries, which may or may not be directly related to this particular industry, would, by and large, be a matter of secondary concern. For examplie, an automobile industry standard for deep drawing quality of steel would primarily take care of auto body building needs but may not suit the needs of some other users of the material. This is where national integration of industry level standards becomes essential. At the same time it must be appreciated that the national task of coordination and integration would be considerably facilitated, if the considered views of all the interests of various industries concerned were readily available, which would certainly be the case if industry level standards existed to a sufficiently large extent. There is, however, always the danger that if in a particular instance national effort of coordination were delayed over a long period of time, industry level standards would become so well entrenched that any belated national level effort at unification might be handicapped by the resistance to change in the already well-established practices. If this happens the national economy would continue to suffer from the adverse effects of avoidable multiplicity of standards.

8.3 In certain industrially advanced countries, industry level standardization activity is highly developed. In the USA for example, there are nearly 500 associations, professional bodies and governmental agencies[1] which have issued and continue to issue thousands of industry level standards on a large number of subjects. On the other hand, national standardization activity in that country is comparatively limited. In most other industrially advanced countries, the balance between industry and national level standardization is fairly well maintained. In contrast, the picture is somewhat the opposite in developing countries. In these countries, standardization movement as such, being of relatively recent origin, is mainly centred around the National Standards Bodies and industry level standardization is often generally conspicuous by its virtual absence. While this situation does facilitate coordination of industry practices on the basis of national standards, it does tend to place considerable burden on the National Standards Bodies.

8.4 The position of the USA in this respect is somewhat unique. There industry level standardization is so well developed that the standards issued by most industry associations are generally considered by those concerned as standards of national importance and used as such. The procedures usually followed by these societies[2 3] are extensive and exhaustive enough to take care of the national need for widespread consultation among almost all interests concerned. Thus the national acceptance of many industry level standards becomes almost automatic, so much so, that many of the industry and association standards are accepted as national standards without any change, when they are formally submitted for such acceptance under the procedure of the American National Standards Institute (ANSI).* Indeed many of the projects for preparing new national standards are sponsored and dealt with by industry associations under the auspices of ANSI and in accordance with the latter's formal procedure which, among other things, requires nation-wide consultation with all interests. Thus, the industry associations and professional bodies play a very important and significant role in standardization movement in the USA. In this respect the position of the American Society for Testing and Materials (ASTM) is particularly noteworthy. While most industry associations deal with subjects of particular interest to their own individual field of industry or trade, ASTM covers all the ramifications of materials and their testing, pertaining to all fields of endeavour. It has so far issued well over four thousand standards,[4] all of which are considered as of national importance, though in the strictest sense of the word they are industry level standards. It may also be noted

* This is the latest title of the USA National Standards Body. Earlier titles have been American Standards Association (ASA, from 1928 to 1966) and United States of America Standards Institution (USASI, from 1966 to 1969).

that quite a few of the ASTM standards have been adopted as American Standards under the ANSI procedure without any change.

8.5 In certain other economically advanced countries also, industry associations are quite active in industry and national level standardization, but nowhere do they exist in such large numbers as in the USA. For example, the technical institutes of industries in the USSR, which are quite numerous, and in France, the research and development institutions of several industries play a somewhat similar role in initiating and developing national standards projects under the auspices of the respective National Standards Bodies, namely the Committee for Standardization, Measures and Measuring Instruments of the USSR (GOST) and the Association Française de Normalisation (AFNOR). Similarly in the UK, professional societies of engineers and technologists take a particularly active interest in the evolution of codes of practice for design, construction and other industrial processes and operations.

8.6 In the newly developing countries, industry level standardization movement, as mentioned earlier, has had little chance to develop. It was only after the conclusion of the Second World War, that a large number of colonial countries acquired control of their own affairs. During the first two decades of independence, though many countries have had so little time, they have had to tackle problems of construction and reconstruction pertaining to every facet of national life. While, as an essential infrastructure activity for industrialization and commercial expansion, standardization has also attracted the attention of many of these countries, their first concern has been with increasing indigenous production of goods and services. In some of these countries National Standards Bodies have been or are in the course of being established, but they have still to build up their cadres to adequate strength and spread the movement to the grass roots of their economies. While this is going on at the national level, the industrialists and entrepreneurs at the industry level are busy tackling problems of securing finance, acquiring know-how, arranging procurement of materials, marketing of products and so on. In most cases they have hardly had an adequate chance of banding together and organizing themselves in industry and trade associations which could address themselves to the problems of standardization. Similarly, learned societies of scientists and engineers are comparatively new and have on hand a good many problems of organizational and professional character to face. Nor are such societies actively engaged in the advancement of the frontiers of knowledge in connection with which the need for standardization might be urgently felt.

8.7 Thus, it is obvious that the burden of standardization work at the early stages of development of a country must be carried largely by its National Standards Body (NSB). This has to be borne in mind from the very beginning and provided for in the initial plans for setting up an NSB. Neither the industry level activity nor even the company level work can be expected to play a very significant role, until the NSB has been fairly well organized. So, the highest of priorities should be accorded to this nation building task. It should, however, be considered incumbent that the NSB when established should not only encourage the advancement of industry level and company level standardization movements, but should also fully recognize that handsome dividends would accrue, if a programme of active promotion of the lower level activities is vigorously pursued.

8.8 In some developing countries of the East, like India and Iran, considerable experience has been accumulated during the past two decades in getting the national standards movements under way. The experience in India is particularly significant in regard to the relationship of the national level standardization to the industry or association level. As in most other developing countries, industry associations and professional bodies in India though existing in large numbers are, by and large, inactive in setting up industry level standards. There are some important exceptions, however, where certain governmental agencies have been rather active in issuing standards for their own use, which could well be equated to industry level standards. The more important of these agencies are:

(1) Directorate of Standardization of the Defence Production Organization
(2) Directorate General of Supplies and Disposals
(3) Railway Designs and Standards Organization
(4) Indian Roads Congress
(5) Directorate of Marketing and Inspection (for grading of agricultural products)
(6) Central Committee for Food Standards
(7) Drug Controller
(8) Director General of Mines Safety
(9) Chief Inspector of Explosives
(10) Textile Committee, and
(11) Export Inspection Council.

8.9 The last seven of the agencies listed above are statutory authorities established with the express purpose of administering definite pieces of legislation under which they exercise prescribed powers. The standards they issue for regulating the particular sectors of industry under their charge, though restricted to safety, health or other specific considerations, serve, by and large, the functions of industry standards in the ordinary sense of the word. They have to be followed by industry as a whole in the normal course of operation as required by law. Furthermore, all such standards must necessarily be taken into account when national level standards are being evolved by the NSB of the country, which incidentally would have to cover a much wider scope and stipulate various other requirements of the products besides those required by law. For example, national standards would have to provide for detailed specifications, service requirements, methods of tests and sometimes even certain elements of process of manufacture, none of which need appear in the standards issued by governmental agencies of the type exemplified by most of the last seven agencies of the above list. Nevertheless, the ground covered by governmental standards is essential and serves as the foundation on which the superstructure of national standards can be built.

8.10 As regards the first four agencies listed in para 8.8, they are more of an administrative character and issue standards, specifications and codes – required for the goods and services dealt with in the course of their day-to-day operations. These standards, though originally evolved to meet the needs of specific departments, are of a more general and broad character and cover most aspects of the subject matter dealt with. Inasmuch as they serve departmental needs only, they could well be likened to company standards. But actually in a developing country where no national or industry level standards otherwise exist, such departmental purchase specifications and codes of practice acquire a somewhat higher status. Governmental agencies constitue large-scale organized buyers and users of goods and services, and, therefore, their standards come to wield an important and significant influence in shaping and modeling the bulk of production of various industries which supply their needs. These industries by themselves usually constitute quite an influential sector in each group and the standards they follow become worthy of being recognized as industry level standards. It is thus that in developing economies governmental specifications and codes acquire the status of industry level standards. The picture is not a great deal different in industrially advanced countries, particularly in certain sectors of industry where national standards may be lagging behind production. Take, for instance, the standards followed by all the private and public sector producers of equipment for the aerospace programme in the USA and the USSR or the atomic power plan equipment in these and several other countries.

8.11 In developing countries, the governmental purchase specifications and codes of practice, which have been widely used, become a highly valuable source of material for the national standardization programme, as soon as a National Standards Body begins to function. These standards, having been tried out for years by producers and consumers alike, serve as a firm foundation for the national structure in developing countries, exactly in the same manner as industry or association standards do in industrially advanced countries where, incidentally, government specifications and codes are also available. In view of the virtual lack of company and industry standards, the developing countries have to make an extra effort in organizing national programmes of standardization.

REFERENCES TO CHAPTER 8

1. Hartman, Joan E.: *Directory of United States standardization activity* (National Bureau of Standards Miscellaneous Publication 288). US Government Printing Office, Washington DC, 1967, p. 276.
2. Gupta, A. K.: Standards activity of National Electrical Manufacturers Association. *Report No 40 of National Productivity Council of India,* New Delhi, 1964, pp. 113-119.
3. Gilbert, J.: Society of Automotive Engineers (SAE) contribution to standardization. *Report of the Secretary and General Manager of SAE, Detroit, Michegan, USA.*
4. *Annual book of Standards* in 33 parts. American Society of Testing and Materials, Philadelphia, 1971.

Chapter 9

National Standardization: Organization

9.0 Next to the industry level on the pyramid of standardization is the national level of standardization, which is perhaps the most important of all levels. Its scope and influence is far-reaching both ways – to all the lower levels already described as well as the next higher level – the international level. It is at this level that the requirements of the individual, company and industry levels are coordinated and integrated, and purposeful national standards emerge which serve as effective instruments for guiding the development of a nation's industry and commerce and bringing about order in the existing pattern of national economy. At the same time, national level standards serve as a basis for forging international agreements on international standards, which help promote worldwide exchange of goods and services, in which every nation is vitally interested. Constituting as it does the key level, national standardization has been treated in two chapters – this one on Organization and the next one on Procedures and Practices.

GROWTH OF NSBs

9.1 The work of national level standardization is dealt with by national organizations, which in some countries are called institutes or institutions and in other associations or societies. In some countries a department or an agency of the government is made responsible for the work. In certain socialist countries the work is entrusted to governmental committees of officials, which are created either under a statute or a government order. For the purpose of our discussion, the various types of organizations will simply be referred to as the National Standards Bodies (NSBs). Most NSBs of the world are members of the International Organization for Standardization (ISO). A partial list of such of the world's NSBs as are presently members of ISO will be found in an ISO publication on the subject.[1] There are a few other NSBs which are not ISO members, but these include mostly those that are in early stages of development, and have not yet acquired either the financial or the technological wherewithal required to make their participation in ISO fruitful. Altogether there may be about 70 NSBs in the world today, including the 55 constituting the present membership of ISO.

9.2 Although all NSBs are responsible for national standardization work within their own national orbit, yet the internal organization of each differs considerably, depending upon the historical, economic and political background of the particular country. Interesting descriptions of the internal structure of each of the ISO Member Bodies is contained in another ISO publication.[2]

The United Nations Economic Commission for Asia and the Far East (ECAFE) has also issued a similar booklet[3] describing briefly the organization and the accomplishments of the NSBs in the ECAFE region countries. Both these publications contain a considerable amount of data on the working of the various NSBs and the present stage of their development, say up to 1969–70. In Table 9.1 is collected some of the relevant information about different NSBs of the world, which includes their abbreviated names, year of establishment as a national body, character of the organization (governmental, private or joint), and the number of standards each one has issued. It also indicates whether the NSB publishes one or more periodicals dealing with standards and standardization and the frequency of such periodicals, whether the NSB has organized a certification marks scheme, and whether it undertakes industrial testing for certification marks and for other purposes.

9.3 Table 9.2 gives the historical sequence of creation of NSBs in various countries, which process, it will be seen, started early in the current century. This table also indicates the total number of NSBs in existence at various epochs of time. The year of creation of an NSB shown in Tables 9.1 and 9.2 is that in which either the present organization, or its predecessor organ closely resembling the present one, was set up. It does

TABLE 9.1

WORLD'S NATIONAL STANDARDS BODIES (NSBs)

(As at the end of Dec 1971)

All full ISO members, except marked (C) – Correspondent members and (N) – non-members

Serial No.	Country	NSB	Established	Character	Standards		Periodical(s)		Certification Marks (CM)		Testing for	
					V or M	Number*	No.	Periodicity (months)	Yes or No	Number	CM	Other Purposes
1	Albania (N)	ASBS	1951	G – † Planning Commission	M	800	–	–	–	–	–	–
2	Argentina (N)	IRAM	1935	P	V	2,300	2	2, 3	Yes	–	–	–
3	Australia	SAA	1922	J	V	1,661	1	1	Yes	100	–	–
4	Austria	ON	1920	P – † Chamber of Commerce	V	1,030	1	1	Yes	–	–	–
5	Belgium	IBN	1919	J	V	1,050	1	1	Yes	–	–	–
6	Brazil	ABNT	1940	J	M	1,100	1	2	Yes	–	–	–
7	Bulgaria	KKCM	1948	G – A: Measuring Instruments	M	7,810	1	1	–	–	–	–
8	Burma (N)	–	1956	G – † Industrial Research	–	–	–	–	–	–	–	–
9	Canada	CSA	1919	J	V	1,000	2	2, 3	Yes	–	Yes	Yes
10	Central America‡ (N)	ICAITI	1962	G – † Industrial Research	V	110	–	–	No	–	–	Yes
11	Ceylon	BCS	1964	G	V	118	–	–	Yes	–	Yes	Yes
12	Chile	INDITECNOR	1944	J	V	539	1	1	Yes	–	–	–
13	China (Taiwan) (N)	NBS	1947	J – A: Weights and Measures	V	2,750	1	1	Yes	1,910	Yes	Yes

Serial No.	Country	NSB	Established	Character	Standards		Periodical(s)		Certification Marks (CM)		Testing for	
					V or M	Number*	No.	Periodicity (months)	Yes or No	Number	CM	Other Purposes
14	Colombia	ICONTEC	1963	J	V	374	1	1	Yes	–	–	–
15	Cuba	NC	1961	G – A: Metrology	M	360	1	2	–	–	–	Yes
16	Czechoslovakia	CSN	1922	G – A: Weights and Measures	M	12,670	3	1, 1, 1	Yes	–	Yes	Yes
17	Denmark	DS	1926	J	V	1,000	2	1, 1	Yes	–	–	–
18	Egypt	EOS	1957	G – A: Metrology	V	985	1	2	Yes	–	Yes	Yes
19	Ethiopia (C)	ESI	1969	–	–	–	–	–	–	–	–	–
20	Finland	SFS	1924	J	V	1,003	1	1	Yes	–	–	–
21	France	AFNOR	1926	J	V	6,449	2	1, 2	Yes	–	–	–
22	Germany	DNA	1917	P	V	11,118	3	1, 1, 2	Yes	–	–	–
23	Ghana	ISIG	1967	G – A: Industrial Testing	V	731	–	–	Yes	–	–	–
24	Greece	NHS	1968	G	V	25	–	–	Yes	–	–	–
25	Hungary	MSZH	1933	G	M	8,650	1	1	Yes	–	–	–
26	India	ISI	1947	J	V	5,724	2	1, 1	Yes	2,196	Yes	–
27	Indonesia	DNI	1954	P	V	57	1	2	No	–	–	–
28	Iran	ISIRI	1960	G – A: Weights and Measures‖	V	527	–	–	Yes	31	Yes	Yes
29	Iraq	IOS	1963	G – A: Weights and Measures	V	37	–	–	–	–	–	–
30	Ireland	IIRS	1946	G – A: Industrial Research	V	153	1	1	Yes	–	Yes	–
31	Israel	SII	1945	J –	V	796	1	3	Yes	–	Yes	Yes
32	Italy	UNI	1928	J – † National Res. Council	V	4,700	1	3	No	–	–	–

Serial No.	Country	NSB	Established	Character	Standards		Periodical(s)		Certification Marks (CM)		Testing for	
					V or M	Number*	No.	Periodicity (months)	Yes or No	Number	CM	Other Purposes
33	Jamaica (N)	JBS	1969	–	–	–	–	–	–	–	–	–
34	Japan	JISC	1921	G	V	6,855	–	–	Yes	14,594	Yes	–
35	Korea, North	CSK	1949	G	M	3,000	2	2, 2	No	–	–	–
36	Korea, South	KBS	1961	G	V	1,545	1	1	Yes	351	–	–
37	Lebanon	LIBNOR	1962	J	V	64	–	–	Yes	–	–	–
38	Malaysia	SIM	1966	J	V	–	–	–	Yes	–	Yes	–
39	Mexico	DGN	1946	G – A: Weights and Measures	V	1,173	–	–	Yes	–	–	–
40	Morocco	SNIMA	1962	G	V	24	–	–	–	–	–	–
41	Netherlands	NNI	1919	J	V	2,200	1	1	Yes	–	–	–
42	New Zealand	SANZ	1932	J	V	1,869	2	1, 3	Yes	1,335	–	–
43	Nigeria (C)	NSO	1970	G	–	–	–	–	–	–	–	–
44	Norway	NSF	1924	J	V	1,276	1	3	Yes	–	–	–
45	Pakistan	PSI	1951	J – A: Weights and Measures	V	760	1	3	Yes	41	–	–
46	Paraguay (N)	INTECNOR	1965	G	V	1	–	–	–	–	–	–
47	Peru	ITINTEC	1959	J	V	167	–	–	Yes	–	–	–
48	Philippines	KP	1947	G	M	108	–	–	Yes	–	Yes	–
49	Poland	PKN	1924	G	M	9,715	2	1, 1	Yes	–	–	–
50	Portugal	IGPAI	1952	J	V	675	1	1	Yes	–	–	–
51	Rhodesia (N)	CASA	1960	J	V	58	–	–	Yes	–	Yes	Yes
52	Rumania	IRS	1948	G	M	6,095	1	1	Yes	–	–	–
53	Singapore	SISIR	1963	J – A: Industrial Research	V	21	–	–	Yes	–	Yes	Yes
54	South Africa	SABS	1934	G	V	1,570	1	1	Yes	–	Yes	Yes
55	Spain	IRANOR	1946	J – † Council of Sc. Research	V	2,660	1	2	Yes	–	–	–

Serial No.	Country	NSB	Established	Character	Standards		Periodical(s)		Certification Marks (CM)		Testing for	
					V or M	Number*	No.	Periodicity (months)	Yes or No	Number	CM	Other Purposes
56	Sudan (C)	SOS	1969		–	9	–	–	–	–	–	–
57	Sweden	SIS	1922	J	V	3,800	1	1	Yes	130	–	–
58	Switzerland	SNV	1919	P	V	2,300	1	1	No	–	–	–
59	Thailand	CTNSS	1967	J – † Appl. Sc. Res. Corpn.	V	4	–	–	No	–	–	–
60	Turkey	TSE	1954	J	V	760	1	1	Yes	–	Yes	Yes
61	UK	BSI	1918	J	V	6,000	3	1, 1, 3	Yes	1,400	Yes	–
62	USA	ANSI	1918	P	V	4,000	3	0.2, 2, 3	Yes	–	–	–
63	USSR	GOST	1925	G – A: Scientific Instruments	M	13,000	3	0.2, 1, 1	Yes	–	–	–
64	Venezuela	COVENIN	1958	G – † Directorate of Industry	V	539	–	–	Yes	–	–	–
65	Viet Nam, South (N)	VIS	1967	G	V	7	–	–	–	–	–	–
66	Yugoslavia	JZS	1950	G	M	6,560	1	1	Yes	–	–	–
67	Zambia (C)	ZSA	1969	–	–	–	–	–	–	–	–	–

– Information not available or vague.
* The number of standards given refers mostly to those published by the end of 1969.
† NSB is organizationally a part of the organization mentioned.
‡ Joint venture of 5 countries – Costa Rica, El Salvador, Guatemala, Honduras and Nicaragua.
|| ISIRI is also responsible for industrial research, weights and measures, industrial testing, pre-shipment inspection and metrology.
A–Additional responsibility of NSB.
G–NSB managed and financed by government.
J–NSB jointly managed and financed by government and industry.
M–Standards issued are generally mandatory.
P–NSB managed and financed by private sector interests.
V–Standards issued are initially or generally voluntary.

not necessarily indicate the year of initiation of national standardization movement in a given country. In many countries, long before the creation of an NSB, national standardization activity began to develop in limited or specialized spheres of work, such as electrotechnology and government purchasing. For example, in the UK, the British Engineering Standards Association, the predecessor of the present British Standards Institution, was incorporated in 1918; yet the earliest record of the national standards movement in that country, which is regarded as the oldest in the world, goes as far back as 1901, when an engineering standards committee was jointly formed by five professional bodies of engineers and architects.

9.4 The data of Table 9.2 concerning creation of NSBs are plotted as a

TABLE 9.2

CALENDAR OF CREATION OF NATIONAL STANDARDS BODIES

(As at the end of Dec 1971)

Year	*Country*	*Cumulative Total*	*Year*	*Country*	*Cumulative Total*
1917	Germany	1	1949	Korea (N)	35
1918	UK, USA	3	1950	Yugoslavia	36
1919	Belgium, Canada, Netherlands, Switzerland	7	1951	Albania, Pakistan	38
			1952	Portugal	39
			1954	Indonesia, Turkey	41
1920	Austria	8	1956	Burma	42
1921	Japan	9	1957	UAR	43
1922	Australia, Czechoslovakia, Sweden	12	1958	Venezuela	44
			1959	Peru	45
			1960	Iran, Rhodesia	47
1924	Finland, Norway, Poland	15	1961	Cuba, Korea (S)	49
1925	USSR	16	1962	Central America, Lebanon, Morocco	52
1926	Denmark, France	18	1963	Colombia, Iraq, Singapore	55
1928	Italy	19			
1932	New Zealand	20	1964	Ceylon	56
1933	Hungary	21	1965	Paraguay	57
1934	South Africa	22	1966	Malaysia	58
1935	Argentina	23	1967	Ghana, Thailand, Viet Nam (S)	61
1940	Brazil	24			
1944	Chile	25	1968	Greece	62
1945	Israel	26	1969	Ethiopia, Jamaica, Sudan, Zambia	66
1946	Ireland, Mexico, Spain	29	1970	Nigeria	67
1947	China (T), India, Philippines	32			
1948	Bulgaria, Rumania	34			

growth curve in Fig. 9.1, which reveals some interesting characteristics. It is seen that after the First World War, up to about the close of the third decade of the century, the rate of creation of NSBs was about 1.8 per year

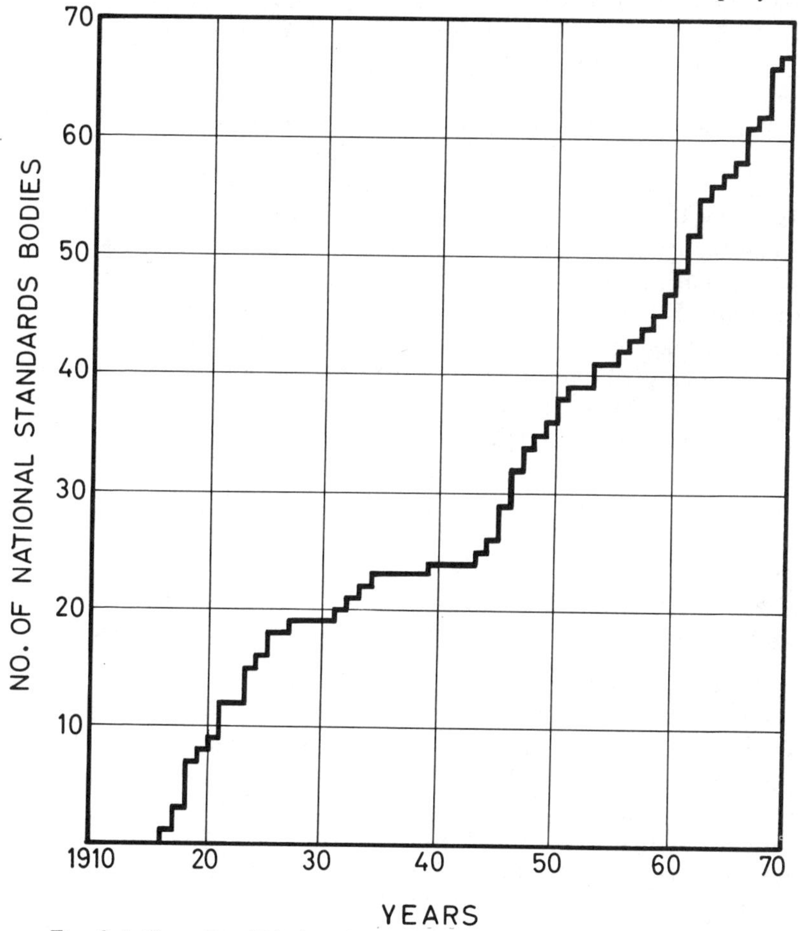

Fig. 9.1 Growth of National Standards Bodies in the world over the years.

on the average, as 18 NSBs had been formally established during the ten-year period 1918–28. All of these were in European countries with the exception of the USA and Japan. During the following years, up to the close of the Second World War, the rate of growth came down to one-fifth this value, when during the 16 years from 1928 to 1944 only 6 NSBs came into being. During the quarter of a century since 1944, the rate of growth has somewhat fluctuated between about one and three new NSBs per year, with an average again of about 1.8. During the fifties, for some reason, the growth was rather slow, that is, about one per year, though during the five or six years immediately following the Second World War and during the sixties the rate has been maintained at an all-time high, that is, roughly two or more new NSBs per year.

9.5 The extraordinary rate of growth of NSBs in recent times may be traced to the fact that during the years following the Second World War, many new countries came into being as a result of the breakdown of the colonial system. In most of these countries, there has been an upsurge of economic development activity which has naturally led to the realization of the importance of standardization. It would perhaps be justified to conclude that, during the period between the two wars, the national standardization movement took a firm root among the countries of Western Europe, North America, the USSR and Japan. As a result of the experience gained and the positive benefits of standardization to industrial development realized in these countries, the newly emerging countries as well as some of the others were led to follow the same path and create National Standards Bodies of their own.

BRIEF SURVEY OF NSBs

9.6 A close study of Table 9.1 would reveal several other interesting features about the national standards bodies of the world. For example:

(1) Of the total of 67 NSBs listed, 36 are in the developing countries and another 31 in advanced countries. This indicates the extent of world-wide interest that has been created in standardization during the recent years, particularly among the newer countries.

(2) While there are only 6 NSBs which are private organizations managed and financed by industry, commerce and other private sector interests, there are 28 NSBs which are jointly run by governments and private interests and 29 NSBs which are government departments or agencies. The character of the NSBs in Ethiopia, Jamaica, Sudan and Zambia, which is not known, is also likely to be governmental. Thus by and large the preference appears to be almost equally divided between joint ventures and government run organizations, while privately run bodies are not much in favour.

(3) Of the 29 government-run NSBs, 17 are in the developing countries and of the 28 jointly-run bodies, 13 are among the developing countries, while among the 6 private bodies, only 2 are so placed. Thus, it would

appear that the preference among the developing countries is somewhat in favour of government-run organizations, though a considerable proportion of them are jointly run establishments. Private bodies, in comparison, appear to be much more unpopular everywhere, especially among the developing countries.

(4) In certain smaller countries, and there are 8 such listed, the NSBs form a part of some other larger establishments, as, for example, the scientific and industrial research organization, the department of industry, the planning commission or even a chamber of commerce. This approach helps to achieve a certain degree of economy through a more efficient utilization of common services, which is an important factor in smaller countries. But this should be balanced against the advantage of independence of action of an NSB, particularly in relation to its work in the international context.

(5) In 13 of the countries listed, most of them comparatively small, the NSBs are entrusted with additional responsibilities other than national standardization. But in all these cases the additional activities are closely related to standardization, as, for example, metrology and measuring instruments, administration of weights and measures, industrial research, pre-shipment inspection of exports and so on. Such a combination is often found useful in smaller countries from the point of view of economy of manpower and finances.

(6) Altogether more than 140,000 national standards appear to have been published by the various NSBs up to the year 1969–70, the individual shares of some being quite high – 13,000 for the USSR, over 12,000 for Czechoslovakia, 11,000 odd for Germany, and nearly 10,000 each for Poland and Hungary.

(7) A large proportion of NSBs, 49 out of 61, issue standards which are initially voluntary in character, while 12 NSBs issue mainly mandatory standards. Ten out of these 12 belong to countries which have centrally controlled economies of the socialist pattern, the remaining two being Brazil and the Philippines. It must, however, be noted that in the countries where NSBs primarily issue voluntary standards, it is common practice for governments to declare as mandatory in public interest some of the critical standards relating to health, safety, export promotion, or other important aspects of national life.

(8) Of the 67 countries listed, 41 regularly publish a total of 59 periodicals on standardization matters, of which 36 are issued monthly, 2 more frequently, 12 bimonthly, and 9 quarterly.

(9) Forty-eight NSBs have responsibility for administering marks

indicating conformity with standards, often referred to as certification marks or standards marks for short. (*see* Chapter 15). Data available regarding the number of certification marks licenses granted by the various NSBs are rather meagre, but Japan appears to have issued perhaps the largest number, that is, nearly 15,000. Among the remaining NSBs, there appears to prevail a general acceptance of the principles of standards marks, but most of them are marking time to reach a suitable stage of development of standardization activity before introducing certification marking schemes.

(10) Altogether, only 19 of the 67 NSBs have industrial testing facilities. Of these 16 carry out tests required for their certification marks operations while 13 undertake industrial testing of a general character as well. Of the latter, 2 NSBs do not operate any standards marks. So it would appear that 29 of the 48 NSBs having standards marks have no testing facilities of their own and depend on other laboratory organizations for this purpose. It is known, however, that most of the NSBs having their own facilities also have to depend for certain specialized tests on outside agencies; only very few could be considered completely self-sufficient.

TWO WINGS OF NSBs

9.7 In view of the great variety of types and sizes of NSBs existing in different countries, it will be appreciated that no generalizations can be attempted as to the pattern of their organization that may be considered ideal or desirable. To quote Hadass of Israel, "there is nothing so non-standard as a standards institution." The prevailing circumstances in a given country – economic, political and social – would largely determine this pattern. One organizational feature, however, is common to all. In each NSB, the totality of responsibility is generally divided between two broad wings of the organization and their interrelationships defined in the light of prevailing circumstances. One of the wings is the central office or the secretariat, which is often called the directorate or sometimes the directorate general, and the other is the committee and council wing. The personnel manning the secretariat is the full-time staff paid out of the funds of the NSB. In some relatively smaller NSBs, part-time paid staff is also engaged for reasons of economy. On the other hand, the committee and council personnel consists mostly of specialists and experts plus a few administrators drawn from outside the NSB, such as industry, trade, technological institutions and government departments. The interests chosen for such representation are those having a stake in standardization and possessing the necessary expertise and experience. These people often work in an honorary capacity and their travel and subsistence expenditure are met by their principals. But there are occasions when such expenses may be met by the NSB itself. In certain countries they are also paid a nominal honorarium for the time spent on NSB work.

9.8 *Directorate Wing* The responsibility of the secretariat or directorate includes chiefly the administration of the affairs of the NSB and its day-to-day functioning, which involves multifarious duties. The most important among these duties are those concerned with the servicing of the committee and council structure; for example:

(1) to organize in accordance with the constitution or statutes of the NSB the relevant committees and councils required to deal with various technical and administrative matters;

(2) to formulate proposals for consideration by the competent authority for securing nomination of representatives on committees and councils to represent the various interests concerned;

(3) to bring new projects for the preparation of standards to the notice of committees concerned along with due recommendations based on enquiries made and investigations carried out;

(4) to collect and furnish the necessary bibliographical and other background data related to the approved projects for the preparation of new standards or the amendment of existing ones;

(5) to follow up decisions of the different committees and councils;

(6) to undertake preliminary drafting of standards in the light of committee decisions wherever called for;

(7) to edit and publish and to organize the sale of standards and other publications of the NSB;

(8) to promote the implementation of standards in different sectors of industry and commerce of the country and generally to act as information centre on all matters concerning standards and standardization;

(9) to organize and control the work of certification marking and arrange for proper inspection and testing of products covered by standards marks;

(10) to propose and process nominations of national delegations to international meetings and conferences on standardization and, in the light of the directions of competent authorities of the NSB, to prepare appropriate briefs for such delegations, which would express the national views on international standardization matters;

(11) to carry out other associated functions as may have been

entrusted to the NSB, such as industrial testing, pre-shipment inspection, weights and measures administration, instrument calibration, etc.

9.9 *Committee Wing* As regards the committee and council wing of an NSB, it is the general rule that the administrative and policy decisions are entrusted to a supreme council or general council or sometimes the general body of its membership. This supreme body, unless it is relatively small, would for detailed work operate through a subordinate executive or working committee which could conveniently meet more often, say every quarter. The technical work on standardization, on the other hand, may be divided among a limited number of high level division councils or industry committees, each being responsible for a fairly large sector of industry. These are appointed by and function under the authority of the supreme body in overall charge of the NSB. The division councils or industry committees determine the technical policies and programmes of standardization for their own sectors of industry. They are constituted by representatives of the industry, trade and other interests concerned with standardization. In government run bodies, the representatives may be nominated by the government or the supreme council, but in jointly operated or privately run NSBs, the membership is defined by naming the industry organizations, companies, government departments and institutions, and the nomination of the personnel is generally left to the interests constituting the membership of the councils. In joint and private NSBs, this practice also applies to other high level administrative bodies and the subordinate level sectional or technical committees.

9.10 The division councils or industries committees appoint sectional or technical committees to undertake the actual work of preparing standards in specific fields, each committee dealing with one or more specializations. These committees, in turn, may create working groups or subcommittees for making a more detailed study and investigations of specified aspects of the problems in hand and for preparing preliminary drafts of standards. In some countries the practice has grown to disband a technical committee soon after the standards assigned to it have been prepared. It is, however, advisable that the technical committees be treated as permanent organs of the NSB, even though some of them may not always be fully active. It is always convenient to have a competent committee available when further proposals of closely allied character are to be dealt with, and then there is always the need for processing amendments to and revisions of published standards. It is generally the responsibility of the technical or sectional committees to accept standards in their final and detailed form, which are subject to the ultimate approval of the division council or industries committees. In NSBs which are governmental agencies, the ultimate approval of standards may be the responsibility of a minister or a ministry of the government or a properly authorized official. This is particularly important wherever national standards are of mandatory character, for after all it is the government's responsibility to ensure that the relevant sectors of economy promptly adopt and strictly follow the standards.

9.11 This general picture of the overall structure of the organization of an NSB is by no means to be regarded as universal. As stated earlier, circumstances in each country differ and more significant differences are perhaps in the countries where the government itself operates the NSB. In the USSR, for example, the various technical institutes which are responsible for the development of different industries assume the role of the so-called division councils or industry committees. Similarly, in France, certain research associations organized by industrial groups assume this role. In certain other smaller NSBs, the division councils may be entirely dispensed with and the work of preparing standards entrusted directly to the technical or sectional committees under the authority of the supreme body and/or the central office directorate. Whatever the detailed picture may be, it may safely be said that any NSB would have its work divided into two more or less diffusely defined compartments – a central secretariat for all administrative, financial and public relations purposes on the one hand, and on the other a consultative wing composed of representative interests exercising varying degrees of responsibility for taking technical decisions on standards.

9.12 *Rationale of Two Wings* This division of responsibility in two wings is the natural consequence of the fact that standardization covers almost all sectors of national economy and it is almost impossible to gather within the framework of a reasonably sized centralized organization all the experts and specialists necessary to deal competently with every branch of industry and trade and to produce standards of requisite quality to guide its operations. Nor would it be economically feasible to do so, because even if the experts were so gathered, they could not be usefully kept engaged on full-time basis, the work-load being rather light and intermittent. Thus it follows that the industrial and other interests concerned should make available to the NSB all the expertise and experience they possess by placing at its disposal the services of their own personnel. This incidentally gives them a sense of participation and a degree of authority in the important nation building work of an NSB, the results of which are of direct interest to themselves. Some NSBs are rather apprehensive that such participation of the vested interest may detract from their own independent status. But to guard against such a contingency, other expedients exist, for example, giving counter-weighing representation to other interests like consumers, users, governments, technologists and scientists.

9.13 Industry's direct participation in the preparation of standards has many other important advantages. When an interest, say a manufacturing unit, is represented by its own official nominee, it becomes incumbent on the latter to consult all other departments of his unit on the contents of a proposed standard and determine the collective viewpoint of his unit as a whole. It is this viewpoint that will determine the attitude of the unit to a given standard and it will be on this basis that the representative will give his consent or otherwise to any compromise that may be called for in finalizing the standard. Thus, all such representatives coming from various interests, some of them intrinsically divergent, constituting a standards writing committee will be able to pool together their experience and highlight their own needs and requirements. This pool of knowledge and experience, representing the overall national background, will determine the shape and content of the national standard. The standard emerging out of this effort will automatically acquire an authoritative character and lead to an automatic and voluntary implementation by the very interests who have had a hand in its shaping. It is also obvious that it will contribute to strengthening the independent status of the NSB.

MEMBERSHIP

9.14 In line with this policy of enlisting the cooperation and support of industry, commerce and other interests in the technical work of the NSB, a good many NSBs, particularly those which are private or jointly run, throw open their memberships to organized bodies and individuals who are interested in the work of national standardization. Certain annual subscription fees along with the privileges to be enjoyed are prescribed for such members. Funds thus collected go to support the work of the NSB. The privileges granted to subscribing members may include the right to receive, free of charge, certain standards and other publications, discount on purchase of additional standards and publications, information on standards and related matters, right to be represented in certain councils and committees of interest, and so on. Most of all, it gives them a sense of participation which is most important in enlisting their cooperation in the implementation of standards.

9.15 Subscribing membership is often divided into various classes intended to cater to the different sizes of enterprises, their capacity to pay, and the extent of their interest. Privileges of members in different classes also vary accordingly. Perhaps a good illustrative example is that of the membership structure of the Indian Standards Institution (ISI), which is briefly as follows:

There are two broad categories of membership.

(1) *Committee members,* who are technical experts and serve on committees and councils and contribute their specialized knowledge and experience and render technical and laboratory assistance as may be required in connection with the investigations concerned with the development of standards. They pay no subscription, but a few of them receive travel and subsistance allowance, in case they serve in personal capacity or represent non-profit making bodies such as educational institutions. They also receive free copies of standards on which they have worked and a discount on purchases.

(2) *Subscribing members* are divided in 5 classes: Patron, Donor, Sustaining, Associate, and Individual Members, each paying annually a subscription of Rs. 25,000, Rs. 10,000, Rs. 500, Rs. 200 and Rs. 50, which roughly amounts to US $3425, $1370, $68, $27 and $7 respectively (US $1.00 = about Rs. 7.30). The first three classes of membership are open to governments of States and Union territories; corporations, municipalities and local authorities, companies and firms; commercial houses and associations. Associate Membership is reserved for smaller firms with limited business and professional institutions, and individuals. Individual Membership is limited to individuals. The different classes of members enjoy graded series of privileges, including limited representation on the General Council which is the supreme governing body, the division councils and other committees, receiving free copies of standards and so on.[2, 4, 5]

9.16 In government-run NSBs, the practice of enlisting members from industry and trade becomes somewhat complicated. Though some such institutions do enlist subscribing members, this is not the general tendency. It would readily be appreciated that for a government department the consideration of financial assistance derived from membership fees is not as important a factor as it is in the case of privately or jointly-run bodies. The technological contribution to be derived from industrial members and their support in implementation of standards are, no doubt, important factors, but even these could be readily forthcoming whenever a government backed body would call for them. Considering all the pros and cons of the situation, however, it would appear that an active, direct and voluntary participation of industry and trade in the policy making and management organs of an NSB would give the NSB an added advantage. It would be desirable, therefore, that even in government-run organizations this feature may be maintained as far as feasible. Institutions jointly run with direct government participation as well as that of private interests would thus have a natural edge both on a purely government-run NSB or a private one. A government-run NSB may suffer from lack of active support from private interests and a private-run NSB from lack of financial support from government.

9.17 There is, however, another important consideration. In a small, newly emerging country, which desires to make quick progress towards industrialization, it may so happen that its industry is practically non-existent, or in such an initial stage of development that it is in no position to assist either financially or by way of supplying expertise to a newly created NSB. Under these conditions, it may well be advisable to decide on creating the NSB under government auspices, which may initially be of modest size but headed by a competent, imaginative and dedicated director. It would, however, be wise in such cases to provide for a certain amount of flexibility in the organization of the NSB so that industry's active participation could be secured when it does develop and become competent to make useful contributions.

FINANCES

9.18 No matter what the structure or the character of an NSB may be, its financial problems are more or less common. Like any other nation building activity, standardization presents problems which must be solved diligently and throws up tasks which must be performed with the greatest of expedition. Finances are, therefore, most vital to the satisfactory working of an NSB.

9.19 The chief sources of income for an NSB may be:

(1) government grants and subsidies;
(2) membership fees from subscribing members;
(3) sale of standards and other publications;
(4) fees from certification mark licenses;
(5) fees for other services rendered, such as participation fees charged for training courses, seminars, symposia and conferences; testing and

inspection fees; levies on pre-shipment and quality control operation; subscription and advertising revenue from periodicals and so on; and

(6) grants, donations and bequests from miscellaneous sources, loans, interest on investments, and the like.

9.20 All these sources may or may not be available to every NSB. Sources listed under item (6) above are casual in nature and may be effective only occasionally, or not at all, depending on circumstances. Income from sources mentioned under item (5) would accrue only if the corresponding activities listed are undertaken by the NSB. Government grants (item 1) would more often than not form the only source of income for most government-run bodies. This source may not be at all available to private bodies, which would depend chiefly on income from industry and trade. On the other hand, membership dues (item 2), while forming the mainstay of private bodies, would not generally be available to government-run institutions, but may represent an important proportion of the income of a jointly-run NSB. Sales income (item 3), however, would be available to all NSBs, no matter what their character. But it may be noted that in government bodies it is the usual practice, with of course some exceptions, that sales as well as any other type of income accruing to the NSB would be credited directly to the government treasury and NSB would only receive the annual budgetary allotment of funds. Certification marks fees (item 4), again, are available to about 80 percent or so of the world's NSBs which have in operation an active programme of standards marking, and such bodies may belong to any one of the three categories – government, private or joint.

NSBs FOR NEW COUNTRIES

9.21 It must be recognized that while in the early days of the twentieth century, national standardization was chiefly aimed at producing order out of the then existing chaos, today it has come to be recognized as a potent instrument for guiding and planning new industries and enterprises. Even in the western countries, where new technologies are developing at a fast rate, it is beginning to be realized that standardization, as far as possible, should precede rather than follow industrial development. In fact, it should become a necessary compliment of all technological research and industrial innovation which are essential prerequisites for all production, industrial as well as agricultural.

9.22 In developing countries, on the other hand, everything by way of industrial and economic development has to be started almost from scratch. The function of standardization, therefore, as an infra-structure for all development is being appreciated progressively all over the world.[6] Most developing countries are adopting as a matter of guiding policy a planned approach for the development of their economies. In the well-meaning rush for quick development, it sometimes happens that the importance of standardization is overlooked or not fully appreciated. But, in order to realize fully the various advantages which planning has to offer, every phase of planned activity should be so designed as to produce the best possible results with the least expenditure of resources – financial, material or manpower. This can happen only if an enlightened programme of standardization becomes part and parcel of the planning effort and is pursued simultaneously along with all the other developmental activity at every level. In order that such activity is meaningfully coordinated at all the essential levels, it is important to provide a central focal point where it can be appropriately guided and coordinated in accordance with the policy of central planning (*see also* Chapter 20).

9.23 The important objectives to be borne in mind to ensure a balanced development of economy through a planned programme of standardization may be summarized as follows:

(1) to secure maximum utilization of all financial, material and human resources in the interest of achieving overall economy;

(2) to ensure an integrated and coordinated development of all industrial, agricultural and other enterprises, so that at all points of transfer of the products and services from one hand to the other, a minimum of effort is involved;

(3) to ensure interchangeability of all products, which should not only include dimensional interchangeability but also functional interchangeability and adequate reliability;

(4) to maintain a cross-check on unnecessary and wasteful growth of variety of products and practices required for guiding design, construction and maintenance;

(5) to maintain a high enough level of quality of products commensurate with the demands of the local as well as the export market, so that their serviceability and marketability are ensured and consumer satisfaction is constantly secured;

(6) to keep a special eye on the ever-changing pattern of overseas markets, tastes and demands in the interest of promotion of export trade; and

(7) to encourage applied and industrial research for developing new

and indigenous technologies of special interest to the country and for the purpose of adapting to local needs the technologies borrowed from abroad to the best advantage.

PLANNING AN NSB

9.24 At this stage it may be worthwhile to discuss briefly the procedure that may be recommended for setting up a new NSB in a country where it does not exist. As mentioned earlier, conditions in each country differ and the peculiar circumstances prevailing there would demand specific and individual approach. It is most difficult, therefore, to lay down any general rules of procedure. Some broad guidelines, however, may be indicated. But necessary discretion should be exercised in pursuing these guidelines, reserving full freedom for making such deviations as may be necessary to meet the local needs. Another thing to be borne in mind is that an objective approach must be ensured. Resources in smaller countries where standards bodies do not yet exist are, generally speaking, quite limited both in terms of manpower and finances. At the same time, competing demands for such resources for nation building works are manifold. The need for caution is, therefore, paramount.

9.25 First of all, it is most important to determine the need for a new organization. Questions to be asked and answered would include:

(1) Whether the industry and trade of the country are sufficiently advanced or are expected soon to develop to such dimensions as to require the services of an NSB.

(2) Whether the country has adopted as a matter of policy a planned approach to its economic and social development, in which case standardization as an infra-structure would assume a special importance.

(3) Whether the existing needs for standards are being adequately met by the prevailing trade standards, the overseas standards, the international standards, or otherwise.

(4) Are any one or more of the existing organizations in a position to furnish the necessary standards, such as the professional societies, research institutions, trade associations or industries organizations and the like?

(5) Whether there is a need for a separate central body like the NSB to coordinate and supplement the efforts of the existing organizations, or whether these tasks could well be entrusted to one of the existing organizations.

9.26 An objective study of the points mentioned above should indicate the appropriate line of action. If the case for the creation of an NSB is established, then the next set of questions to be answered would be those concerning its size, shape and character, like the following:

(6) Whether the NSB could be made a part of an existing organization with a view to making it an independent body later on, or whether it would be more desirable to create a separate body to begin with.

(7) What should be the duties entrusted to it besides the preparation of standards and certification marking?

(8) What should be the size of its staff – technical and others – and the magnitude of its financial requirements to begin with and how fast it may be expected to expand during the immediate future, say in the next five years or so?

(9) What should be the character of the NSB – should it be a government department or agency, a private sector enterprise or a jointly run organization in which both government and private sector would participate in its financing and management?

(10) How would its finances be secured from various possible sources from year to year in the foreseeable future?

(11) What steps need be taken or provisions made in its constitution to ensure its national character, independence of policy and action, technical capability, and financial stability?

9.27 In financing answers to these questions and deciding the various issues involved, the brief survey data given in paragraph 9.6 would be of some value. But the aim of the survey should be to study the local climate and estimate its possibilities in the light of experience of other countries. The next step would be to proceed with the drafting of the constitution of the new NSB in the requisite form. The exact form would be determined by the prevailing legal practices of the country. For example, it could be drafted in the form of a government decree, a bill for legislation, a memorandum of association to be registered under a societies registration act, a prospectus for organizing a non-profit making company under the companies act, or in some other form, as may be considered most advisable in the light of decisions arrived at in answer to the questions listed in paragraphs 9.25 and 9.26. For this purpose it would be most helpful if a close study is made of the constitutions and byelaws of some of the NSBs, which have achieved a measure of success. It is important that the NSBs chosen for such close study with the possibility of being used as model in mind are from countries having a somewhat similar industrial structure, stage of development and economic system as the country seeking to establish a new NSB. Some assistance may also be derived from the literature of the United Nations' organs such as the Economic Commission for Asia and the Far East (ECAFE),[7] [8] and from the Cunliffe Committee Report on BSI.[9]

9.28 In conducting this survey and preparing the basic groundwork for the establishment of a new NSB, it is highly desirable to ensure from the very beginning active participation of all the important parties concerned with standards in the country and take into account their views in arriving at the various decisions. The success of the venture would depend a great deal on this factor.

LOCATION

9.29 In finding a suitable location for the headquarters or the central office of an NSB, several considerations may arise. In resolving this issue, the basic needs of the organization would have to be borne in mind. As repeatedly stated earlier, cooperation of all the groups concerned in standardization is a prerequisite for the success of any NSB. The location of the headquarters, therefore, should be so chosen as to facilitate such cooperation from the largest number of interests concerned. While it may be possible in some countries to find a location where both the governmental and industry organizations may be concentrated, it may so happen in other countries that such a location is not possible. Even in the former case, other important centres of industry and commerce may exist scattered over a large area of the country, in locations far removed from the central point of concentration, as for example, port towns. It becomes imperative in such cases to locate the headquarters at the central point of concentration and to rely on organizing branch offices in remotely placed important centres. These branches could be entrusted with the responsibility for carrying out certain functions of the NSB, which are of particular interest to the respective areas covered by them.

9.30 In locating the central headquarters and branch offices care should be taken that the locations chosen are well served by a communications and transport network. In certain developing countries it has been observed to be extremely beneficial to locate the headquarters of the NSB at the seat of government, where communication facilities are generally available and which the business and industrial leaders often visit in connection with their normal work with government agencies. This facilitates their participation in the work of the NSB, without burdening their financial commitments too much. Thus, whether or not a capital city does have a concentration of industry and trade, it is worthwhile considering it as a possible seat for the NSB headquarters in a developing country. The proximity to the seat of government itself would prove of great advantage for various other reasons as well. For example, consultations with the highest authority in the government would be facilitated and plans and policies of the NSB could readily be coordinated with the latest official thinking on the development plans and industrial and trade policies. Furthermore, the financial and other needs of the NSB requiring government assistance could be better explained to the authorities concerned and their active cooperation enlisted for meeting them.

BODIES OTHER THAN NSBs

9.31 In several countries – advanced as well as developing – it so happens that some organizations which cover specific fields of industrial, trade or professional activity, prepare standards for use in their own particular fields. Such standards should, strictly speaking, be treated as industry level standards, but sometimes do assume national importance by virtue of the authoritative backing they receive from most national interests concerned with their application. This happens in all countries with or without an NSB. The case of the former type of countries has been dealt with in Chapter 8 (*see* paras 8.4–8.5). In the latter type of countries having no NSB, the services rendered by these specialist organizations fill a very useful need and may be considered to be of great advantage to national economy. But the situation should not be allowed to lead to complacency of attitude that, under the circumstances, an NSB is perhaps not necessary at all.

9.32 Wherever several bodies, besides the officially recognized NSB, happen to be interested in national standardization, many problems are bound to arise in respect of coordination of their activities and overlapping of their jurisdictions. No generalized solution can be suggested to resolve these problems, except to say that the NSB should, in such cases, be suitably strengthened so as to be able to exercise an effective influence to bring the various organizations to work together in harmony in national interest. This requires a purely objective and impartial approach on the part of the NSB. In addition, a great deal of tact and psychological understanding of human relationships is most essential. The leadership of the NSB should, therefore, be entrusted to a man who possesses not only the necessary knowledge and experience of as many branches of industry and technology as possible, but who also has the proven ability to deal with the complex human relationships and who is endowed with a broad vision of economic and social affairs of the nation.

REFERENCES TO CHAPTER 9

1. *ISO Memento 1971* (an annual publication). International Organization for Standardization, Geneva, 1971, p. 48.
2. *General information on the ISO member bodies.* Internationl Organization for Standardization, Geneva, 1970.
3. *National standards bodies in the ECAFE region.* Economic Commission for Asia and the Far East, Bangkok, 1970, p. 11.
4. *ISI Constitution.* Indian Standards Institution, New Delhi, 1966, p. 21.
5. *ISI Byelaws.* Indian Standards Institution, New Delhi, 1970, p. 37.
6. Verman, Lal C.: *Standardization as infrastructure for the development of ECAFE region* (Asian Conference on Industrialization Doc. No E/CN. 11/I & NR/Ind. Conf. 2/L.10. May 1970). ECAFE, Bangkok, p. 31.
7. *General considerations and problems in the setting up of national standards bodies* (Asian Industrial Development Council Doc. No E/CN. 11/I & NR/AIDC. 2/L.2) ECAFE, Bangkok, 1967, pp. 29-45.
8. *A draft bill – an act to create the X-land Institute of Standards and Industrial Research* (Asian Industrial Development Council Doc. No AIDC/ASAC (2)/8). ECAFE, Bangkok, 1969, pp. 3-18.
9. Cunliffe, Geoffrey: *Report of the committee on the organization and constitution of the British Standards Institution.* Her Majesty's Stationery Office, London, 1950, p. 44.
10. *Cinquante ans de normes françaises* (Fifty years of French standards). Association Française de Normalisation, Paris, 1970, p. 134 (*in French*).

Chapter 10

National Standardization: Procedure and Practice

10.0 In spite of all the diversity existing in the internal structure and organization of the various NSBs in the world as described in the previous chapter, one cannot help noticing the basic uniformity of approach everywhere. Similarly, in respect of the procedures and practices adopted for the preparation and promulgation of national standards, there exists some degree of diversity, but a much greater degree of uniformity is quite evident. The basic principles on which these procedures and practices are based seem to have had their origin in the United Kingdom, where the first national standardization movement in the world began to be organized as early as 1901. From then on the procedures have been continuously developed, modified and improved from time to time, until now a considerable similarity is discernible between the procedures of various countries.

PRINCIPLES

10.1 The underlying principles for the preparation of national standards as universally acknowledged today include the following:

(1) National standards shall fulfil a generally recognized need of industry, trade, technology and other sectors of national life.

(2) They shall be in accordance with the current and immediate future needs of the economy of the country.

(3) They shall safeguard the interests of both the producer and the consumer.

(4) They shall represent the largest possible national consensus of opinion between all the interests concerned.

(5) They shall keep in view the latest scientific advancements, but remain technologically and economically practical for application to various sectors of activity to which they relate.

(6) They shall ensure optimum overall economy of the nation.

(7) They shall be so designed as to encourage the development of more efficient economical practices, and leave the way open for devising new ways and means for carrying out operations more efficiently and effectively.

(8) They shall be subject to periodic revision and amendment and be kept up-to-date in respect of the latest advancements of technology and the progressively changing conditions of the nation's economy.

10.2 The validity of most of these principles would be self-evident, but a few words of explanation in regard to each may be useful. Firstly, it is clear that, unless a national standard fulfils a generally recognized want and is in line with the needs of the economy (items (1) and (2) above), it would mean a waste of effort, for it may not receive the ready acceptance necessary for its nation-wide implementation. In extreme cases such an unwanted standard may even prove harmful to the smooth flow of trade in goods and services, within the country or in outside markets. It is important, therefore, that before a new project for standardization is approved, and assigned to the appropriate technical committee, a thorough investigation is carried out to determine whether it is really justified in the context of the prevailing conditions and wanted by those concerned. This investigation may take the form of a nation-wide consultation with the concerned interests through a questionnaire. But wherever division councils composed of the representatives of large sectors of industry have been established (*see* Chapter 4), it might be adequate to consult the particular council involved. In certain cases, for subjects of extreme importance or of great complexity and where a division council may find it necessary, a special conference of producers, consumers and technologists concerned may have to be called. Such a conference would not only establish the need or otherwise for the standard proposed, but could also indicate general lines of approach, involving the programme, the priorities, the coverage and so forth. In addition, it would help to create an interest among the community which constitutes the ultimate user of the standard.

10.3 Safeguarding the interests of the consumer as well as the producer (item (3) above) implicitly covers the interests of the trader, the exporter, the distributor and others concerned with the use and handling of goods and services. In this regard all parties concerned, except perhaps the common consumer, could readily be consulted and their interests determined. In most countries all these interests, including the industrial consumer, are fairly well organized and in a position to determine their collective point of view, but the common consumer – the man in the street or a housewife – is not so well organized or well enough informed, except in a few advanced countries. How to get him and her to express their views on the technological implications of standards is, therefore, a serious problem. This problem becomes particularly difficult in developing countries. Some aspects of the consumer's problems are separately discussed later in Chapter 18.

10.4 The consensus principle (item (4) above) is of great importance, particularly in countries where the majority of standards are to remain voluntary instruments. Even in other countries where standards may be considered mandatory by law, the consensus principle, if followed, would enable their implementation to be secured much more expeditiously than if they were to be based on majority acceptance or on authoritarian decisions. The success of standardization movement in the world appears in a great measure to be due to the adoption of the consensus principle for taking decisions in preference to the 51 percent majority rule, which generally prevails in most other walks of human affairs. It stands to reason that when diverse groups of interests, who are not in any way responsible to an electorate or a nominating authority, participate in the standards making process of their own free will, it would hardly do to take decisions on a majority basis and expect them to be accepted by all concerned including the minority. In political and other circles, the 51 percent rule tends to ignore the minority view, however large the minority or however justified the view itself may be. But in standards making it becomes important to seek solutions which would try to meet almost all points of view within a wide circle of interests, however divergent, so as to ensure a national consensus.[1] This is not to say that unjustified views of any given party motivated by narrow selfish interests, should be accommodated. But when independent representatives of all the interests directly concerned with a subject come together in a meeting, it is not impossible to find compromise solutions to any technical controversy, which would meet the various legitimate viewpoints based on theoretical reasoning and experimental verification. Experience in standardization all over the world has proved this to be so in almost every case. The process may be time-consuming but it has its rewards. It is sometimes argued that this may lead to a highly watered down standard of little economic value, which is not quite unlikely. But this is exactly the situation in which the authority of an independent NSB can play a very important part, by insisting on actual investigations and research before any decisions are arrived at.

10.5 As regards ensuring the practical character of standards (item (5) above), it is obvious that a highly academic approach, no matter how scientific and how up-to-date, could be wasteful and futile. It is important that the engineer at the design table, the technologist on the production line, the foreman and worker on the shop floor, the scientist in the testing and development laboratory should all be able equally to comprehend and implement the standards with the help of the tools, machines and apparatus placed at their disposal. Otherwise neither could the manufacturer hope to make any profit nor the consumer get his money's worth. An impractical standard may indeed do little more than collect dust on the shelves of the NSB.

10.6 The question of optimum overall economy (item (6) above) has already been touched upon in Chapter 2 (*see* para 2.12). Suffice it to emphasize here that the overall economy of the nation in some exceptional cases may require solutions contrary to the economy of a particular sectional interest. This may be illustrated by an example or two. During the Second World War, when the USA found copper in short supply, silver was drawn from Fort Knox for use as bus-bars in electrical power plants and as transformer windings.[2] From the cost point of view such a step could hardly be justified, but from that of optimum overall national economy of war days, the prevailing standard practice was considered worthy of being modified. Similarly, though the use of coking coal as locomotive fuel in India in pre-independence days was an accepted practice, no doubt because of its efficiency and economy, yet the practice had ultimately to be given up in the larger national interest of having to preserve the coking coal resources so urgently required for the development and expansion of the steel industry.

10.7 In regard to item (7) above, suffice it to say that standards would be doing a disservice to economy, if they tended to block the way for improvements and innovations. For example, if a national standard attempted to freeze the detailed design of a machine, it would mean that every manufacturer who wished to comply with the standard would have to produce exactly the same machine. Any improvement in design involving a variation would be considered as non-standard practice, even if it performed its task more efficiently. Such design standards belong properly to company level and should remain relegated to that level, where improvements and changes when called for can readily be introduced, and where they have constantly to stand the test of competition. National standards, however, may profitably deal with important and interchangeable machine parts and components, with methods of testing the machines, and with their performance, efficiency, safety and reliability. This approach would give the engineer a free hand to employ his ingenuity and originality in working out and improving the details of design to achieve the best possible results within the framework of the vital factors which may be covered by national standards.

10.8 The question of the desirability and the need for continuous review of standards (item (8) above) has already been discussed in Chapter 3 (*see* paras 3.9 and 3.10), where the circumstances have been listed, under which standards would have to be revised and amended. This would apply to standards on any subject, for any aspect and at any level, but the NSB would have to exercise, as a matter of principle, a special vigilance in regard to keeping the national standards up-to-date, for they affect standardization at all other levels.

PROCEDURE

10.9 The principles enunciated and briefly explained in the foregoing paragraphs would provide the guidelines for laying down the procedure in every country. Since the principles are universally accepted, the procedures based upon them may differ only in detail and these differences may arise largely from the differences in the organizational set-up of NSBs and other economic and social factors peculiar to each country. Generally speaking, the procedure of an NSB for preparing national standards would include a number of steps, namely:

(1) emergence and receipt of proposals,
(2) preliminary scrutiny of proposals,
(3) approval of projects,
(4) allotment of work,
(5) appointment of joint responsibility committees,
(6) preparation of draft standards,
(7) wide circulation of drafts,
(8) compilation of comments,

(9) finalization of drafts,
(10) approval of standards, and
(11) publication and publicity.

Each of these steps is briefly discussed individually in the following paragraphs.

10.10 *Emergence and Receipt of Proposals* Proposals for preparing new standards or revising or amending existing standards may arise from the needs of any sector of economy. An NSB would welcome them from all authentic sources, such as:

(1) organizations of industry, trade, technologists, consumers and users;

(2) individual industrial units, commercial houses, technologists, professional engineers, industrial and other users and consumers;

(3) government departments and agencies, both central and state, city corporations, municipalities and other local authorities;

(4) councils, committees, sub-committees and panels of NSB or their constituent members;

(5) any other organized body having an interest in standardization in the proposed field.

Personnel employed in the NSB itself sometimes has important proposals to make. While there could be no basic objection to accepting proposals from the NSB staff for further processing, certain NSBs would prefer that such suggestions may be put through the relevant committee organs in the first instance, thus securing support of the interests concerned from the very beginning.

10.11 *Preliminary Scrutiny of Proposals* Each one of the proposals for new work received in this manner is first of all examined by the central directorate to determine:

(1) whether the proposal is such as may be considered consistent with the principles enunciated in paras 10.1, 10.2 and 10.7;

(2) whether related national standards in the field already exist in the country and the extent to which they could meet the need for standards envisaged to be prepared under the new proposal;

(3) whether ISO or IEC recommendations, or standards or recommendations issued by any other authoritative international body exist, which should be or could be used as a basis for the proposed national standards;

(4) whether assistance could be derived from national or other standards from other countries;

(5) whether entirely original standards would have to be prepared;

(6) whether any survey, investigation or research work would have to be undertaken;

(7) whether in making a decision for approval of the project it would be necessary to call a conference of the interests concerned, or to make a postal enquiry among such interests, or whether the relevant division council or industry committee could perhaps act directly;

(8) whether a competent technical or sectional committee exists which could be allotted the work, if it were approved, or whether a new committee would have to be created; and

(9) in case a new committee is called for, what would be its desirable composition.

10.12 *Approval of Projects* The result of the preliminary scrutiny carried out by the directorate, together with its recommendation for further action, is placed before the divisional council or the industry committee concerned for deciding whether the proposal be approved and a project for new work set-up, or rejected or its consideration postponed.

10.13 *Allotment of Work* In case the proposal, after due consideration by the division council or industry committee, or after a conference of the national interests, is decided to be approved, the work is allotted to an existing technical committee or sectional committee, giving such directives and allotting such priority as may be considered necessary. In case a component committee for undertaking the particular task does not exist, the authority concerned takes the necessary action to create such a committee, for which purpose the relevant recommendations of the directorate would be useful. Some other aspects of the allotment of subjects to division councils and industry committees have already been discussed in Chapter 4, which it would be worthwhile to review at this stage (*see particularly* paras 4.1 to 4.8).

10.14 *Joint Responsibility Committee* If it so happens that the subject matter of the proposal falls within the scope of more than one division council (*see also* Chapter 4, para 4.7 (6)) action is initiated by any one of these division councils to appoint a joint responsibility technical committee to deal with it. Approval of the other divisions concerned is obtained before creating such a committee. Alternatively, in certain NSBs and under certain circumstances, the matter of appointing such a committee and allotment of the work may have to be referred to the supreme body of the organization or its executive committee, as the case may be.

10.15 *Preparation of Draft Standards* On due allotment of the subject, the technical committee proceeds with the preparation of draft standards, with such help as the directorate secretariat can render. This step is by far the most important and most time-consuming element of the whole procedure. The success or failure of any standardization project would depend on how well this part of the work is organized and carried out. Each committee has to plough its own furrow and acquire its own experience in handling the projects assigned to it. The responsibility of the chairman of the committee is such that he has to be able to judge the capabilities of each of the members and have the tasks allotted to them accordingly.

10.16 In economically advanced countries and under more favourable circumstances in the developing countries as well, it is often possible to rely on technical committee, subcommittee and panel members to undertake the preliminary drafting work. In such a case the assistance of secretariat personnel is called for only in supplying the background material, editing the drafts and bringing them in line with the standard format or style which the NSB might have adopted (*see* Chapter 22). The great advantage of this approach is that the persons concerned with preliminary drafting are not only experts in the field but also have a representative character. They are in a position not only to take into account the experience of sister countries from their own knowledge of industrial practices prevailing elsewhere, international standards and standards of other NSBs, but also directly to take care of the requirements of their own country with which they are personally concerned in their daily work.

10.17 It sometimes happens, particularly in the developing countries, that no committee member may be available with experience enough to undertake the drafting of a national standard; or that competent members may otherwise be so heavily committed to their normal duties that they cannot find the time required for the purpose. Under such circumstances the burden of preparing preliminary drafts devolves on the secretariat personnel, whose actual field experience is bound to be limited, and the drafts thus produced have usually to be based on library research of available standards and other literature. But when such preliminary drafts are put up for acceptance of the technical committees, they are subjected to close scrutiny by the more experienced representatives of industry and others familiar with the prevailing conditions in the country. This process ultimately leads to a fairly well balanced draft incorporating both the experience of other NSBs and the requirements of the country itself.

10.18 When competent personnel is available among committee members to undertake the task of drafting, it is often convenient for the technical committee to put them together in a subcommittee or a panel so that they could divide the work among themselves according to their own convenience. It is also possible to nominate as subcommittee members other knowledgeable persons who may not be members of technical committees in their representative capacity, but who could usefully contribute to its work in their personal capacity. Care should, however, be exercised in appointing such a subcommittee or panel that it is an effective group, which can meet formally or informally as required at short notice. In some NSBs the practice has thus grown to appoint subcommittees and panels preferably of one or two members only.

10.19 When it so happens that the data available in standards from outside sources are inadequate or unsatisfactory for the purpose of drafting national standards, it becomes necessary to gather original data particularly relevant to the nation's own enterprises under consideration. In such a case the committee or subcommittee responsible may initiate action along any one or more of the following lines of approach whichever is appropriate:

(1) Issue a questionnaire to all the known interested parties in the country to collect the necessary information.

(2) Issue a similar enquiry to the countries with which the country may have trade relations in the field under study; as also to other countries which may be thought to be in a position to assist with relevant data.

(3) Initiate surveys, investigations and research projects in cooperation with institutions which are competent and adequately equipped and manned for the purpose.

10.20 The organization and follow-up action on these projects naturally devolves on the central secretariat of the NSB, but the planning and basic approach of the enquiry or investigation must be settled by the relevant committee initiating the work. However, the financing of such projects often presents problems, particularly those requiring extensive field investigations and laboratory research. In such cases, the NSB has to mobilize all the possible resources and avenues available to it. For example:

(1) A national laboratory or a cooperative research organization of the industry may be able to underwrite the expenditure as a part of its own activities.

(2) The industrial and other interests participating in the NSB project may collectively arrange to share the work and/or the expenditure.

(3) A special grant may be obtained from a national or international foundation or from the government.

(4) Failing all other resources, the NSB itself may have to carry the financial burden.

The last contingency would seldom arise but the first two are most common. The third option, though most suited to developing countries, is usually a time-consuming affair, unless the project happens to have been anticipated by making provision in the current budget.

10.21 The draft standards thus prepared by subcommittees and/or panels, after being duly edited by the secretariat, are fully reviewed by the technical committee or sectional committee as a whole, with the object of determining their technological soundness and their compliance with the principles enumerated in paras 10.1 to 10.8. This review is generally carried out in one or more meetings of the committee, where all differences of opinion could be freely ventilated, discussed and resolved, so as to arrive at the largest possible consensus of opinion on the largest possible number of the clauses of the draft. It is possible sometimes to dispense with an actual meeting and carry out this part of the process by correspondence, especially when the subject is simple or non-controversial. The draft at this stage is again reviewed by the secretariat with the object of incorporating the decisions of the committee and further editing, if need be.

10.22 *Wide Circulation of Drafts* The draft standard at this stage is ready for wide circulation. It would be appreciated that the membership of the technical committee producing the draft would at best be limited compared with the number of parties interested in the subject matter of a national standard. The object of wide circulation is to inform every interest in the country that may be affected and those abroad about the contents of the draft and invite their critical review and comments, with a view to modifying it suitably in such a way as to make it more generally acceptable. The NSB secretariat arranges for copies of the draft to be sent to as many parties as are known to it to be concerned and invites their close study and suggestions, for modification, if they have any to offer, or to indicate their acceptance of the draft as such. At the same time, it is most advisable to give wide publicity in the technical and also sometimes in the daily press, inviting the attention of all those interested to the fact that such and such draft standard has been circulated for public opinion and anyone interested may receive on request a free copy for review from the NSB. This step helps to approach a much wider audience than could possibly be covered by any mailing list available with the central secretariat, no matter how exhaustive it may be.

10.23 It is the normal practice to give a time limit for the receipt of comments, and this may be as long as two or three months, to facilitate coverage of overseas interests. Sometimes, at the request of an important interest, the originally stipulated time limit may have to be extended. But, in case of non-controversial drafts and those that are considered most urgent, the limit may even be reduced. Nevertheless, it is important to adopt a rule, say, of three months and reduce the period only by a decision of such authority as a division council.

10.24 In some countries it is the practice that, while circulating a draft, an indication is given that if no comments are received it would be assumed that the party addressed is in agreement with the contents of the draft standards. This indication helps to stimulate some parties, who may otherwise be indifferent, to at least read the draft in their own interest. In spite of all efforts some drafts fail to receive the attention they deserve from a few of the potential users of standards. This is perhaps more frequent in developing countries than in industrially advanced countries, largely because, in the former, familiarity with standards is not so widespread and their importance not so generally understood. For this reason, it is not uncommon to receive belated complaints about the inadequacy of a national standard soon after it is published and begins to be implemented. If the interest raising the objection is important enough to the economy of the country, it may become necessary to consider a suitable amendment. Such a situation is unfortunate, but has to be faced in the early stages of development.

10.25 The need for circulating drafts to overseas interests must be emphasized, particularly in relation to standards of interest to the export or import trade of the country. It is not so easy to reach all such interests directly. A few of course may be on the mailing list of the NSB and some may be reached through the export and import houses of the country. Some NSBs make a practice of circulating drafts to commercial counsellors and attachés of their embassies abroad with a request for securing comments from interested parties, but most NSBs find it satisfactory to invite other NSBs of the interested overseas countries to assist in this matter. It must be admitted that an NSB is the best agency in any country to render this service, for it would have its own representative national committees on most of the important subjects, which could quite readily be consulted. This practice of commenting on one another's drafts has indeed become quite common among the NSBs of the world and is proving quite useful in harmonizing standards of various countries, which, indeed, is the first step towards the evolution of international standards.

10.26 *Compilation of Comments* The comments received as a result of wide circulation should be so collected and collated that on presentation to the technical committee they could be systematically examined and corresponding decisions conveniently recorded. It is always helpful for the secretariat to append its own suggestions, wherever pertinent, indicating the manner in which comments on certain clauses of the draft could be dealt with in the light of the earlier thinking of the committee. In doing so, it would be well worthwhile for the secretariat to adhere to a purely objective viewpoint and attempt to suggest solutions which, while meeting the needs of the national economy, would also comply with the suggestions of the commentators to the extent possible.

10.27 *Finalization of Drafts* It is advisable at this stage to circulate to the technical or sectional committee the comments so collated together with secretariat's suggestions to meet them, so that they may all be studied at leisure by the respective members before being discussed at a meeting. This could conveniently be done by sending the collated comments together with a notice for the meeting well in advance of the due date. In taking decisions on the various comments, the committee will again have to rely on the basic principles enunciated in the early paragraphs of this chapter but it must always keep in view the desirability of adopting such compromises as would enable maximum possible implementation of the standard to be secured on its publication. In the light of committee decisions, the final version of the draft is compiled by the secretariat for the next few stages of processing, which are largely formal.

10.28 *Approval of Standards* The final version of the draft is now ready for being accepted as the national standard. The procedure for this acceptance varies in different countries. But wherever there exist division councils or industries committees, the draft is first presented to the relevant body for acceptance or, on behalf of the body concerned, to its chairman. Then it goes to the supreme body, or its chairman on its behalf, for final approval as a national standard. The main objective of these rather formal steps is to ensure that the procedure as prescribed in the rules or byelaws for the NSB has been scrupulously adhered to and that due weightage has been given to the views of most interests involved. Aggrieved parties can bring their viewpoint for consideration at this stage. The secretariat is expected to make available to the accepting and approving authorities the full background of the processing of the standard, so as to enable them to adjudge the issues involved.

10.29 Usually no changes are made in the draft at the acceptance or approval stage; but if changes are considered necessary at any one of these stages, the draft standard is normally referred back to the committee concerned for review in the light of the directives that may be issued by the higher authority. It is obvious that this is the only way, because it is the committee producing the original draft, which is technically competent, possesses all the background information and data, and has the necessary representative character to be in a position to make adjustments, if any such are called for by the higher administrative body.

10.30 *Publication and Publicity* The final draft thus emerges as the duly approved national standard, which is now ready for publication, except that it has to be readied for press complete with drawings and illustrations in the form required by the printer. At this stage the number of copies to be printed is also to be estimated, which must include those required for free distribution to committee members, subscribing members, and others, and those that are expected to be sold before the next possible revision of the standard. It is hardly possible to estimate the latter with any degree of accuracy and an *ad hoc* rule is generally adopted, which with experience may be amended from time to time.

10.31 Directly on publication of a standard, or sometimes even before it is actually made available for sale, it is necessary to publicize widely the fact that such and such standard or a related group of standards has been issued by the NSB. The press note issued in this regard generally contains a brief description of the ground covered and highlights the important points. Besides publishing such information in the NSB's own periodical journal (in case there is one), it is important to issue it to all the technical and trade press concerned with the subject matter. In the case of a standard the subject matter of which is of general interest to the public, for example, a standard on a common consumer item, it is well worthwhile to cover it also in the popular daily and weekly press.

10.32 While such press notes are readily accepted and published by the technical and trade press, in view of their specialized interest in the subject matter of the standard, they are not so readily accepted for publication by the popular press. In order to make these press notes acceptable to this section of the press, it is necessary to write them especially for the purpose, taking every care to make them interesting to the layman. The notes written for the technical and trade press would not ordinarily meet this criterion. For this reason certain NSBs employ special staff having some experience and aptitude for popular writing. But very few NSBs can really afford this. If, however, an NSB is large enough and active enough to issue, say, a couple of hundred standards or more a year, it may be well worth its while to consider such a step, for it is quite likely that effective publicity will bring its own returns, not only by way of increased sale of standards, but also in other ways by making the NSB more popularly known as an important national development agency.

JUSTIFICATION FOR THE PROCEDURE

10.33 The lengthy and complex procedure described in the foregoing paragraphs (10.9 to 10.32) is by no means adopted by every NSB in the form it has been described here. But most of its essential elements are considered important and are provided for in most procedures in one way or another. It would not be incorrect to say that, in general, the procedures of all NSBs are quite long drawn-out and complicated. Why should this have to be so? The question may simply be answered by asserting that if the underlying principles (paras 10.1 to 10.8) are to be accepted as valid, there is little one can do to abbreviate or simplify the practice that has been evolved over several decades in several countries.

10.34 The main objective of the underlying principles enunciated is two-fold: firstly, to ensure the standards being sound from technological as well as economical points of view and secondly, to secure its acceptability on the widest possible scale. It may be argued that if the first condition of soundness is fulfilled, acceptability should naturally follow. To a large extent this may be true, but the process of trying to make a standard technologically and economically sound itself requires a wide-scale consultation with producers, users and technologists, so as to ensure that it reflects the latest trends of the market demand, advancement of knowledge and accumulation of experience. Besides ensuring soundness of a standard, this consultation process helps achieve, to some extent, the objective of acceptability also. But no matter how large a group of specialists is invited to participate in this process, it can never replace the wide circulation element of the procedure, which is considered indispensible in maximizing its acceptability.

10.35 In countries where standards are voluntary, it will be admitted that most of the elements of the procedure as described need be as fully retained as possible. But even in countries where they serve as mandatory instruments, the procedure has got to be more or less the same, because the underlying principles have necessarily to be adhered to everywhere. It is of course possible to simplify it to a certain extent depending on the organizational and technological structure of the economy.

COMPLEXITY OF PROCEDURE

10.36 In view of the very lengthy and rather complex procedure involved in the preparation of national standards it is natural that the time consumed in the process becomes a source of anxiety. In the context of the rapid advancement of technology, it is highly important that if a standard is going to be of real service, the interval between the demand for a standard and its availability be reduced to the barest minimum. It has not been an infrequent experience that a standard on its publication has required immediate amendment, because of the technological developments registered during the course of its preparation. On the other hand, none of the underlying principles adopted could be sacrificed or even compromised without detracting from the value of a standard and its effectiveness. The only possibility of reducing the inherently lengthy time involved is for each NSB to take a good look at its own procedure and discover ways and means to simplify and rationalize it.

10.37 It is to be noted that the various steps involved in the procedure may be divided into two broad categories – one requiring to be dealt with by the secretariat or the directorate staff of the NSB, and the other by the various councils and committees. It may be reasoned that the former part of the work could readily be expedited by strengthening the staff complement and improving its efficiency through training and experience, by providing attractive incentive schemes, by directing their work with the help of modern management organization and methods (O & M),[3–5] and so on. But there is a limit to this approach, not only because of the finances involved and the limitations of trained personnel that may be available, but also because the major delays involved take place in steps of the second category concerned with committee work.

10.38 It is in the committee part of the procedure where particularly time-consuming processes are sometimes involved, which must necessarily be put up with, as in the following cases:

(1) A standard may involve research or investigation of one or more problems concerned with the collections of data concerning the requirements to be specified or an agreed method of test to be evolved, or a national survey to be conducted.

(2) Certain interests concerned may have widely differing points of view. This situation arises not only between distinct categories of interests, such as consumers and producers, but also between two interests in the same category; for example, two manufacturers producing somewhat differing products. To resolve such differences, a great deal of discussion

may be necessary, in which the secretariat may have to play an important mediatory role.

(3) Sometimes a strong interest represented on the committee finds it useful to delay the preparation of a standard for reasons which may or may not be explicitly stated, while other interests find it expedient to push ahead with the project. Such conflicts sometimes arise between small and large-scale industries and have to be patiently and tactfully resolved.

(4) The size of a project may be quite large requiring the collection of a considerable amount of data, such as a glossary of terms for a given industry. It is sometimes found expedient to break up such a project into smaller parts and tackle each part separately.

TIME STUDIES

10.39 With a view to determining how and where the delays take place and to what extent, and whether they could be mitigated or eliminated, the Indian Standards Institution (ISI) initiated, in 1962, a detailed statistical study[6] of the time consumed at each step of the procedure. This study covered a period of 5 years, 1956–61, and came to the most unexpected conclusion that it took, on the average, a period of 70 months or nearly six years to prepare an Indian Standard from the date of receipt of the proposal to its final publication. The spread was from a minimum of 2 years to a maximum of 14 years. This was all the more amazing when it was considered that ISI published on the average 200 standards annually during this period. Stage-wise analysis showed that an average of 29 months were taken from the receipt of the proposal to the initiation of work and that only 41 months were involved to go through the whole of the remaining procedure from this point to the publication stage. Even if the time taken for the preliminary formalities prior to starting the actual work were ignored, the period of three years and five months for completing the project was considered alarming enough to warrant further and closer examination.

10.40 At this stage other countries were invited to make similar studies and to join forces in helping to pinpoint the lacuna and speed up the process of preparation of national standards. Through the initiative taken by the Commonwealth Standards Conference[7 8] and the Standing Committee for the Study of Principles of Standardization of ISO (STACO), a number of NSBs undertook studies on lines similar to those followed by ISI and contributed their findings to STACO for further deliberation.[9] In the meantime ISI had carried out another study, this time spread over a period of 4 years,[10] 1961–65, the objective of which was to discover whether any improvement had been registered by introducing certain modifications in the procedure, which had been adopted as a result of the first study. The results of this and all the other studies available were analyzed on behalf of ISO by S. K. Sen of ISI.[11] Table 10.1 summarizes the average time taken for preparing a national standard in the eleven countries, which participated in these studies.

TABLE 10.1

AVERAGE TIME TAKEN FOR THE FORMULATION OF NATIONAL STANDARDS (DURING 1959–63)

	Time in months up to publication stage	
Country	*From receipt of proposal*	*From initiation of work of drafting*
France	32	28
India*	52	33
Israel†	–	32
Japan	–	42
New Zealand	56	47
Poland	32	25
Portugal	45	41
South Africa	77	70
Turkey	29	24
UK	–	37
USSR	28	24
Average	44	37

* Period of study 1961–65
† Period of study 1946–66

10.41 Another way to present the same data would be to indicate the percentage of the total number of standards issued during the period under study which took different durations of time, say one year, two years, and so on. Table 10.2 presents the data in this manner. The average of the 8 countries covered by the data are plotted in Fig. 10.1 in a cumulative as well as a distributive manner.

10.42 It will be seen from the data given in Tables 10.1 and 10.2 that the studies conducted in all the eleven countries show a remarkably similar trend. The time taken in all countries is quite long and differs much less from the average time than might have been expected in the light of the prevailing differences in organizational structures and procedures. The length of time taken to issue a national standard indicated by the averages

TABLE 10.2

PERCENTAGE DISTRIBUTION OF STANDARDS ISSUED ACCORDING TO TIME TAKEN (DURING 1959–63)

	Time taken in years										
Country	*1*	*1-2*	*2-3*	*3-4*	*4-5*	*5-6*	*6-7*	*7-8*	*8-9*	*9-10*	*>10*
France	3	43	27	11	5	4	5			2	
India*	2	11	26	23	13	9	6	1	2	2	5
Israel†	2	23	34	19	10	7			5		
Japan	20	39	24	13				4			
Poland	2	26	42	18	9	2	1	0	0	0	0
Portugal	0	0	39	19	16	10	4		12		
UK	7	31	23	14	9	7	4	1	2	1	1
USSR	14	28	24	17	10	7	0	0	0	0	0
Average	6	25	30	17	10	6	3	0	1	1	1

* Period of study 1961–65
† Period of study 1946–66

amounting to 44 months from the receipt of proposal and 37 months from initiation of work (*see* Table 10.1) is quite long – 3 to 4 years. This represents a real challenge to the standards engineers of the world. So does the fact that some 60 percent of the total number of standards takes as much as 3 years on the average to issue (*see* Fig. 10.1). In view of the fast pace with which technology is progressing today, this challenge has to be met as soon and as effectively as possible.

POSSIBILITY OF SIMPLIFYING PROCEDURE

10.43 The standards engineer is thus faced with a serious dilemma. While he knows that the various time-consuming steps of his procedure are essential by virtue of the very nature of the standardization process, he realizes that he is also expected to meet the demand of the fast developing technology which cannot await his convenience. Indeed, the pressure of this demand may one day bring about a radical change in the whole approach to standardization. But, in the meantime, some measures have been suggested as a result of the studies referred to above, which are being tried out in several countries. These measures are more in the nature of palliative medication rather than a specific cure. This state of affairs will have to continue for the time being because there is hardly any possibility

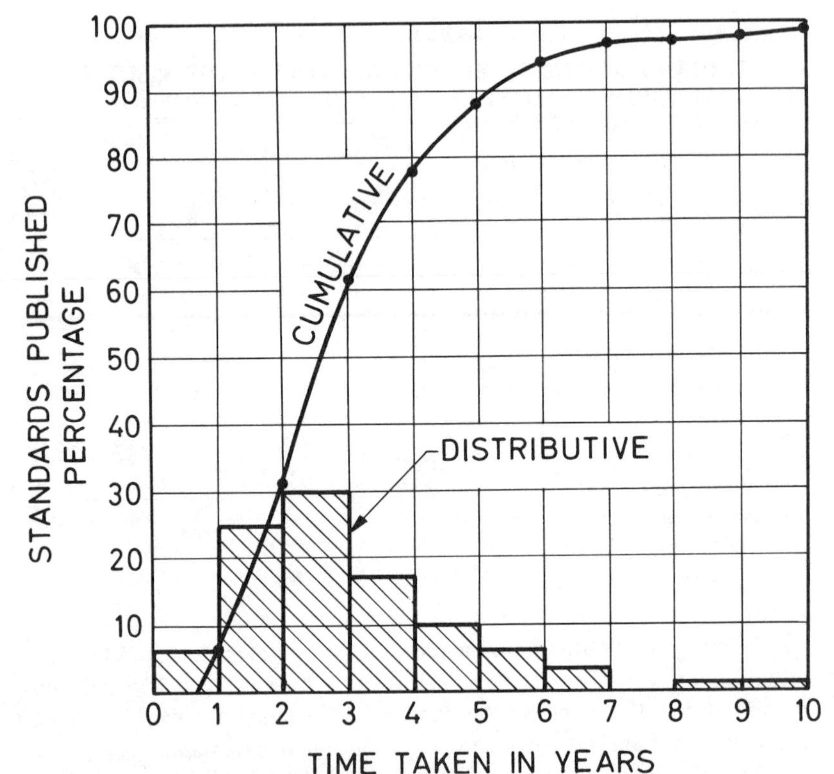

Fig. 10.1 Percent distribution of standards published according to time taken for their preparation. Average of 8 NSBs (*see* Table 10.2).

at present of departing appreciably from the basic underlying principles of standardization as accepted all over the world.

10.44 Some of these palliative measures have already been tried out with positive results, as in Turkey[9] and India.[10] It may be useful to list them here for ready reference:

(1) Minutely analyzing the various operations of the procedure involved, fixing time limits for the different stages and reviewing the progress from time to time; in other words, applying PERT and other management techniques[3] to all the operational stages in meticulous detail.

(2) Carrying out a really exhaustive examination as a preliminary to the acceptance of a proposal for standardization, which may enable the new project being approved by the head of the NSB, subject to later

approval by the authority concerned, namely, a division council,[12,13] and which would certainly assist the technical committee in making quicker progress.

(3) Preparation of the preliminary drafts by the staff of the directorate secretariat.

(4) Failing to be able to follow this course, assigning the drafting work to a one-man panel instead of a subcommittee or a larger group.

(5) Adopting schemes for incentives and awards for individuals contributing well prepared preliminary drafts leading to speedy production of standards.[14]

(6) Putting the preliminary draft in wide circulation before its consideration by the technical committee.

(7) Bringing together and arbitrating between two or more parties amongst whom serious differences of opinion might have arisen.

(8) Adopting, if feasible, the practice of issuing tentative standards for a prescribed period of time, which, though not complete in all details, are sufficiently comprehensive for a brief period of use, pending the availability of research data on some of the characteristics that may be considered time-consuming.

10.45 There is, of course, the possibility of bypassing some steps of the process of preparation of national standards by adopting an international standard issued as recommendations by ISO, IEC, a specialized agency of UN, or some other authoritative international organization. But even here there is the need to examine closely the relevant international standard before accepting it as such or after making certain modifications or additions. This process of examination, essential for safeguarding national interests, would itself require the application of the essential steps of the formal procedure for preparing national standards, but, of course, in a considerably modified form, which is bound to save a great deal of time. On the other hand, if the NSB concerned has already actively participated in and contributed to the preparation of the international standard in question, the time element involved in accepting it as a national standard would be further shortened very appreciably.

10.46 However, there is another side to this approach, namely, that the process of preparation of international standards itself is a far more time-consuming affair than that for the national standards. Furthermore, the number of international standards available today is quite limited as compared to the total needs; neither do they cover many of the subjects and aspects for which national standards are required in several countries.

This is particularly the situation in the context of the developing nations, all the needs of which ISO and IEC cannot hope to meet within a foreseeable future. The position is further complicated because both ISO and IEC are expected simultaneously to keep pace with the demand for new standards to serve the newly developing technologies in the advanced countries.[15] This emphasis on new technologies is fully justified in view of the important place they are expected to occupy in the future economy of the world as a whole, including the developing countries.

10.47 It is, therefore, important for every NSB to examine critically its own procedure and adopt such measures as may help save an amount of valuable time, which ultimately means not only the saving of both manpower and financial resources, but also creating a more prompt and efficient national standardization service in the interest of speeding up development. For new NSBs the studies described and the suggestions given in the above paragraphs would be of special value in designing their own procedures from the very beginning on the basis of experience gained elsewhere.

EMERGENCY PROCEDURES

10.48 In the case of a national emergency such as war, the normal procedures become quite unsatisfactory, because of the time element involved. In all matters of standardization, where strategic materials have to be conserved, productivity pushed to the limit and every resource stretched to the extreme, emergent action is called for. Every existing standard has to be re-examined for its adequacy and amended where necessary. Certain new standards are called for to be made available overnight. Emergency procedures for such contingencies have been adopted in most countries which have had to face a war, both for meeting defence as well as civilian needs. In this connection no detailed recommendations can or should be made except to refer to the experience of some other countries. Riebensahm[18] has given an interesting account of the German experience of the Second World War, while many accounts of the utility scheme of the UK[19] are also available.

REFERENCES TO CHAPTER 10

1. Verman, Lal C.: *Standardization – a triple point discipline* (Indian Science Congress Presidential Address, Calcutta, 1970). Indian Science Congress Association, pp. 6-7.
2. Cochrane, R. C.: *Measures for progress – history of National Bureau of Standards.* US Department of Commerce, Washington, 1966, p. 416.

3. See references 17, 18, 20, 21, 23, 24 and 27 of Chapter 5.
4. Rangan, R. K.: Time factor in standardization – an O & M approach. *Ninth Indian Standards Convention (Bangalore). Doc S-1/6.* Indian Standards Institution, New Delhi, 1965, p. 5.
5. Jain, V. C.: A graphic technique to control the time taken in the preparation of Indian Standards. *Ninth Indian Standards Convention (Bangalore). Doc S-1/10.* Indian Standards Institution, New Delhi, 1965, p. 9.
6. Time taken for the formulation of an Indian Standard. *ISI Bulletin.* 1965, vol 17, pp. 108-110.
7. *Third Commonwealth Standards Conference, 1957 – Report.* Indian Standards Institution, New Delhi, p. 45.
8. *Fifth Commonwealth Standards Conference, 1962 – Report.* Standards Association of Australia, Sydney, p. 38.
9. *Document ISO/STACO (India-11) 290,*
   Document ISO/STACO (UK-14) 314,
   Document ISO/STACO (Poland-21) 337,
   Document ISO/STACO (France-15) 338,
   Document ISO/STACO (Turkey-1) 346,
   Document ISO/STACO (Turkey-2) 372,
   Document ISO/STACO (Israel-1) 394,
   and others. International Organization for Standardization, Geneva, 1964–67.
10. Time taken for the formulation of an Indian Standard – second study. *ISI Bulletin,* 1965, vol 17, pp. 271-273.
11. Sen, S. K.: Time taken to develop national standards. *Document ISO/STACO (Sen-1), February 1967.* International Organization for Standardization, Geneva.
12. Subramu, S.: Expediting preparation of Indian standards through procedural changes. *Ninth Indian Standards Convention (Bangalore). Doc S-1/5.* Indian Standards Institution, New Delhi, 1965, p. 4.
13. Rao, A. Sangameswara: Expediting formulation of Indian standards with special reference to the standards on chemical analysis. *Ninth Indian Standards Convention (Bangalore). Doc S-1/7.* Indian Standards Institution, New Delhi, 1965, p. 5.
14. Vaidyanathan, S. S.: Suggestions for speedy formulation of Indian standards. *Ninth Indian Standards Convention (Bangalore). Doc S-1/8.* Indian Standards Institution, New Delhi, 1965, pp.3.
15. Frontard, R.: Normes de papa, normes de demain (out-dated standards, stanards of tomorrow). *Courrier de la Normalisation.* Jan-Feb 1970, no 211, pp. 5-10 (*in French*).
16. Ene, I.: Scurtarea ciclului de elaborare a standardelor. *Standardizarea.* 1966, vol 18, pp. 19-23 (*in Rumanian*).
17. Binney, H. A. R.: Standards and efficiency – speeding BSI's work to help export trade. *BSI News,* Oct 1966, pp. 6-7.
18. Riebensahm, Hans E.: Contribution of standardization to the substitution of scarce materials. *Tenth Indian Standards Convention (Ernakulam). Doc S-1/2.* Indian Standards Institution, New Delhi, 1966–67, p. 16.
19. *Final Report of the Committee on Consumer Protection.* Her Majesty's Stationery Office, London, 1962, p. 331 (*see* clauses 242-250, pp. 76–78).
20. *Procedure manual for the development of CSA standards.* Canadian Standards Association, Rexdale, Ontario, 1971, p. 15.

Chapter 11

International Standardization

11.0 The ultimate and the highest goal of standardization effort is to achieve international accord on all technical questions which are related to the exchange of goods and services between one nation and another. It is at the international level that standardization begins to contribute to the deepest aspirations of mankind by helping to bring about understanding among nations and to transform the one world concept into a reality.

BRIEF HISTORY

11.1 The first organized effort ever made to achieve international accord on standardization matters was perhaps just about a century ago, when in 1870 the French government initially called together a number of nations to a conference to consider the adoption of the metric system as the international system of measurement (*see* Chapter 13). The second occasion arose with the opening of the present century, when electrical technology was beginning to make rapid advances. Leaders in the field at the time felt that an immediate need existed for an international accord on the various new concepts, terms and units which were being evolved in different parts of the world. Accordingly, by a resolution of the International Electrical Congress[1] of 1904, the nucleus of an idea was formed for the creation of what has now come to be known as the International

Electrotechnical Commission (IEC). The foundation was thus laid for international standardization long before the concepts of national standardization became really crystallized. Another closely allied international effort was that which brought about the establishment of the International Association for Testing Materials in 1897, which had for its objective the exchange of information and views on development in testing and unification of standards for testing.[23] While this organization became inactive with the outbreak of the Second World War and went out of existence, the Metric Convention is now well on its way to becoming universally adopted and the IEC has come to be recognized as the premier international authority for standardization in two of the most important fields of industry, namely the electrical and the electronics industries.

11.2 As mentioned earlier (*see* Chapter 9), during the decade that followed the First World War, a number of national standards organizations had come into being in several countries. Soon after their creation, these bodies began to realize the need for cooperation to achieve international accord on standards. In 1921, seven of these countries, namely, Belgium, Canada, Netherlands, Norway, Switzerland, UK and USA, came to a mutually agreed arrangement for regular exchange of information on standards. This led the way in 1926 to the creation of the International Federation of the National Standardizing Associations (ISA), with the initial participation of 14 NSBs of the following countries:[2]

Austria	Japan
Belgium	Netherlands
Canada	Norway
Czechoslovakia	Sweden
France	Switzerland
Germany	UK
Italy	USA

11.3 The stresses and strains of the Second World War brought ISA activites to a standstill. From 1943 to 1947, its place was temporarily taken by the United Nations Standards Coordinating Committee (UNSCC) which had a membership of the following 18 nations:

Australia	Denmark	Norway
Belgium	France	Poland
Brazil	India	South Africa
Canada	Mexico	UK
China	Netherlands	USA
Czechoslovakia	New Zealand	USSR

At the UNSCC's own initiative, the International Organization for Standardization (ISO) was brought into being in 1947, which inherited the legacy of both the ISA and the UNSCC. Soon after the creation of ISO, an agreement was reached with IEC which made the latter the Electrical Division of ISO, while maintaining its full identity and independence of status. Today ISO and IEC constitute the only two organizations at the international level, which are exclusively occupied with standardization, covering not only the field of industry but also, to an extent, agriculture and other economic spheres. A number of specialized agencies of the United Nations, and some other organizations as well, do undertake a certain amount of work on international standards, but this is limited to their own respective fields of specialization. In addition, there are certain regional groupings of countries, which deal with international questions of standardization from various viewpoints.

INTERNATIONAL ELECTROTECHNICAL COMMISSION (IEC)

11.4 As mentioned earlier, the germ of the idea for creating an international body to undertake standardization at the international level in the electrotechnical field was born at the 1904 session of the International Electrical Congress which had been meeting occasionally during the preceding quarter of a century and whose object was to provide opportunities to the electrical engineers and scientists to exchange ideas on the rapid developments taking place in the new science. At its 1904 session, the Chamber of Government delegates adopted the key resolution reading as follows:

> That steps should be taken to secure the cooperation of the technical societies of the world, by the appointment of a representative commission to consider the question of standardization of nomenclature and ratings of electrical apparatus and machinery.

11.5 After a preliminary meeting of the organization in 1906, the statutes of the newly created International Electrotechnical Commission (IEC) were adopted at the first meeting of its Council in 1908, at which 15 countries were represented.[1] These statutes provide representation to member countries through the so-called National Committees of the IEC which were to be formed in each country wishing to become a member. Today the following 41 countries are members of IEC through their respective National Committees:

Argentina	Germany	*Pakistan
*Australia	Greece	*Poland
Austria	*Hungary	*Portugal

Belgium
*Brazil
*Bulgaria
*Canada
China (People's Republic of)
*Czechoslovakia
*Cuba
Denmark
Finland
France
*India
*Indonesia
*Iran
*Israel
Italy
*Japan
*Korea (Republic of)
*Korea (Democratic People's Republic of)
*Netherlands
Norway
*Rumania
Spain
*South Africa
Sweden
Switzerland
*Turkey
UAR
*UK
*USA
USSR
*Venezuela
*Yugoslavia

(National Standards Committees of the countries marked (*) are associated with the respective National Standards Bodies constituting ISO Member Bodies *See also* para 11.20).

11.6 The objectives of IEC are defined in the following terms:[3]

> To facilitate the coordination and unification of national electrotechnical standards and to coordinate the activities of other international organizations in this field.

The supreme governing body of IEC is the Council consisting of the Presidents of the National Committees, and the management is entrusted to a smaller body of 9 elected countries constituting the Committee of Action. Each member of the Committee of Action is elected for a term of 6 years, one-third of whom retire every two years. The standardization work is entrusted to Technical Committees on which every National Committee is represented. At present there are 68 such committees which together with 98 Subcommittees cover the whole field of electrotechnology in almost all its branches, including power engineering and electronics. As mentioned earlier, IEC became the Electrotechnical Division of ISO in 1947, and thus established the close ties which enable its work to be effectively coordinated with the ISO work in other fields of industry in which electricity plays an essential part, as for example, in acoustics, refrigeration, computers and nuclear energy. The work of the organization has been competently described by several of its leaders in a number of excellent articles[1, 3-6] and other publications,[7, 8] to which reference may be made for detailed information.

11.7 The impact of this work on the development and expansion of electrotechnology and its associated industries has been far reaching throughout the world and this trend is likely to continue in the years to come. One of the chief reasons for this extraordinary result would appear to be that the IEC came into being at a time when electrotechnology was in its early stages of development and when its work could help avoid the possibility of diverse practices developing in different countries. Another important effect that the pioneering work of IEC had, was to inspire, by its example, other industries to undertake standardization programmes in an organized manner and to provide the leadership for the formation of National Standards Bodies all over the world.

INTERNATIONAL ORGANIZATION FOR STANDARDIZATION (ISO)

11.8 At the close of the Second World War, the UNSCC (*see* para 11.3 above) decided that time was then ripe to create a new and permanent international body which could take over the work of international standardization from where the ISA had left it and carry it forward from the point which the UNSCC itself had reached. Accordingly, in October 1946, a conference[9] of all the UNSCC and the ISA members together with representatives from some other countries was arranged in London. At this organizing meeting delegates from the following 25 countries participated:

Australia	France	Poland
Austria	India	South Africa
Belgium	Italy	Sweden
Brazil	Mexico	Switzerland
Canada	Netherlands	UK
China	New Zealand	USA
Czechoslovakia	Norway	USSR
Denmark	Palestine (now Israel)	Yugoslavia
Finland		

11.9 At this conference, the constitution of the new organization – the International Organization for Standardization (ISO) – was drafted and duly adopted subject to later ratification by the participating countries and others. ISO's objectives were laid down and the rules of procedure were agreed to in general terms. Subsequently, from time to time, the Constitution and Rules of Procedure[10] have been modified and now the objectives of ISO read as follows:

> To promote the development of standards in the world with a view to facilitating international exchange of goods and services and to

developing mutual cooperation in the sphere of intellectual, scientific, technological and economic activity.

As means to the attainment of these objectives, the following activities are listed:

(1) to take action to facilitate coordination and unification of national standards and issue necessary recommendations to Member Bodies for this purpose;
(2) to set up international standards provided, in each case, no Member Body dissents;
(3) to encourage and facilitate, as occasion demands, the development of new standards having common requirements for use in the national or international sphere;
(4) to arrange for exchange of information regarding work of its Member Bodies and of its Technical Committees; and
(5) to cooperate with other international organizations interested in related matters, particularly by undertaking at their request studies relating to standardization projects.

11.10 Soon after ISO had begun to function with its first meeting in June 1947, it was recognized officially as a non-governmental consultative organization to the United Nations Economic and Social Commission and the various UN specialized agencies. More recently, in 1970, it was granted the consultative status to the United Nations Industrial Development Organization (UNIDO).[11] The top level governing body of the organization is the General Assembly (GA) consisting of representatives of all of its Member Bodies, which is required to meet at least once every three years. At the time of its first meeting in 1947 there were in all 24 National Standards Bodies (NSBs) which had formally joined the ISO as its members; at the end of 1971, the membership had risen to 56 NSBs. These are listed in Table 9.1 of Chapter 9. Only one standards organization from each country may be recognized as a representative body competent to join the ISO membership. This is extremely important, because in certain countries, there are a number of bodies claiming to issue standards of national importance (*see* Chapter 9, paras 9.31–9.32). If every such organization were made eligible for ISO membership, it can be appreciated how much more difficult it would become to reach international agreements which are already quite time-consuming. In recognizing one body as the country's representative, the resolution of all differences in viewpoints prevailing within the national boundaries is made the responsibility of this body and to that extent international work is facilitated.

11.11 The management of ISO is carried out by its Council in accordance with the policies and plans adopted by the General Assembly. The Council consists of representatives of 14 Member Bodies duly elected from among its members, of whom approximately one-third retire every year but are eligible for re-election. By convention, special preference is accorded to the re-election of four Member Bodies, namely those of France, UK, USA and USSR. As a result, these Member Bodies have been repeatedly elected to the Council. This provision is intended to make the valuable experience of these advanced countries continuously available to the international standardization work, without giving them a permanent status or a veto on the pattern of the UN Security Council. By still another convention, it is also provided that new Member Bodies which may not have served on the Council in the past might be given special consideration for election to the Council. This convention enables Member Bodies of the newly developing countries to gain first-hand experience and direct insight into the working of the international standardization process, and by so doing to help promote the standardization movement among newly independent nations.

11.12 The actual work of standardization is carried out by the various Technical Committees (TCs), operating on the basis of Directives[12] which formally lay down the detailed procedure required to be followed in the preparation of ISO Recommendations for standardization and of international ISO Standards. Each of the 146 TCs that exist today (January 1972) and on which each Member Body is entitled to be represented, is created by the Council to deal with one or more approved projects. A TC is created only after it has been ascertained that a majority of the Member Bodies are not opposed to undertaking the proposed project and that at least 10 Member Bodies would be prepared actively to participate in the TC, if one were appointed. Under special circumstances, where the work is considered to be of special international importance, the Council is empowered to authorize the creation of a TC, even if the number of willing participants is as low as 5. The scope[13] of each TC is subject to approval by the Council and liaisons are established with other ISO and IEC Technical Committees and outside international bodies interested in the subject. In establishing these liaisons, particular care is taken that no authoritative body having important interests is overlooked. Such co-operation helps to avoid possible overlapping and duplication of work and leads to better coordination and implementation of standards.

11.13 Within the framework of the directives and the constitutional provisions, a TC may set up its own Subcommittees (SCs) consisting of Member Bodies which may or may not be participating members of the parent TC. Incidentally, it may be mentioned that apart from the participating (P) category of Technical Committee members, two other categories are recognized, namely the observers (O) and not-interested (N). Both the TCs and SCs are competent to set up Working Groups (WGs) consisting of individual specialists and experts drawn from various countries on the basis of their merit, to undertake a prescribed part of the work in hand. At present nearly 450 SCs and 500 WGs[14] are operating under the existing TCs. One of the Member Bodies among the participating members of a TC or SC is designated as its secretariat, with responsibility to process and carry on the work at its own expense. The secretariat also has a major voice in the appointment of the chairman of the TC or SC as the case may be. The enormous load of detailed work which would otherwise devolve on the Central Secretariat is thus distributed among several countries and so is the financial burden involved. However, the travel and subsistence expenses of delegates attending the various meetings are met by the respective countries appointing them.

11.14 After a TC has approved a Draft Recommendation, which may be by majority vote, the draft is circulated to all Member Bodies for approval or comments. After all comments have been duly taken into account and a majority of 60 percent of Member Bodies have approved the draft, it is sent out to the Council members for acceptance as an ISO Recommendation. So far some 2,000 ISO Recommendations[15] have been issued on a variety of topics. The rate of publication of ISO Recommendations has now reached the level of 350 per year, which is commendable, considering how much longer it takes to finalize an international document on standardization than a national one (*see* Chapter 10, paras 10.39–10.42). Until recently ISO statutes have required that for an ISO Recommendation to become an ISO Standard, no dissenting vote should have been cast by a Member Body during its circulation for acceptance by the organization. Under this rule no ISO standards had been issued. More recently, however, the rule has been changed to the effect that ISO Recommendations will be replaced by ISO standards and that 75 percent of the Member Bodies voting would be required to cast an assenting vote for an ISO Standard to be issued.[22] This move is designed to help project a more authoritative image of ISO, especially among those who are not closely connected with its internal working.

IMPLEMENTATION OF INTERNATIONAL RECOMMENDATIONS

11.15 Whether it is an ISO Recommendation or an international ISO Standard, or whether it is an IEC Recommendation or report, each of the Member Bodies and National Committees is free to adopt it as its own national standard without any change or after altering it suitably to meet its own peculiar needs. The underlying idea is to attain the maximum possible coordination and alignment of national standards, consistent with the right of each sovereign country to make its own final decisions. As a result, many small, and some large countries also, have been able to adopt ISO and IEC Recommendations as national standards without any modification. But some of the Member Bodies, even though they may not be able to adopt ISO or IEC Recommendations word for word *in toto,* find it most useful to follow them closely in preparing their own national standards. This is largely because of the element of international agreement they reflect and the consolidated world-wide experience they represent. While the adoption of ISO and IEC Recommendations as national standards without any change may be the most desirable thing to do to achieve full international harmony, this course of action sometimes does become rather difficult to follow because of one or more of the following reasons:

(1) The relevant recommendation may not have covered all the different aspects of a given subject or all the various requirements desired to be covered in a national standard. Such a situation may have arisen out of the difficulty experienced in arriving at international accord on all the important aspects and requirements, or even from a lack of general interest in such matters at the international level.

(2) Certain parts of the recommendation while reflecting the majority view of those participating in the Technical Committee's work, may not meet the needs of some of those within the minority group, who may have important national considerations which could not be ignored, as for example, climatic factors, legal restrictions, raw material supply and market conditions.

(3) It sometimes happens that a Member Body has not been able to take active part in discussions at the meetings of the Technical Committee and its views otherwise expressed by correspondence may not, for some reason, have been duly reflected in the international recommendation. Under these circumstances, it may become unavoidable for the member country to depart from some of the provisions of a given ISO or IEC Recommendation.

(4) Some Member Bodies or National Committees may have adopted a particular style of presentation for their national standards which, according to their practice, may invariably have to be adhered to for the sake of uniformity and which may require departure from the actual text and/or the form of presentation of an ISO or IEC Recommendation.

11.16 In the case of developing countries, as also those with relatively limited resources, it is most advantageous to adopt the ISO and IEC Recommendations as such or with such minor modifications as are consistent with their contents. It must be generally admitted that no country which is keenly interested in expanding its export and import trade can afford to introduce in its national standards requirements which may be contradictory to those in international recommendations, or which may otherwise tend to exclude its products from international markets or to seriously restrict the markets from which it procures its own needs.

RELATIONSHIP BETWEEN ISO AND IEC

11.17 As mentioned earlier, through a formal agreement between the two organizations entered into in 1947, IEC began to act as the Electrotechnical Division of ISO, but it continued to maintain its own technical, administrative and financial independence. The IEC collects its own dues and administers its own affairs. But, in order to ensure coordination, the Central Office of IEC is housed in the same building in Geneva as the Central Secretariat of ISO. The two offices work as closely together as humanly feasible. In the technological sphere, the scopes of the subjects dealt with by their Committees are respectively defined by the Council of the ISO on the one hand and the Committee of Action of the IEC on the other. The representation on the Technical Committees is secured through the so-called Member Bodies of ISO and the National Committees of IEC, and the two groups of Committees carry on their work more or less independently of each other with occasional cross-representation through liaison officers.

11.18 However, electrotechnical matters concern many industries which are not basically electrotechnical and, therefore, do not fall under the purview of IEC. On the other hand, several industries, in fact most, depend on electrical devices and some are rather closely related to electrotechnology as such. It is thus quite obvious that the two sets of Committees have to work in close liaison and harmony on many fronts, if contradictions in their respective recommendations for standardization are to be avoided. Apart from taking note of each other's prior work, numerous contacts are established between the various Technical

Committees. Attempts are made to resolve questions of overlap and duplication, if any, through the intermediary of the two Secretaries General in consultation, if need be, with the secretariats of the Committees concerned. While most such questions are readily resolved at this level, the issues that may remain unresolved are referred to a higher authority created collectively by the two organizations, namely the ISO/IEC Joint Committee for Coordination and Consultation, which reports to both the organizations.

11.19 Thus for dealing with the internal problems of financial, administrative and technical character, the two organizations have worked out a satisfactory enough *modus operandi.* There are, however, other external issues which have also to be tackled collectively; for example, their relationships with the community of nations outside their membership and with other international organizations. On these and many other matters of common interest, the two organizations arrange to consult each other and try to arrive at a common course of action. Despite this prevailing harmony, questions are sometimes asked, particularly among the developing countries, whether it is essential that the two organizations should remain so distinct as they presently are; why the two sets of subscriptions should have to be paid; and whether it would not be more efficient and economical to have one consolidated international body to deal with all the standardization problems, a good many of which are closely interrelated.

11.20 At the national level, it may be observed that electrotechnical and other fields of standardization are usually the responsibility of one National Standards Body. For example, among the 41 National Committees of the IEC; as many as 25 constitute integral parts of the National Standards Bodies of the various countries which are ISO Members. These countries are indicated by an asterisk mark in the list given in paragraph 11.5 above. On the other hand, in the remaining 16 countries the National Committees of IEC constitute organizations which are distinct from the respective NSBs. In addition, there are over 20 other NSBs in the world (*see* Chapter 9, Table 9.1), which may be considered as prospective IEC members, and in which the electrical and other standardization work is carried out in an integrated manner. Thus in some 75 percent of the countries of the world having standardization activity, the national organizational apparatus is unitary in character.

11.21 But it must be realized that the IEC was created some 40 years earlier than ISO, when there was no NSB existing anywhere. National Committees of IEC, therefore, were the forerunners of the present day

NSBs. Later on, in some countries the National Committees of IEC became associated with their own NSBs. In only 16 countries do they still continue to maintain their independent status. It will also be noted that these 16 are relatively older and more advanced countries with their own technological heritage and industrial tradition. By combining all standardization activity under one NSB, most of the newer countries appear to be establishing their own tradition by learning from the experience of the older ones and justifiably wondering why similar steps could not be taken at the international level.

11.22 The answer to this rather debatable question has really to be found in the historical sequence of events and the fact that changing any tradition, even if considered desirable, takes time. In the meanwhile, it must be recognized that both the ISO and IEC are doing their best continuously to improve their working efficiency by developing an effective pattern of cooperation in all matters of common interest. In due course, when more of the remaining 16 countries find it feasible and desirable to unify their standardization effort at the national level and the movement for similar unification at the international level gathers further strength, a change may still take place. But whether it would lead to any better or more economical working still remains to be seen. In this connection, it may be of interest to quote the following passage from the report of the Conference on Standardization in the Developing Countries,[16] organized jointly by ISO and the UN Centre for Industrial Development, the predecessor of UNIDO, in Moscow in 1967.

> A question was raised as to the division of work and collaboration between the IEC and ISO. It was emphasized that the ISO and IEC collaborate closely and work hand-in-hand thus avoiding duplication of work. It was mentioned that there has never been a case where these two organizations did not reach agreement as to the distribution between themselves of subjects for technical work.

OTHER INTERNATIONAL ACTIVITY

11.23 Since the IEC came into being as early as the opening of this century, almost at the beginning of the development of electrotechnical industries and began to take care of the basic needs for standardization in the field from the very start, there did not appear to be much justification for other international bodies to undertake similar work. But in certain specialized sectors, in which the IEC was either not prepared to or was not equipped to meet the needs of the industry, certain other international organizations became interested. The following examples may be cited:

CIGRE – International Conference on Large Electric Systems
CIRM – International Marine Radio Association
ITU – International Telecommunication Union and its two Commissions, namely:
CCIR – International Radio Consultative Commission, and
CCITT – International Telegraph and Telephone Consultative Commission
UNIPEDE – International Union of Producers and Distributors of Electrical Energy
URSI – International Scientific Radio Union

11.24 While the international bodies exemplified by the above list do undertake a certain amount of standardization work in their own restricted fields of specialization, they are in addition concerned with a large number of other problems concerning development, research, organization and personnel and so forth. Besides, their standardization work is carried out in close liaison with IEC and, in turn, IEC depends on liaison with them for its own work. In addition, there are about a score of other international bodies[3] the work of which is related to electrotechnology, and IEC has a similar close liaison with them. This liaison is essential in the interest of uniformity of approach and avoidance of divergence in the standards requirements of the different international authorities. They are most effective, for they operate at the working level of the Technical Committees through cross-representation of observers of one organization to the other. Though such observers do not carry power of voting as do the official delegates from member countries, they do wield a considerable influence in framing the final decisions, because their submissions invariably are given due weight and considered extremely important for achieving a truly international accord.

11.25 A similar situation exists in the case of ISO, but with this difference that a much wider field, comprising every other industry and commercial and economic spheres, is required to be covered. It must also be borne in mind that ISO, or for that matter even ISA, came into being much later than IEC and for this reason also many other international bodies had begun to take active interest in standardization in the specialized fields of their own. This had indeed happened long before ISO could be organized well enough to be able to meet all their needs. Neither can it be said that ISO is yet in a position to meet immediately the whole of the worldwide demand for standards in all fields of interest. But whatever work it is doing has of necessity to be carried out in close collaboration with all the other international bodies having interest in the subjects in hand.

11.26 In some rather important sectors, certain international organizations have fairly well developed programmes of standardization under way, for example:

CGPM – International Conference on Weights and Measures defines and standardizes the units of weights, measures and other physical quantities.

OIML – Organization for Legal Metrology is engaged on the unification of practices and procedures for the administration of metrological legislations, which covers standardization of methods of verification of weights and measures and measurement of other physical quantities, as also specification of certain types of measuring instruments.

CAC – Codex Alimentarius Commission of the FAO and WHO[17] has for its objective compiling or collection of the internationally adopted food standards presented in a unified form.

ILO – International Labour Organization deals with standard codes of practice for safety of personnel and evaluation of injuries, as also with standards for certain safety devices.

Many others which deal with relatively more restricted fields of standardization could be mentioned, such as:

BISFA – International Bureau for the Standardization of Man-made Fibres
CIE – International Commission on Illumination
FID – International Federation for Documentation
IATA – International Air Transport Association
ICHCA – International Cargo Handling Coordination Association
ICRU – International Commission on Radiation Units and Measurements
IIR – International Institute of Refrigeration
IIW – International Institute of Welding
IRU – International Road Transport Union
UIC – International Union of Railways

11.27 Altogether there are some 250 international bodies with which ISO has established cooperative working relationship through mutual liaison between them and the respective Technical Committees.[18] In order to distinguish between the various organizations as regards the extent of their interest in a given Technical Committee of ISO, three categories of liaison have been recognized, namely:

Category A: Where an international organization makes an *effective contribution* to the work of a technical Committee for the greater part of the subjects dealt with by it.

Category B: Where an international body makes an *effective contribution* to the work for *only one particular subject* on the work programme of a Technical Committee.

Category C: Where an international organization wishes to be *kept abreast* of the work of a Technical Committee.

This categorization, which is made on the basis of requests of the international organizations, helps to simplify not only their own work but also that of the ISO Committee Secretariats.

11.28 Of special interest among the ISO's international relationships are those with the various United Nations organs including the specialized agencies and the Economic Commissions.[18] Of these the following may be mentioned:

ECOSOC – Economic and Social Council of UN
ECA – Economic Commission for Africa
ECAFE – Economic Commission for Asia and the Far East
ECE – Economic Commission for Europe
ECLA – Economic Commission for Latin America
FAO – Food and Agriculture Organization
GATT – General Agreement on Tariffs and Trade
IAEA – International Atomic Energy Agency
ICAO – International Civil Aviation Organization
ILO – International Labour Organization
IMCO – Inter-Governmental Maritime Consultative Organization
ITU – International Telecommunication Union
UNESCO – United Nations Educational, Scientific and Cultural Organization
UNIDO – United Nations Industrial Development Organization
UNCTAD – United Nations Conference on Trade and Development
UPU – Universal Postal Union
WHO – World Health Organization
WMO – World Meteorological Organization

With the specialized agencies the relationship is mainly at the working level of the Technical Committees, but with the ECOSOC, UNESCO, UNIDO and the regional Commissions, the liaison extends to several other spheres of activity including promotion and implementation of standards, technical and financial aid for training and seminars, research and investigations, assistance to developing countries for establishing and strengthening NSBs, and so on. In these matters, the Special Fund of UN Development Programme plays a very significant part by allotting funds to various projects, particularly those aimed at the advancement of standardization activity in the newly developing nations.[19-21]

11.29 With a view to creating effective links at all levels, including the technical, and providing readily available channels of communication, particularly with certain important international organizations, such as those of the UN family, the Secretary General of ISO maintains a continuous liaison with their highest administrative organs and top level officials. In this task he is assisted, when necessary, by the chief executives of ISO's own Member Bodies and reports to the Council on all important matters of policy. It is through such liaison activity that coordination of work at the committee level is reinforced and collaboration at the administrative and staff levels ensured, leading to overall harmonious working. ISO and, through ISO, IEC as well get all the opportunities to make known to a very wide circle of organized interests the progress that international standardization is making under their own aegis and the services that they are capable of rendering to the international community. This does not mean, however, that every problem finds a ready or immediate solution through liaison channels. Nevertheless, the mechanism provides an excellent means to bring about mutual understanding and cooperative working at the international level among the various independent organizations of sovereign status.

DEVELOPING COUNTRIES AND INTERNATIONAL STANDARDIZATION

11.30 With a view to facilitating the development of standards bodies in the newly emerging countries, the ISO provides a special class of membership at a concessional rate of subscription, namely the correspondent membership. This is open to those countries where there may not exist a National Standards Body or where it may be in a rudimentary stage of formation. Every five years such membership is reviewed to determine whether the country has advanced far enough to join the ISO as a full member. At present there are 13 correspondent members in the following countries:

Barbados	Madagascar
Cyprus	Malta
Ethiopia	Nigeria
Hongkong	Sudan
Iceland	Syria
Kenya	Zambia
Kuwait	

11.31 Correspondent membership of ISO is made available to any national organization or government department, which may be actively interested or engaged in setting up a national standards body. Every correspondent member receives useful literature, advice and assistance. A correspondent member is also entitled to send observers to attend meetings of such Technical Committees as may be of interest, at which they may express their national views and participate in deliberations, but may not vote. A special section within the ISO Central Secretariat is devoted specially to cater to the needs of developing countries and in this work it receives full support of UNIDO and UNESCO. There is also a Development Committee of the ISO Council (DEVCO) which determines the policies of ISO concerning developing countries and deliberates on their problems.[16]

11.32 Among the active full membership of ISO, 25 out of 56, and in IEC 13 out of 41 countries, may be classed as developing. These, being duplicated, constitute 25 nations and together with the 13 correspondent members, they have a special obligation in respect of international standardization towards the whole group of developing countries in the world, of which there are about 80 within the membership of the United Nations. Thus, only 45 percent of the total number of developing countries appear to be associated in any way with international standardization. On the other hand, all the developed countries in the world may be considered to be closely associated with the formulation of international standards and recommendations. It has, in fact, become largely their concern, which is indeed quite understandable, for they are the ones advanced far enough, both in technology and trade, to be vitally interested in international standardization, and they are capable, both financially and manpower-wise, of actively participating in ISO and IEC work. In contrast, the developing countries are handicapped in many ways to wield similar influence, although their interests in international standardization are equally vital, if not more so. Their actual difficulties and possible means of dealing with them will be discussed in Chapters 12 and 19.

11.33 In spite of the general handicap from which a majority of the developing countries suffers in making worthwhile contribution to the work of ISO and IEC, some relatively better organized countries among them have already succeeded in influencing the advancement and shaping of certain international standards of direct interest to them; for example, Iran and Turkey, in respect of grading of dry fruits and test methods for handmade carpets. India has also contributed significantly in respect of standardization of test atmosphere; rationalization of sizes of electric motors and steel sections; measurement of fluid flow in open channels; pictorial marking for handling of goods: specifications for spices, condiments and stimulant foods; and so on. It is within the reach of each member country, or under special circumstances even a non-member country through the intervention of a member nation, to bring to the notice of ISO and IEC their special needs and requirements which they consider should be reflected in the international standards under active consideration of the respective international committees, or to bring new subjects of vital interest within their purview.

11.34 It must be emphasized that active participation of all the Member Bodies of ISO and IEC in the international discussions on standards should be considered a matter of high priority, because it not only enables them to influence final decisions but also helps them to have their own peculiar needs duly reflected in the international recommendations. Even if the representatives of the developing countries do not happen to be in a position to make such positive contributions to the international work as those from the developed ones, they stand to gain considerably in other indirect ways by merely attending the international technical committee meetings; for example, by developing and widening their contacts with their counterparts from other countries, from whom they stand to learn about the latest advances in technology and the important lessons being learnt from experience everywhere. Such direct contacts are bound to open up new avenues of communication and channels for the exchange of valuable data not ordinarily available to developing countries. Another aspect to be borne in mind is that the very fact of having been present at the discussion leading to international agreements on standards, would help such representatives from developing countries to improve their own national practices and procedures, and enable them to incorporate the international recommendations in their national standards on the basis of a thorough understanding of their genesis.

11.35 In order, therefore, to be in a position to reap the benefits of such participation and to have an active hand in the advancement of world standardization, it is important that every country, developing or otherwise, should become an active member of ISO and IEC. Only then will international standards begin to wield a powerful influence on international trade and bring about the universal understanding which they are purported to do.

REFERENCES TO CHAPTER 11

1. Ruppert, Louis: 56 years of IEC. *ISI Bulletin,* 1960, vol 12, pp. 278-281.
2. Coonley, Howard: "The international standards movement" pp. 37-45, *in* Reck, Dickson (Ed.): *National standards in a modern economy,* Harper, New York, 1956, p. 372.
3. *The International Electrotechnical Commission – what it is, what it does, how it works.* International Electrotechnical Commission, Geneva, 1958, p. 30.
4. Herlitz, Ivar: The relation between IEC recommendations and national standards. *ISI Bulletin.* 1960, vol 12, pp. 286-288.
5. Dunsheath, Percy: International standardization in the field of cables and measuring instruments. *ISI Bulletin.* 1960, vol 12, pp. 284-285 and 336.
6. Series of the following Charles le Maistre Memorial Lectures delivered at Annual IEC meetings and published by International Electrotechnical Commission, Geneva:
   (a) *Charles le Maistre – his work, the IEC,* by Andre Lange, 1955, London.
   (b) *Standardization – virtue and necessity,* by Clarence H. Linder, 1956, Munich.
   (c) *The national acceptance of international standards,* by Reginald O. Kapp, 1957, Moscow.
   (d) *Measuring – standardizing – producing,* by Richard Vieweg, 1958, Stockholm.
   (e) *International standardization and the development of economic relations between countries,* by A. M. Nekrasov, 1959, Madrid.
   (f) *The IEC in an expanding electrical technology and a contracting world,* by Gijsbertus de Zoeten, 1960, New Delhi.
   (g) *What the IEC means to the developing countries,* by Mohammed Hayath, 1961, Interlaken.
   (h) *The IEC yesterday, today, and tomorrow,* by Ivar Herlitz, 1962, Bucharest.
7. *Catalogue of IEC publications – 1971.* International Electrotechnical Commission, Geneva, p. 80 (describes 364 publications up to 31 Dec 1970).
8. *General directives for the work of the IEC.* International Electrotechnical Commission, Geneva, 1958, p. 20.
9. *Report of the conference of United Nations Standards Coordinating Committee together with delegates from certain other national standards bodies.* UNSCC, London, 1946, p. 35.

10. *Constitution and Rules of Procedure.* International Organization for Standardization, Geneva, 1971, p. 25.
11. UNIDO grants ISO consultative status. *ISO Bulletin,* June 1970.
12. *Directives for the technical work of ISO.* International Organization for Standardization, Geneva, 1963, p. 43.
13. *Information on ISO Technical Committees* – (a) *Titles,* (b) *Scopes,* (c) *Liaisons.* International Organization for Standardization, Geneva, 1970, p. 71.
14. *ISO Memento 1971.* International Organization for Standardization, Geneva, p. 48.
15. *ISO Catalogue 1971.* International Organization for Standardization, Geneva, p. 132.
16. *Report on the conference on standardization in the developing countries, 16-21 June 1967.* Moscow. International Organization for Standardization, Geneva, p. 24 + 2 + 1 + 2.
17. *Report of the joint FAO/WHO Conference on food standards, 1-5 October 1962.* Geneva. Food and Agricultural Organization, Rome; World Health Organization, Geneva. Issued by FAO, 1962, p. 68.
18. *Draft general survey of ISO liaisons with other international organizations* (Document ISO/Council 1970–11/1: ISO/General Assembly 1970–10/1), p. 88.
19. *Promotion of industrial standardization in developing countries* (Report of the UN inter-regional seminar of Helsingör, Denmark, Oct 1965). United Nations Publication. Sales no 66.11. B.12, 1966, p. 30.
20. Industrial development and standardization. *Industrialization and Productivity* (published by UNIDO). 1969, no. 13, pp. 109-117.
21. *Assistance to developing countries* (Document ISO/DEVCO (Secretariat-17) 24, Feb 1969). International Organization for Standardization, Geneva, pp. 6 + 2.
22. International standards. *ISO Bulletin.* 1971, vol 2, no. 10, p. 4.
23. Campus, F: *Tr.* Sinclair, D. A.: *International organizations in the field of testing of materials.* National Research Council of Canada, Ottawa, 1966, p. 12.
24. ISO progress in 1971. *ISO news service (no.139/01/19 dated 19 January 1972).* International Organization for Standardization, Geneva, p. 1.

Chapter 12

Regional Level Activity

12.0 Reference has been made in paragraph 3.5 of Chapter 3 to the emergence of standardization activity at the regional level during the past two decades or so. It was also stated that it would be inadvisable to attribute a formal level to this activity in the generalized presentation of standardization space. Nevertheless, the activity at the regional level does exist and as stated earlier, it is motivated by varying objectives. Mainly, however, it is designed to meet the peculiar economic needs of a group of countries involved. As such, it does contribute to the advancement of international standardization, though the manner in which each regional group may be making its contribution may differ. The activity, therefore, must be taken note of as being useful, though not quite indispensable from international viewpoint. It is important that its various motivations should be fully understood and its operations properly guided, so as to ensure against its assuming a character which might adversely affect the international movement.

12.1 First of all, it is essential to distinguish clearly between the regional level standardization activity and the multi-national standardization activity. Groups organized to undertake either type of activity consist of several nations. But while a regional group may be organized on the basis of the National Standards Bodies, existing in the various member countries of the group, a multi-national organization would itself constitute a National Standards Body of the group of member countries involved, in none of which there may be any other standards body of national character. Thus a multi-national standards body would serve as a collective NSB of a number of countries and be entitled to ISO membership on behalf of all its member nations. On the other hand, a truly regional group consisting of the NSBs of several countries, would be represented independently in the ISO by several of the NSBs of its member countries and thus carry very much more influence as compared to a multi-national body. The number of multi-national standards bodies are few in the world – two have been included in Tables 9.1 and 9.2 of Chapter 9, namely the ICAITI (Instituto Centroamericano de Investigacion y Tecnologia Industrial) consisting of five Central American republics, namely Costa Rica, El Salvador, Guatemala, Honduras, and Nicaragua, and the SACA (Standards Association of Central Africa) consisting of Southern Rhodesia, Zambia and Malawi. But the number of truly regional groups is much larger; some of the more important ones will be briefly reviewed in this Chapter.

12.2 It may also be noted that the term "regional" has been used to describe this grouping merely as a matter of convenience and for lack of a better alternative. While most of the groups may indeed consist of countries lying within a given geographical region, others may consist of countries having common political or ideological or even linguistic and ethnical ties. The most fundamental underlying consideration which brings each one of these groups together appears to be the common economic interest, which after all is the chief motivation for all standardization activity. For these very reasons, it is quite difficult to classify the various groups in any systematic manner, except perhaps as the geographical groups and non-geographical ones.

12.3 The more important of the geographical groups may be mentioned as under:

(1) CEN – European Committee for Standardization (Comité européen de normalisation)
(2) CENEL – European Electrotechnical Standards Coordinating Committee (Comité européen de coordination des normes dans le domaine electrotecnique)
(3) CEE – International Commission on Rules for the Approval of Electrical Equipment (Commission internationale de réglementation en vue de l'approbation de l'équipment electrique)

(4) ASAC – Asian Standards Advisory Committee to the Asian Industrial Development Council of the UN Economic Commission for Asia and the Far East
(5) COPANT – Pan American Standards Coordinating Committee (Comisión Panamericana de Normes Técnicas) earlier known as PASC – Pan American Standards Committee

It is also understood that the UN Economic Commission for Africa (ECA) is planning to establish a standards advisory committee within its framework,[30] presumably on the pattern of ASAC. The regional groups which may be classed as non-geographical would include:

(6) ABC – America, Britain and Canada Group
(7) CSC – Commonwealth Standards Conference
(8) CMEA – Council of Mutual Economic Assistance of the Socialist Countries
(9) AOSM – Arab Organization for Standardization and Metrology

In addition, there is a consultative group among the four Scandinavian countries, Denmark, Finland, Norway and Sweden, which is perhaps less formal than all the others. Comparatively very little information is published about this group, but it is known that, besides consulting on collective approach to the international standardization problems, the member countries regularly exchange draft standards and undertake mutual discussions with a view to achieving coordination of their national standards.

12.4 *CEN* Soon after the European Economic Community (EEC) was established, it was realized that unification of standards of the various member countries was essential for a healthy development and consolidation of the Common Market. A standards committee was, therefore, set up for the purpose in 1957. Soon after, in 1960, the European Free Trade Association (EFTA) or the common market of the so-called outer seven countries came into being, which set up a similar committee of its own for achieving unification of standards. Both these committees independently agreed to operate on the basis of ISO Recommendations as far as possible. This was quite understandable, since all the countries involved were ISO members and already committed to such a course of action. Soon afterwards, that is, in 1961, it was agreed to amalgamate the two committees of the EEC and EFTA to bring into being what is now known as the European Committee for Standardization (CEN). Thus the membership of

CEN at present consists of the National Standards Bodies of the following countries:

Austria	Germany (Federal Republic)	Norway
Belgium	Greece	Portugal
Denmark	Iceland	Sweden
Finland	Italy	Switzerland
France	Luxembourg	UK
	Netherlands	

12.5 The aim of CEN is to establish standards common to the member countries in order to promote commerce and interchange of services between them. These common standards are intended to be used as unification documents to bring about maximum possible coordination of national standards.[1] More recently, at the direct initiative of the governments involved, serious consideration is being given to issue what may be called European Standards for direct adoption as national standards.[2] Consisting, as it does, of the NSBs of the various countries, only a few of which are governmental agencies, CEN cannot deal with the unification of official regulations of a technical nature which may constitute non-tariff barriers to the free flow of trade. This function is discharged by the official government representatives on the EEC and EFTA themselves with such consultation with CEN as may be necessary. The standardization activities of CEN, on the other hand, are related on most fronts to those of the ISO – it promotes the implementation of ISO Recommendations in Europe, it determines the collective European attitudes to some of the ISO projects under way, and finally it makes proposals for the initiation of new ISO projects on technical matters of interest to Europe, suggesting such priorities as it may deem fit.[3] It is to be appreciated that for achieving unification of national standards of Europe, CEN largely "works on the basis of ISO Recommendations, for the intention is to avoid creating yet another series of standards which would erect more technical barriers to global trade."[4] At the same time, it must also be admitted that through the collective action of CEN, the European countries are enabled to exert a much more effective influence on the working of ISO, in which they had already been playing a leading role. Being industrially and technologically well advanced and being in a position substantially to finance the expensive operations of ISO, it is but natural that the European countries should assume its leadership, and in so doing bear a proportionately larger share of the burden for preparing and propagating international standards.

12.6 *CENEL* This organization, the European Electrotechnical

Standards Coordinating Committee may be considered quite parallel to that of CEN for dealing with all standardization problems in the electrotechnical field in the EEC and EFTA countries. The membership, the objectives and the working of CENEL are more or less similar to those of CEN.[5] The reason why a separate organization has been considered necessary is perhaps that many of the European countries have separate National Committees of IEC for dealing with electrotechnicological standards at the national level (*see* Chapter 11, paras 11.7–11.22). It is, therefore more convenient for CENEL to operate independently and deal with IEC directly as CEN does with ISO.

12.7 *CEE* This International Commission on Rules for the Approval of Electrical Equipment is not, as its name may imply, truly international in character but is confined only to the European countries. As a regional organization in the field of standardization, it is perhaps the oldest body, having its origin in 1926 under the name of IFK – Installationsfragenkommission (Committee on Installation Matters). IFK was reorganized after the Second World War under its present name as CEE in 1947. Its membership is open to organizations, one in a given country, which are concerned with specifying rules for safety of electrical equipment and its testing. The following countries are its members:

Austria	Greece	Portugal
Belgium	Hungary	Spain
Czechoslovakia	Italy	Sweden
Denmark	Netherlands	Switzerland
Germany (Federal Republic of)	Norway	UK
Finland	Poland	Yugoslavia
France		

12.8 Its objectives are to issue safety standards for electrical equipment for protection against fire and risks to life and property, with a view to achieving unification of government regulations of member countries, which would facilitate freer import and export of electrical equipment intended for domestic, factory and other uses. By a formal agreement CEE works in close cooperation with IEC and tries to avoid duplication of work. But the existence of this agreement has had the effect of retarding for some years the initiation and progress of IEC standards dealing directly with the safety of electrical equipment, which would enjoy worldwide authority outside Europe. Safety matters were of course covered in most of the specifications which IEC issued on electrical equipment in general, and in this respect CEE work provided a very useful guidance. It was only in recent years that IEC took up the work on matters pertaining to safety of domestic equipment, for which worldwide demand has existed for many years. In the meantime, the non-European countries have been deriving benefit from the efforts of the European countries as represented in the publications of CEE, of which roughly 29 have already been issued, covering most fields of domestic safety.[6]

12.9 *ASAC* The Asian Standards Advisory Committee established in 1967 being an organ of the Asian Industrial Development Council (AIDC) of the UN Economic Commission for Asia and the Far East (ECAFE) is not truly an organization in the same sense as CEN and CENEL are with their own independent secretariats and well-defined mandates from the parent bodies. Its membership is open to all the governments of the 25 regional countries which are members of ECAFE, namely:

Afghanistan	*India	*Nepal
*Australia	*Indonesia	*New Zealand
Brunei	*Iran	*Pakistan
*Burma	*Japan	*Philippines
Cambodia	*Korea (Republic of)	*Singapore
*Ceylon	Laos	*Thailand
*China (Taiwan)	*Malaysia	*Viet Nam (Republic of)
Fiji	Mongolia	Western Samoa
Hong Kong		

(NSBs exist in * marked countries)

Though it is the governments of the ECAFE region countries which nominate delegates to the ASAC meetings, their choice usually falls on the chiefs and senior executives of the NSBs, wherever they exist. The main burden of participation in ASAC's work, therefore, falls on the National Standards Bodies of the countries (see * marks in the list above). Being an advisory body to AIDC and ECAFE, ASAC makes recommendations on standardization matters to member governments and the United Nations. But more importantly, it constitutes a regional organ which also takes decisions and makes recommendations to the NSBs of the region for collective and cooperative action. Nevertheless, according to an explicitly enunciated policy it may not undertake the preparation of regional standards, though it would do everything needful to secure coordination and alignment of national standards of the member countries among themselves as also with ISO and IEC Recommendations.

12.10 ASAC is comparatively a new body and so far has had only two formal meetings[7, 8] during which it has made many recommendations for the advancement of standardization movement in the region. They include, among others, those for encouraging the creation of new NSBs in countries wherever needed, for strengthening the NSBs which are in early stages of formation, for all NSBs to join the membership of ISO and IEC, for organizing training programmes for standardization personnel, for rendering mutual assistance, for securing technical and financial aid from the UN agencies, and so on. The most important action that ASAC has so far taken is perhaps the adoption of a regional strategy[9] which would enable the countries of the region to overcome some of the serious handicaps from which they usually suffer in being able to make positive contribution to the international standardization effort of ISO and IEC. By removing these handicaps, the strategy may help accelerate the pace of coordination of national standards among themselves and with those of ISO and IEC. This strategy, being of interest to other developing regions of the world besides Asia and the Far East, is proposed to be briefly outlined later in this Chapter. But it may be added that in the pursuit of this strategy, ASAC has already organized mutual consultations on eight or nine of the subjects of direct interest to the countries of the region, which are currently being dealt with by ISO and IEC Technical Committees or which may especially be proposed to be taken up.

12.11 *COPANT* The Pan American Standards Committee (PASC) was originally organized in 1947 at the initiative of the Organization of American States (OAS) and the Pan American Union (PAU), and consisted of NSBs of the Latin American countries.[10] but in spite of many attempts at initiating its activities it could not start functioning[11] until it was reorganized in 1961 on a wider basis with US participation and with the inclusion of other Latin American countries having no NSB as associate members.[12, 13] Its name was subsequently changed in 1965 to *Comision Panamericana de Normas Tecnicas* and abbreviated as COPANT.[14] The objective of PASC was and that of COPANT continues to be to prepare regional standards to assist the development of the economy of the Southern and Central American countries. These so-called Pan American standards are based whenever possible on ISO and IEC Recommendation, but, whenever necessary, assistance is also derived from standards of USA and certain European countries.[15, 16] Even Indian Standards have sometimes been used as a starting point, especially in the realm of steel.[17, 18] The Latin American Free Trade Association depends on COPANT to develop standards to facilitate trade between Latin American countries and establish the Latin American common market on a firmer footing.[19-21] Membership of COPANT in 1968 consisted of:

Argentina	Venezuela
Brazil	ICAITI, representing:
Chile	Costa Rica
Colombia	El Salvador
Mexico	Guatamela
Paraguay	Honduras
Peru	Nicaragua
Uruguay	

It is understood that as many as 234 Pan American standards had been issued by May 1970.

12.12 *ABC Group* During the Second World War, the USA, the UK and Canada came across certain serious handicaps in planning and conducting war operations, which could be directly attributed to the lack of uniform standards prevailing among the allied countries. Considerable time and resources were expended, for example, in the maintenance and repair of equipment because of the non-uniformity of screw thread standards. Avoidable delays took place in the planning and tooling of production of equipment of one another's designs, because of the differences in drawing office practices. The so-called ABC Group consisting of the representatives of the three English-speaking countries (America, Britain and Canada) came to be formed to iron out some of the existing differences in certain engineering standards of urgent importance. The unified screw thread that emerged out of these discussions led the way to the subsequent agreement on the ISO thread profile.

12.13 The usefulness of the work of the ABC Group was appreciated not only for war-time conditions but also for the progressing of the post-war deliberations within ISO. It has, therefore, continued to operate as an active group even after the end of the war.[29]

12.14 *CSC* The Commonwealth Standards Conference is another body, like the ABC Group, consisting of countries which geographically occupy no particular region but are literally spread over all the continents of the world. It started off in 1946 at the initiative of the British Standards Institution as a group of eight countries having NSBs, out of a total of some 15 countries which belonged to the then British Commonwealth. These included:

Australia	New Zealand
Canada	Palestine
Eire	South Africa
*India	United Kingdom

(*India's NSB was in its formative stage)

In recent years, many other countries of the then British Empire have progressively gained independence but have chosen to remain a part of the Commonwealth, and the appellation "British" has been dropped from its name. Some countries like Eire and Palestine (now Israel) have opted out of the Commonwealth. Commonwealth countries, which now number some 33, are listed below:

*Australia	Jamaica	Sierra Leone
Barbados	*Kenya	*Singapore
Botswana	Lesotho	Swaziland
*Canada	Malawi	Tanzania
*Ceylon	*Malaysia	Tonga
Cyprus	Malta	Tobago
Fiji	Mauritius	Trinidad
Gambia, The	Nauru	Uganda
*Ghana	*New Zealand	*UK
Guyana	Nigeria	Western Samoa
*India	*Pakistan	*Zambia

(Of these, the 12 countries marked * having NSBs of their own constitute the core of CSC.)

12.15 The objective of CSC since its inception has been to develop closer cooperation between the various NSBs and seek an alignment of their procedures and practices, and to achieve coordination of their standards, with a view mainly to maintaining and developing smooth flow of trade and technology among its members. It may be noted that the Commonwealth has constituted a sort of common market for many years, based on tariff preferences. The meetings of CSC, which have been quite informal, have taken place at irregular intervals of two to six years in different capitals. Numerous recommendations made by CSC from time to time have had a positive influence on the alignment of policies and procedures of member countries and brought about an active exchange of drafts and comments on one another's standards.[22] Although at one or two conferences, such as those held in New Delhi in 1957, and in Ottawa in 1959, some technical discussions did take place on the contents of certain standards, with a view to arriving at a common approach, this practice was soon given up in favour of the ISO and IEC forums which were considered more appropriate for dealing with such matters.

12.16 *CMEA* The Council of Mutual Economic Assistance organized by the socialist countries under the leadership of USSR has had under way, among other things, a very active programme of standardization since 1957. In 1962, CMEA decided to appoint a Permanent Commission for Standardization which determines all policy issues and prepares all the basic standards including those for metrology. There is also a CMEA Institute for Standardization to undertake planning and investigations on standards problems leading to the preparation of standards. In addition, there are a number of permanent commissions of CMEA for different industries, which, while dealing with the problems of their own industries, undertake standardization work pertaining to their specialization.[23] In other words, CMEA, through its various commissions, operates more or less like a multi-national standards body, except that it cannot strictly be so regarded because all its member countries have their own NSBs to prepare and promulgate their own national standards.

12.17 The standards issued by CMEA are intended not only to help the coordination of national standards and to facilitate international trade but also to assist in developing interchangeability, specialization and cooperation between the member countries, aimed at achieving an economic integration of production and distribution within the region as a whole. The so-called Standards Recommendations of CMEA (RS-CMEA) are not considered obligatory for every member country but members are expected to follow them under their basic obligation to CMEA. The activity is so intense and so extensive that between 1957 and 1970 more than 2,600 RS–CMEA were issued and the present rate of production is well over 500 per year. Member countries, it is estimated, have saved well over the equivalent of 350 million US dollars as a result of adopting these standards. An extensive plan for the 1970–80 decade has been prepared which will further help integrate the economy of all the socialist countries within CMEA membership.

12.18 *AOSM* The original idea for the creation of the Arab Organization for Standardization and Metrology was mooted in 1958 at the first Middle East Standards Conference held in Beirut,[24] and AOSM was brought into being in 1967 at an organizing conference held in Cairo attended by delegates and representatives of the following countries together with those of the Arab League:

Iraq	Kuwait	Libya	United Arab Republic
Jordan	Lebanon	Syria	

The objectives of AOSM were stated to be the unification of technical terms, methods of testing, measurements, and specifications among Arab states, aimed at raising the productivity and quality of goods to facilitate exchange, trade and cooperation in the fields of economy, industry, agriculture, science and culture.[25-27] In the pursuit of these objectives, AOSM has already started tackling, among others, some of the problems which may be considered of interest specifically to the Arab countries, such as the evolution of common technical terminology in the Arabic language and the fixation of the metric equivalents of the old measuring units prevailing in the different Arab countries. As a matter of policy, AOSM would be guided by ISO and IEC Recommendations, encourage the formation of new National Standards Bodies, and have them join ISO and IEC. It will, nevertheless, have special problems of the Arab world to resolve, for which specifically Arab standards may be required.

SIGNIFICANCE OF REGIONAL ACTIVITY

12.19 From the variegated character of the nine regional organizations described briefly in the above paragraphs, it will be clear that, though these organizations differ a great deal among each other in their structure, organization and methods of working, they do have a great many points in common also; for example, they all:

(1) attempt coordination of standards of the various member nations of the group with the object of regulating and facilitating the intra-regional trade and exchange of services, keeping in view the requirements of international trade;

(2) owe allegiance to and use the Recommendations of ISO and IEC as the preferred basis for coordinating their own standards; and

(3) promote the implementation of international standards wherever they are available.

12.20 Of the differences to be noted among the various groups, the following may particularly be mentioned:

(1) Organizations, like COPANT and CMEA, prepare regional standards which may directly be adopted as the national standards by the member countries; and CEN may soon join this group.

(2) Some of them, like AOSM, are concerned among other matters

with the problems which are peculiarly their own and which cannot be dealt with at the international level (e.g. linguistic terminology).

(3) Others, like CSC, place a great emphasis on the alignment of the procedures and practices for standardization among the member countries and the exchange of information.

(4) Still others, like CEN, CENEL and ASAC attach particular importance to bringing to the ISO and IEC forums collective and coordinated views of the member countries with the object of influencing international decisions.

(5) In contradistinction to others, one of them, namely the CEE, specializes not so much in standardization in general as such, but concentrates on activities aimed at the unification of official rules and regulations for the approval of electrical appliances from the safety point of view.

12.21 It will be admitted that all these are worthy objects and have the potential of helping to expedite the creation of truly international standards. It is important, however, that parochial interests are not allowed to harden the attitudes of the regional bodies to such an extent as to stand in the way of the spirit of give and take which is so very essential in all international dealings. The danger of such a thing happening exists only in cases where regional standards for national adoption are attempted to be prepared in areas where no ISO or IEC Recommendation is available for guidance. It is obvious that if such standards were intended to serve only the intra-regional trade and the exchange of services between the member countries themselves, there can be no apprehension in regard to the progress of international standardization. But, if they were concerned with a commodity of interest to inter-regional trade as well, then it is important that they be fully coordinated with any possible international standards that might emerge. The existence of un-coordinated regional standards in such cases may, therefore, present real difficulties in arriving at international standards unless, of course, the regional interests concerned are prepared for such compromises as may be called for at that stage.

12.22 In certain cases where the regional groups concerned happen to be quite influential like CEN and CENEL, it may not be always possible to strike the necessary compromise. The danger is, therefore, that if such cases begin to arise often enough, these powerful groups may gradually develop a tendency to circumscribe themselves and to begin to depend more and more on their own efforts and less and less on ISO or IEC for global agreements. thus the very advantage of having the regional groups may be lost and indeed turn into a disadvantage, leading to a weakening of ISO and IEC and the development of an unhealthy isolation in the world of standards. The consequences of such a development are bound to be retrograde in the context of the need of the day for more and more unification and togetherness in world affairs.

12.23 An international standard, in order to be really effective, has to be such as would meet the needs of every country concerned, but such an ideal can hardly be achieved in view of the prevailing diversity of practices and the large number of countries that are involved. There is bound to be, therefore, a certain amount of alteration and addition to the texts of international recommendations before they are adopted as national standards in some of the countries. Though it does not always necessarily happen, yet it is quite likely that such modification of detail may have the effect of throwing a national standard so much out of harmony with the international one that it may lead to real difficulty in the way of trade. Thus the basic advantage of having an internationally agreed standard would be lost at least in such cases. The existence of regional groups in providing forums for prior discussions on international matters, therefore, has a great advantage in minimizing the number of such cases arising. Not only that, but has also the possibility of expediting international decisions by the very fact that some of the local differences of point of view could be more easily resolved in the relatively smaller regional groups.[28]

12.24 Considered from this viewpoint, one may be tempted to advocate the creation of regional groups on a much wider scale than at present. Nevertheless, it will be observed that most of the countries of the world except perhaps those of the African continent have been covered by the existing groups. Some of the countries indeed belong to more than one group. While a geographical group on the African continent might, therefore, well be envisaged,[30] it would be important to emphasize that there should exist a community of economic interests among the prospective members, before such a step may be considered feasible.

DEVELOPING COUNTRIES AND REGIONAL ACTIVITY

12.25 In view of the limited resources, both financial and in manpower, available to an individual developing country, its capacity to contribute to its own development through standards and to the international standardization effort in general is of necessity quite limited. However, through joint efforts of several countries at the regional level this capability could be considerably improved. Standardization activity at the regional level has thus a special significance for the developing countries. The various handicaps from which such countries suffer in being able to make worthwhile contribution to the development and utilization of international standards may include:

(1) Total absence of an NSB in a country, early stages of development of activities of a recently formed NSB, or the existing NSB's non-membership of ISO and IEC.

(2) Paucity of available technical data relating to production, utilization and consumption of items of interest, which are essential for the formulation of standards; or an inadequacy of the means available for organizing the collection of such data and their analysis.

(3) Insufficiency of the available resources to enable properly briefed delegations to be sent for presenting the national viewpoint at the international meetings of ISO and IEC Technical Committees, Subcommittees and Working Groups, which are often held in Europe, sometimes in the United States, but very seldom in the more convenient proximity of the developing countries; and

(4) Even when represented, the NSBs from developing countries would usually find themselves relatively in a minority position and may not thus be able to influence the decisions in their own favour to a significant extent.

For all these and other reasons, NSBs of the developing countries are not often in a position to propose for international standardization new subjects of direct interest to their own economies, nor can they readily offer to undertake the secretariat responsibilities for pursuing such objects.

12.26 Any regional level effort that could help overcome these handicaps through cooperative action should, therefore, be welcomed not only from the national viewpoint but from the international viewpoint as well. As mentioned earlier (*see* para 12.10), ASAC has adopted a planned strategy in this regard for collective action[9] by the ECAFE region countries of Asia and the Far East. The various elements of this regional strategy for coordination of standards may briefly be reproduced as under:

1. *Status of NSB*
   - 1.1 Create an NSB in each country of the region.
   - 1.2 Develop its activities with the active participation and cooperation of all national interests concerned.
   - 1.3 Each NSB to enrol as an ISO and/or IEC member.
   - 1.4 Failing the above, create a standards cell in a government agency or department.

1.5 Enrol the cell as correspondent member of ISO.

1.6 The cell may derive assistance, through ASAC, from sister countries.

1.7 The cell may enrol as regular member of an NSB of a neighbouring country.

1.8 The cell may aim ultimately to develop into an NSB.

1.9 As an alternative to the cell, create a multi-purpose institution for standardization, industrial research, weights and measures, testing, preshipment inspection and/or related activities.

1.10 Depending on the development of specific branches of activity, the multi-purpose institution may give rise to the creation of more specialized institutions.

1.11 Yet another alternative is for a group of countries to form a multi-national standards body, which may also be multi-purpose, if necessary.

1.11.1 A new country may enter in joint partnership with an existing NSB, bringing into being a multi-national standards body.

2. *Collection of technical data*

2.1 Whatever the stage of development, countries should pursue the study of their production, distribution and utilization patterns through technical and economic investigations and research.

2.2 Assistance in this respect may be derived from sister countries through:

(a) bilateral exchanges of personnel,

(b) study tours,

(c) training courses.

2.3 International fellowships could be utilized, if available.

3. *Delegations to international meetings*

3.1 Organize adequate representation at all international technical meetings where subjects of primary interest to the country are under discussion.

3.2 Briefs for the national delegations should be prepared on the basis of research and economic data collected and the views expressed should represent the national consensus.

3.3 Enrol financial assistance of the United Nations, if necessary and possible:
   3.3.1 Utilize UN fellowships where available, by combining them with training programmes, if need be;
   3.3.2 The ECAFE secretariat could perhaps organize representations on behalf of ASAC countries on request;
   3.3.3 Other sources of United Nations assistance may be explored.

3.4 Alternatively, seek representation through proxy.

3.5 In the last resort, send, through committee secretariats, the national views and comments for presentation and discussion at the international meetings.

3.6 Insist on a larger number of meetings of interest being held in the ECAFE region.

3.7 Be prepared to invite and host such meetings.

3.8 Agitate for Western countries to relax their attitude in regard to more meetings being held in the countries of the region.

4. *Minority position of developing countries*

4.1 To off-set the disadvantage of minority position, prepare a sound and strong case based on factual and scientific data.

4.2 Formulate collective views in ASAC, for presentation on behalf of several countries by delegations carrying the proxies of countries not attending.

4.3 Review the participation status in each of the ISO and IEC Technical Committees and subcommittees and ensure one level of participation or the other in each group dealing with subjects of interest to the country.

5. *New projects*

5.1 Having developed the national activity on standardization in the fields of primary export or import interest to the country, be prepared to propose new subjects for international standardization in such fields.

5.2 Be also prepared to accept secretarial responsibilities, if it is agreed to set up a committee to deal with the subject proposed.

5.3 More countries of the region should come forward to set up additional secretariats of ISO and IEC Committees, particularly those countries which are specially equipped to do so, financially, industrially, and manpower-wise.

6. *Coordination of regional standards*
   6.1 Create within ASAC, groups of experts to study collectively subjects of special interest to more than one country, with a view to formulating regional views:
   (a) for presentation to ISO and IEC, and
   (b) as guide to national standards.
   6.2 The pattern of organization of such groups may generally be that of ISO and IEC.
   6.3 In respect of certain products, where two or three countries are interested and agreed, national standards may be developed through mutual collaboration between the NSBs concerned.

12.27 This strategy for coordination of standards in the form as now adopted by one of the important developing regions of the world under the UN-ECAFE umbrella may be applicable to other developing regions as well with such minor modifications as may be required to suit local conditions. Unfortunately, however, little experience is available in regard to the actual implementation of the strategy, being as it is of very recent origin. But given the UN support, through UNIDO or other agencies, it has the potential of going a long way towards assisting the developing nations in the achievement of their economic goals to the maximum possible extent within the limits of their own resources.

REFERENCES TO CHAPTER 12

1. How CEN works to coordinate European national standards. *Magazine of Standards.* 1963, vol 34, pp. 309-310.
2. CEN developments confirmed. *BSI News,* July 1970, pp. 17-18.
3. *Catalogue des normes françaises.* Association Francaise des Normalisation, Paris, 1970, p. 519.
4. CEN meeting in Lisbon. *ISO Bulletin,* Dec 1970, p. 2.
5. The work of the European Electrical Standards Coordinating Committee. *BSI News,* Aug 1970, pp. 12-13.
6. *Aims, structure and results* (Publication 8). International Commission on rules for the approval of electrical equipment, Arnhem (Netherlands), 1961, p. 22.
7. *Report of the Asian Standards Advisory Committee (First Session) to the AIDC* (Document AIDC (3)/9. Jan 1968). Economic Commission for Asia and the Far East, Bangkok, p. 66.

8. *Report of the Asian Standards Advisory Committee (Second Session) to the AIDC* (Document AIDC (5)/2. Jun 1969). Economic Commission for Asia and the Far East, Bangkok, p. 27.
9. *Regional strategy for coordination of standards* (Document AIDC/ASAC (2)/9. Mar 1969) (prepared by Lal C. Verman). Economic Commission for Asia and the Far East, Bangkok, p. 18.
10. Betz, Herbert: Latin America: Industrial development and standards. *Magazine of Standards,* 1960, vol 31, pp. 132-135.
11. *Pan American Standards Committee's report: Origin of Pan American Committee.* COPANT Provisional General Secretariat, Buenos Aires, 1963, p. 12.
12. Ciaburi, Beatriz Ghirelli de: How Argentina develops standards. *Magazine of Standards,* 1961, vol 32, pp. 68-70.
13. Townsend, J. R.: New program for Pan American Standards. *Magazine of Standards,* 1961, vol 32, pp. 207-210.
14. Ainsworth, Cyril: Progress in Pan-American standardization. *Magazine of Standards,* 1964, vol 35, pp. 323-326.
15. Townsend, J. R.: Technical committees work towards Pan-American standards. *Magazine of Standards,* 1963, vol 34, pp. 41-42.
16. Eden, Jack F.: Latin America at the crossroads. *Magazine of Standards,* 1963, vol 34, pp. 305-308.
17. Steel products standardization and simplification. *ISI Bulletin,* 1963, vol 15, pp. 232-233.
18. Standardization of steel profiles in Latin America. *ISI Bulletin,* 1970, vol 22, pp. 340-341.
19. Marshall, Jr, Thomas A.: Accomplishments and trends in Pan American standardization. *Magazine of Standards,* 1967, vol 38, pp. 43-44.
20. Copant meets in Lima. *Magazine of Standards,* 1968, vol 39, pp. 111-113.
21. *Six years of COPANT* (GS Document No 4-67. 1967). COPANT Provisional General Secretariat, Buenos Aires, p. 7.
22. Reports of the Commonwealth Standards Conferences (CSC):
    1st CSC. 1946. BSI, London, p. 23.
    2nd CSC. 1951. BSI, London, p. 18.
    3rd CSC. 1957. ISI, New Delhi, p. 46.
    4th CSC. 1959. CSA, Ottawa, p. 31.
    5th CSC. 1962. SAA, Sydney, p. 38.
    6th CSC. 1965. BSI, London, p. 35.
    (Copies available from any of the host National Standards Bodies noted above.)
23. Stepanenko, S. I.: Collaboration among the members of CMEA in the field of standardization. *Standarti i Kachestvo.* 1970, no 10, pp. 17-19 (*in Russian*).
24. *Transactions of Middle East Standards Conference.* The Industry Institute, Beirut, Lebanon. 1958, p. 109 (*see* p. 97).
25. Arab Organization for Standards and Measures. *UAR Standardization Bulletin,* 1966, no 10, pp. 14-15.
26. Arab Organization for Standards and Measures starts its activities. *UAR Standardization Bulletin,* 1968, no 17, pp. 4-7.
27. Activities of the Arab Organization for Standardization and Metrology. *Bulletin of Iraqi Organization for Standards.* 1968, no 3, pp. 5-8.
28. Tarrant, F. G.: Civil international standardization. *Standardization* (Directorate of Standardization, Ministry of Defence, UK). Dec 1970, no 84, pp. 13-23.
29. ABC conference on certification of engineering standards. *BSI News.* Dec 1970, p. 19.
30. Africans plan standards coordinating committee. *ISO Bulletin,* 1971, vol 2, no 2, p. 2.

Chapter 13

Systems and Units of Measurement

13.0 Standardization of units of measurements may be considered as the most basic function of the discipline of standardization in its generalized sense, for measurements pervade almost all walks of human endeavour. In order that measurements be adequately accurate for the various purposes for which they are intended, it is important that they be carried out in terms of a well-defined and well-understood system of units. All scientific, technological and industrial pursuits and procedures depend on such measurements. Furthermore, they are most essential in the daily life of the man in the street or the housewife in her home. Above all, their importance in connection with the preparation and utilization of standards at all levels must be emphasized. With the exceptions of a few standards such as those concerned with abstract matters like terminology, most of the other standards dealing with any concrete subject or aspect must necessarily define quantitatively the characteristics of the product or process covered. Quantitative definition of a characteristic requires to be given in terms of a specific system of measurements consisting of scientifically and precisely defined units for each quantity. This is essential because in the realm of standardization, measurements to be made in different places, by different operators at different times, must be consistent in themselves and comparable with one another. Otherwise, their value as criteria for judgement would be considerably lost.

PRE-METRIC SITUATION

13.1 At the time when organized standardization began to be developed as a systematic activity, there prevailed a large number of systems of measurements all over the world, which still exist to a great extent.[1] But those in use in the industrially advanced countries where standardization originated could be divided into two broad groups namely the British foot-pound system and the metric system. Both these systems were well entrenched – the former in the English-speaking world and the latter in the continental countries of Europe. In other areas associated with these two groups of countries, one or the other system prevailed together with the various indigenous systems. Each of these groups felt that its own system was so well suited for most applications that it deserved worldwide recognition and universal adoption as a standard. Almost every attempt at the establishment of internationally standardized quantitative values for characteristics of products was invariably met by the most protracted discussions about the particular system of units to be adopted for expressing the characteristics to be covered in the standard. Quite often these difficulties led to an agreed conclusion for specifying two sets of values of characteristics – one in terms of the foot-pound system and the other in the metric system.[2 3]

13.2 This practice was considered by some as a virtual failure of the international system of standardization, but others welcomed it as an achievement, claiming that two agreed world standards represented a definite advancement on the prevailing situation in which one had to deal with several standards. To an extent both points of view were justified, but the urgent need for worldwide unification of measuring systems was acutely felt by all those concerned with the international standards movement.

13.3 Like most other less well-known systems, the origin of the foot-pound system had *ad hoc* foundations. For example, the length of a human foot was originally taken to be the basis of one of the units of measurement of length and so was a barleycorn and the human digit.[4 5 6] There was no rational relationship between the units of various quantities, for example, the length the volume and the mass. In spite of this lack of interrelationship and its other inconsistencies, the foot-pound system in due course came to be quite precisely defined for modern needs, both in the United Kingdom and the USA. Small differences between the UK inch and pound values and the USA inch and pound magnitudes were reconciled in due course.[7] Other English-speaking countries and those connected with the British Empire followed the general practices of the

UK and the USA. For standardization purposes, therefore, this system was considered to be equally as effective and precise as the more scientifically evolved metric system. But for a slight fractional difference between one inch and 25 millimetres (1 in = 25.4 mm), the two systems might have found a workable basis for co-existence,[8] at least in regard to length measurements. However, even this could not have met all the needs of standardization, for length happens to be only one of the quantities involved in standardized measurements. There are several other important quantities like mass, area, volume, capacity and others, which are equally involved in standardization and which could not be reconciled even to the approximation mentioned above in the inch-millimetre case. It is fortunate that what looked like almost impossible to achieve about two decades ago has become quite within reach today,[52] for the world now appears to be poised for a universal acceptance of the metric system within a foreseeable future. Without, therefore, discussing any other systems of measurements in any detail we may well confine ourselves to a brief review of the metric system.

METRIC SYSTEM

13.4 The metric system was conceived and developed during the closing decade of the eighteenth century by a French team of scientists.[9, 10] The chief motivation was to rationalize the then existing variety of measurement systems which prevailed on the European and American continents. At that time there appeared no chance of any one of the known existing systems of units being universally accepted, which could help promote freer intercourse between the various countries, and indeed between certain regions of the same country. So, the French scientists, encouraged by the revolution, addressed themselves to the task of devising a system, using nature as a model and natural phenomena as a guide, to which no national susceptibilities could possibly attach. It goes to the credit of the French Constituent Assembly which took the initiative in 1790 to entrust to the French Academy of Sciences the task of establishing a measuring system which held the promise of being accepted by the world as a whole.

13.5 It was thus that, after as careful measurements as could be made at that time, one ten-millionth part of a quadrant of the earth's meridian was adopted as the unit of length – the metre. The unit of mass was derived from this unit of length by defining the kilogram as equal to the mass of water having a volume, under certain conditions of measurement, equal to a decimetre (one-tenth of a metre) cube. Based on these measurements, two physical prototype standards, both in platinum, were constructed – one for the metre and one for the kilogram – and deposited in the Archives of the French Republic in 1799.

13.6 Another significant step which was also taken at the same time may be considered to be quite novel in the history of systems of units of measurements. Most systems that had prevailed in the world had been based on multiples and submultiples other than ten. These multiples included 8, 12, 16 and others. They had their merits in the sense that the numbers could be readily divided by 2, 3 and 4 etc.; nevertheless, they were inconvenient because of their non-consistency with the normal system of decimal counting and decimal place values. Evidence,[11] however, is available for the existence of a decimal length measure in the Indus Valley civilization some 4,000 years ago, but apparently neither this measure nor its decimal character acquired any degree of popularity during the subsequent ages. It goes to the credit of the originators of the metric system to have, for the first time, completely decimalized all the multiples and submultiples of the basic as well as the derived units and to have confined the system only to the positive and negative integral powers of 10. This innovation, which should be considered a great step forward in the realm of measurements, considerably simplified all calculations and held promise of daily saving millions of man-hours for the users at all levels of human activity. In passing, it may, however, be noted that today's computer technology is based on the binary system of calculation. But this has its own reasons and own uses and is not likely in any way to influence the prevalence of the decimal system of counting in other walks of life.

13.7 *Metre Convention* In spite of the well-founded expectations of its originators for its unquestioned and prompt acceptance by all the then advanced nations of the world, the metric system continued to remain dormant for several decades and indeed its universal acceptance even within France was not as spontaneous as it had originally been expected to be. Voices in favour of a worldwide unification of measuring systems on the basis of the metric system, however, continued to be raised from time to time by various learned societies in France as well as other European countries. There arose a general feeling that an international approach to collective action was called for. It was thus that the French government, in 1870, invited representatives of several countries to meet in Paris. Twenty-four countries responded to this invitation, of whom only 15 could send their delegates, because of the outbreak of the Franco-Prussian War. These delegates constituted the *Commission Internationale du Mètre,* but they could take no decisions. The work of this Commission could, however, be resumed in 1872 with the participation of delegates of 30 countries, 11 of whom were from the American continent. About 40 resolutions were passed dealing with the preparation of new prototypes of kilogram and metre and related matters. The creation of an International Bureau of Weights and Measures was also recommended to the interested governments.

13.8 But the members of this international commission, who were all scientists, had no authority to commit their governments. Hence some years later, in 1875, another conference attended by full-fledged representatives of the governments, was held again in Paris. It was called the *Conférence Diplomatique du Mètre.* This time positive results were achieved. On 20 May 1875, a *Convention du Métre* was signed by 18 states.[12] By this Convention, the signatory states bound themselves to set up and maintain at common expense a permanent scientific body of weights and measures at Paris. It was given the name of *Bureau International des Poids et Mesures* (BIPM).

13.9 *Conférence Générale des Poids et Mesures* The governing authority of the Bureau was the *Conférence Générale des Poids et Mesures* (CGPM), made up of officially appointed delegates from all the member countries, which were signatories to the *Convention du Mètre* and those which might join the Convention later. The duties of the General Conference of Weights and Measures were defined briefly as follows:

(1) to discuss and adopt necessary measures for the propagation and improvement of the metric system;

(2) to sanction the results of new fundamental metrological determinations and various scientific resolutions of international importance; and

(3) to take important decisions concerning the organization and the development of the International Bureau of Weights and Measures.

13.10 The CGPM, which meets every six years, being the supreme authority, takes all the major decisions in regard to the new and revised definitions of metrological standards and all policy matters including finances and programmes for future developments. It also appoints members of the implementing body called the *Comité International des Poids et Mesures* (CIPM), consisting of a maximum of 18 specialists chosen from the signatory countries. The CIPM is expected to meet at least every two years or more frequently, if need be. It is charged with the functions of following up the decisions of the Conference and looking after the operation and management of the Bureau. The CIPM appoints its own specialist consultative committees, of which there are seven at present, one each dealing with the definition of the metre, the definition of the second, thermometry, electricity, photometry, ionizing radiation and the *Système International d'Unités* (SI). Up to the end of 1968 altogether 40 states had signed the Metre Convention, which now constitute the membership of CGPM. A list of the member countries in alphabetic order is given in Table 13.1, indicating the year of their joining the Convention.

13.11 *Bureau International des Poids et Mesures* The Bureau International des Poids et Mesures (BIPM) is the laboratory wing of the organization, where all the standards of weights and measures of the world authority are maintained as also all other prototype standards of various physical quantities.[13, 14] Besides the maintenance of these standards, the functions of the BIPM include the carrying out of research on further refinements of standards and of methods of measurements of ever-improving attainable accuracy. The BIPM also serves the member countries as the central authority for all matters connected with scientific metrology, including such periodic verification of national standards in terms of the international prototypes as may be requested. These services are rendered free of charge to member nations; from other countries requiring such services, an appropriate charge is made. Many countries, particularly among the developing countries, find it more economical to pay for these services fees than to join the convention and pay the annual subscriptions.

13.12 *World Cooperation* The international work on physical metrology is concentrated not only in the BIPM but also spreads over many of the important national laboratories dealing with standards of measurements, which include, among others, the following:

National Standards Laboratory, CSIRO (Australia)
National Research Council (Canada)
Conservatoir National des Arts et Métiers (France)
Physikalisch-Technische Bundesanstalt (Germany)
Deutsches Amt für Mass und Gewicht (Germany)
National Physical Laboratory (India)
National Research Laboratory of Metrology (Japan)
National Physical Laboratory (United Kingdom)
National Bureau of Standards (USA), and
D. I. Mendèleev Institute of Metrology (USSR)

13.13 With the central coordinating role of the BIPM and the co-operative effort of the national laboratories, considerable advances have been made in extending and refining the metric system so that it now goes

TABLE 13.1
MEMBERS OF CGPM AS OF 1968

Serial No.	Name of the Country	Year of Entry
1.	*Argentina*	1875
2.	Australia	1947
3.	*Austria*	1875
4.	*Belgium*	1875
5.	Brazil	1954
6.	Bulgaria	1911
7.	Canada	1907
8.	Chile	1908
9.	Czechoslovakia	1922
10.	*Denmark*	1875
11.	Dominican Republic	1954
12.	Finland	1921
13.	*France*	1875
14.	*Germany* (East and West)	1875
15.	*Hungary*	1875
16.	India	1957
17.	Indonesia	1960
18.	Ireland	1926
19.	*Italy*	1875
20.	Japan	1885
21.	Mexico	1890
22.	Netherlands	1929
23.	*Norway*	1875
24.	Poland	1925
25.	*Portugal*	1875
26.	Rumania	1881
27.	South Africa	1964
28.	South Korea	1959
29.	*Spain*	1875
30.	*Sweden*	1875
31.	*Switzerland*	1875
32.	Thailand	1912
33.	Turkey	1933
34.	UAR	1962
35.	United Kingdom	1884
36.	Uruguay	1908
37.	*USA*	1875
38.	*USSR*	1875
39.	Venezuela	1960
40.	Yugoslavia	1879

Note: Italicized countries are the original signatories of the Metre Convention.

far beyond the original standards of length and mass. At its eleventh session in 1960, the CGPM adopted the so-called International System of Units (SI),[15] based on the six basic units, namely, the units of length, mass, time, temperature, electric current and light intensity, which have by now been as precisely defined as the present-day advances in science and technology make it possible to do so. Incidentally, it may be noted that the ultimate length standard is no longer the prototype metre; it is now defined in terms of a given number of wavelengths of certain light. Similarly, the standard of time, the second, depends no longer on the movements of heavenly bodies, which have been found to be somewhat variable. It is now represented by the time duration of a given number of cycles of vibrations of light of a particular spectrum line.[16] Likewise, all other standards of measurements are constantly being defined and redefined more and more accurately through the work of the BIPM and the associated national laboratories. New fields in which new standards of measurement are now being developed include those of ionizing radiation and nuclear physics.

INTERNATIONAL ORGANIZATION OF LEGAL METROLOGY

13.14 It has always been conceded that so far as scientific aspects of standards of measurements are concerned, the Metre Convention complex of the CGPM, CIPM and BIPM, together with the cooperating national laboratories are organized, equipped and endowed well enough to serve all the national and international needs of measurements. But metrological standards developed by science have ultimately to be applied also to everyday economic activity of man related to his industrial and commercial requirements, namely, those arising from the exchange of goods and services. This requires the highly precise scientific standards to be translated into everyday standards of commerce and industry and the related machinery – legal, technological and administrative – to ensure the availability and utilization of the latter standards at all levels of human activity. This link is provided by what has now come to be known as legal metrology. Until recently, legal metrology problems were dealt with by individual countries according to their own genius and regardless, more or less, of the practices prevailing in sister countries. But more recently, with the unprecedented growth of world trade and travel, a need has been felt for international coordination in the formulation of rules and regulations governing the use of metrological standards. To meet this need, an inter-governmental organization for legal metrology was created in 1956 under the title *Organisation Internationale de Métrologie Légale* (OIML), the headquarters of which are also located in Paris.[17, 18]

13.15 The 43 countries which have so far joined this organization in the two categories of membership include:

Full Members

Australia	Guinea	Norway
Austria	Hungary	Poland
Belgium	India	Rumania
Bulgaria	Indonesia	Spain
Ceylon	Iran	Sweden
Cuba	Israel	Switzerland
Czechoslovakia	Italy	Tunisia
Denmark	Japan	UAR
Dominican Republic	Lebanon	United Kingdom
Finland	Monaco	USSR
France	Morocco	Venezuela
Germany	Netherlands	Yugoslavia

Associated Members

Greece	Nepal	Pakistan
Jordan	New Zealand	Turkey
Luxembourg		

13.16 Dealing with the whole field of legal metrology at the international level, the OIML attempts to bring about a unification and coordination of the legal practices prevailing in member countries, through the issue of recommendations on methods of measurements and standards for instruments used in measurements, on model laws and regulations for the control of weights and measures, on the pattern of services required to be organized for exercising such controls and so on. An important aspect of its work is to set up a central documentation and translation service to collect and disseminate information on all legal metrological matters from different countries. Above all, by promoting closer relations between departments responsible for legal metrology in the various member countries, the OIML serves the cause of facilitating international exchange of goods and services. Thus its work supplements that of the ISO in many ways. The close collaboration established between the ISO on the one hand and the CGPM and OIML on the other has already been referred to in Chapter 11 (*see* para 11.26).

13.17 *Spread of the Metric System* Though 18 countries had signed the Metric Convention in 1875, not all of them had adopted the metric system as an exclusive system of units of weights and measures. Among them the notable exceptions were the UK and the USA. In fact a recent report of the BIPM indicates that of the 81 countries, where use of the metric system is mandatory, only 34 countries are signatories to the metric convention.[21] The remaining six countries of the total of 40 signatories are among those in which the use of the system is optional. Another survey[1, 19] published by the United Nations in 1966 classified 126 countries as metric and 70 as non-metric, with six where both metric and non-metric usage prevails. It has been estimated[20] that the metric countries account for well over 80 percent of the world's population, approximately 60 percent of the world's gross national product, and about 75 percent of the international trade.

13.18 In spite of this preponderance of metric usage, it has been observed that in some countries where the metric system is required to be used universally by law, the rigour of its enforcement varies. This may in some cases be due to force of habit or the established tradition, in others it may be dictated by the demands of that component of the international trade which is based on non-metric standards. Until the last decade or two, it was the general practice to place reliance mainly on the legislative mechnisms for introduction of the metric system in a nation's economic life. Japan's experience in this respect is perhaps typical where it is well known that its factual all-round adoption took some decades to be accomplished.[53] In other countries the associated metrological services and the administrative enforcement machinery has not always been satisfactory enough to ensure its universal usage. Even in certain advanced metric countries pockets may exist where non-metric units may still prevail, such as the use of the "morgan" as an area measure in some parts of Europe.

13.19 India was perhaps the first country where a systematic approach was made for the introduction of the metric system according to a planned programme of action, spread over a stipulated period of time.[21, 22] Outline plans had been prepared even before the legislation of 1956 was passed. This was followed by a period of intense educational and propaganda effort, aimed at preparing the minds of the population and imparting the necessary training to the technical and skilled personnel. Then came in succession the decimalization of coinage and preparation of industrial standards in metric terms, accompanied by industry-wise conferences for planning the sectorial conversion programmes and their interrelationships. All commercial and industrial sectors were dealt with one by one, area by area, and simultaneously metrological services were introduced for the supply and verification of commercial sets of weights and measures and the associated instruments and equipment. By the end of the ten-year plan period, the country had gone metric practically in all spheres of its economic activity and the pockets that remained continued to be identified and tackled.

13.20 Most of the English-speaking countries, namely the USA, Canada and the UK, together with several of the associated countries, remained outside the metric circle until 1965 when the UK declared its intention to go metric and adopted a planned programme of conversion extending to 1975.[23–28] Organized efforts according to detailed plans of conversion for each economic sector are presently well under way.[29] The whole operation is being conducted under the guidance of a high level Metrication Board, with the active participation of the British Standards Institution, which was originally responsible for making the initial move.[30] The prognosis is quite optimistic.[56]

13.21 Countries of the Commonwealth and those associated economically with the UK soon followed suit and have, after due study, either decided already or are about to decide to adopt the metric system. These include Australia,[19] Canada,[31] Ceylon,[32] New Zealand,[33] Malaysia, Singapore, South Africa[34] and others. Many other countries have also joined the metric fraternity in recent years.[20,55] But the most significant event of recent times is perhaps the seriousness with which the USA has taken up the question. In August 1968, the US Congress passed a legislative measure authorizing a comprehensive study of the feasibility and implications of adopting the metric system in the country. The first of a series of reports on this study which is now approaching conclusion is already available and deals with the importance of internationally accepted standards to the economy of the USA. It outlines the role that the USA could play in the preparation of world standards and their wider application.[35] Since most international standards are based on the metric system and the preponderance of international trade is going more and more in that direction, the US concern, as revealed by the first US metric study report, can be readily appreciated. The final report of the US metric study has also now been submitted to Congress,[57] in which a strong case is made out for adoption of the metric system according to a planned programme. Two bills have also since been introduced in the Congress for the implementation of these recommendations. If one or other of these bills or a consolidated one is passed, of which there is little doubt in anyone's mind, it may safely be said that the whole world would soon have adopted a universal measuring system – a real triumph for the originators of the metric system and a blessing for the human race. This would perhaps be the first really international standard of the world which would be legally accepted everywhere.

13.22 Among the non-metric countries,[1, 19] with the exception of very few, all are in the category of developing countries. These countries together with those metric countries where the system has not yet become fully entrenched would do well to watch the planned introduction of the system in the UK, which is presently under way,[29, 36–38] and also to follow closely the programme that might be evolved in the USA.[35, 39, 40] A more useful guide for the developing countries would perhaps be the methods followed in another developing country, namely India, which has recently gone metric and whose experience has been collected in a book[21] published by the Indian Standards Institution in 1970, and also presented from time to time in certain journals.[41] Assistance may also be derived from some UN publications and documents.[42]

SI SYSTEM OF UNITS

13.23 Apart from the need for precisely defining and standardizing the basic units of measurements from which other units may be derived, it is important that a self-contained systematized set of units should be devised and adopted within the orbit of which each and every quantity used for scientific, technological and commercial purposes could uniquely be expressed. Since the origin of the metric units, several systems of such units have been proposed and used in different walks of life, of which the following are the outstanding examples:

cgs – centimetre-gram-second system
MKS – metre-kilogram-second system
MTS – metre-tonne-second system
MKSA – metre-kilogram-second-ampere system
RMKSA – rationalized metre-kilogram-second-ampere system.

For various reasons, which may not be detailed here, these systems have not been found adequate.[43, 49] The latest system which has found universal favour is the *Système International d'Unités* (International System of Units) abbreviated as SI in all languages.

13.24 The basic units of the SI system were defined and adopted at the Tenth Session of the CGPM[44] and the system itself was completely defined at the Eleventh Session of the CGPM.[45,46,47] The six basic units of the system are:

Quantity	*Unit*	*Abbreviation*
Length	Metre	m
Mass	Kilogram	kg
Time	Second	s
Electric current	Ampere	A
Thermodynamic temperature	Degree Kelvin*	K
Luminous intensity	Candela	cd

* Originally adopted abbreviation °K was later changed to K.

In addition, two supplementary units were also adopted, namely the measure of the plain angle radian (rad) and the solid angle steradian (sr). To facilitate the expression of very large and very small quantities compared to the size of the basic and derived units (*see* para 13.25), a set of standard multiples and submultiples applicable to all units were designated together with their names to be prefixed to the name of the units, and the accompanying abbreviation symbols were also specified. These are listed below:

Factor by which the unit is multiplied	*Prefix*	*Symbol*
1 000 000 000 000 = 10^{12}	tera	T
1 000 000 000 = 10^{9}	giga	G
1 000 000 = 10^{6}	mega	M
1 000 = 10^{3}	kilo	k
100 = 10^{2}	hecto	h
10 = 10^{1}	deca	da
0.1 = 10^{-1}	deci	d
0.01 = 10^{-2}	centi	c
0.001 = 10^{-3}	milli	m
0.000 001 = 10^{-6}	micro	μ
0.000 000 001 = 10^{-9}	nano	n
0.000 000 000 001 = 10^{-12}	pico	p
0.000 000 000 000 001 = 10^{-15}	femto	f
0.000 000 000 000 000 001 = 10^{-18}	atto	a

13.25 Furthermore, for expressing various other important physical quantities defined in terms of the basic units, an additional 27 derived units were also adopted, namely:

DERIVED UNITS

Quantity	*Name of the Unit*	*Unit Symbol*
Area	square metre	m^2
Volume	cubic metre	m^3
Frequency	hertz	Hz (s^{-1})
Density	kilogram per cubic metre	kg/m^3
Speed	metre per second	m/s
Angular velocity	radian per second	rad/s
Acceleration	metre per second squared	m/s^2
Angular acceleration	radian per second squared	rad/s^2
Force	newton	N ($kg.m/s^2$)
Pressure	newton per square metre	N/m^2
Viscosity (dynamic)	newton second per square metre	$N.s/m^2$
Viscosity (kinematic)	metre squared per second	m^2/s
Work, energy, quantity of heat	joule	J (N.m)
Power	watt	W(J/S)
Quantity of electricity	coulomb	C (A.s)
Electric tension potential difference electromotive force	volt	V (W/A)
Electric field strength	volt per metre	V/m
Electric resistance	ohm	(V/A)
Electric capacitance	farad	F (A.s/V)
Magnetic flux	weber	Wb (V.s)
Inductance	henry	H (V.s/A)
Magnetic flux density	tesla	T (Wb/m^2)
Magnetic field strength	ampere per metre	A/m
Magnetomotive force	ampere	A
Luminous flux	lumen	lm (cd.sr)
Luminance	candela per square metre	cd/m^2
Illumination	lux	1x (lm/m^2)

13.26 As may be noted from the list of derived units given above, the great merit of the SI system of units is its coherence.[48, 49] This means that whenever a given quantity is the product or quotient of two or more quantities, its unit in a coherent system will be the product or quotient of the units of the latter quantities. For example, in the SI system, the unit of force, newton, would be derived directly by the product of mass unit, kilogram, and the length unit, metre, divided by the square of the time unit, second, in other words:

$$1\ [\mathrm{N}] = \frac{1\ [\mathrm{kg}] \times 1\ [\mathrm{m}]}{1\ [\mathrm{s}] \times 1\ [\mathrm{s}]},$$

in which [ ] denotes the unit of the bracketed quantity. It may also be noted that it is not always necessary to give a special name to the unit of every quantity expressed in terms of the derived units, in the same manner as in the case of newton, joule, farad and several others. For example, there is no special name for the unit of speed, it is known simply as metre per second. Similarly, density is expressed as kilogram per cubic metre, and so on for many other quantities, as may be seen from the list given in the previous paragraph. It may be emphasized here that the strongest point and the most important feature of the SI system is its coherent character. It would also perhaps be correct to say that the unit of force, newton, may be considered as its core. It gets rid of the gravitational unit kilogramforce which involved the use of earth's gravity that varies from place to place. Though the magnitude of a newton is about one-tenth of a kilogramforce, it is quite within the realm of practical units and simplifies calculations considerably. A systematic account of the SI system has been recently published by the BIPM under the authority of CIPM.[54]

13.27 Another equally, if not more, important feature of the system is that it eliminates the use of different units for measuring the same quantity in different disciplines of science and technology. For example, it has been the common practice to measure energy and work in terms of kilowatt-hours in electrical engineering, horsepower-hours or kilogram-force-metres in mechanical engineering and calories or kilo-calories in thermal engineering. In the SI system all these would be replaced by the joule which is one newton-metre. It will no doubt take some time to have this change universally adopted, for not only will the existing literature have to be rewritten and a good many of the measuring instruments replaced, but what is even more significant is that the thought-processes of the present-day scientists and technologists would have to be readjusted.

Only when the new generations have been taught their science and technology in SI terms, will the change become fairly complete.

13.28 *ISO Selection of SI Units* In view of the coherent and rational character of the SI system of units,[49] and the fact that it represents a significant advance on any one of the previous systems of metric units proposed or used, its official adoption by the CGPM was appreciated by all scientists and technologists and widely welcomed by metrologists and weights and measures authorities in most countries. But it was soon realized that its practical implementation either in science or technology or in trade was not going to be a straightforward process. If left to individual choice, it was bound to lead to the use of quite a variety of multiples and submultiples of the basic and derived units for expressing the same quantity, which may give rise to a new type of confusion. For example, the six basic, the two supplementary and the 27 derived units add up to 35, and any one of them could be used with any one of the 14 officially approved prefixes for multiples and submultiples, which would lead to as many as 35 X 14 or 490 possible units. All these would be strictly within the orbit of the CGPM definition of *Systeme International* (SI).[50] Furthermore, there would still remain the need for devising several other units for expressing those derived quantities which have not been covered by the CGPM list of the 27 units, as for example the units for impact strength, momentum, specific heat and so on.[58] These derived units too could be devised on the basis of SI units in many different ways, in different countries and by different authorities.

13.29 Furthermore, it would be necessary to regulate within the orbit of each discipline the use of specially agreed units for expressing the quantities which have not been covered by the CGPM list or covered in a manner not quite convenient for use in all disciplines. For example, to express the density of concrete or concrete aggregate, the SI derived unit kg/m^3 would be quite convenient and acceptable. But for a chemist it would hardly do to talk in terms of kilograms and cubic metres when he is concerned with expressing the density of fine chemicals. He would no doubt prefer to use a more manageable and comprehensible unit, say g/cm^3, which is not in the CGPM list of derived units. In order to comply strictly with the CGPM decision, g/cm^3 will have to be expressed as milli-kg/m^3, because kg/m^3 and the prefix milli happen to be both contained within the SI definition. This, to say the least, would hardly be acceptable to a chemist, being as it is not only awkward but much more difficult of comprehension as compared with g/cm^3, that is, in the context of the size of quantities dealt with by a chemist.

13.30 It was felt extremely necessary, therefore, to settle these and many other questions of the choice of SI units, their multiples and sub-multiples, if industry, technology and trade were to be properly guided for adopting the SI units in a universal and uniform manner in all countries of the world. The ISO recognized its responsibility in this respect at quite an early stage and produced in record time a document[51] on the subject which represented the agreement of a preponderant majority of its membership. This international standard, ISO/R 1000-1969, has appeared at the most opportune time when several countries of the world, especially the inch-pound countries, are in the midst of adopting the metric system as their national system of measurement. It will enable them to go directly to the most up-to-date system of metric units and to convert their standards and practices accordingly without having to go through an intermediate stage, where it might have been necessary to adopt the practices of the erstwhile metric countries as the first stage, as in the case of India. It has now become necessary for certain far-reaching changes in the standards of all the countries where the metric system has prevailed for many years, to be made at the earliest possible opportunity, if full compliance with the SI units in conformity with the ISO/R 1000 is to be secured.[51] In any case the day does not appear to be far off when the whole world will not only be metric but also follow a uniform pattern of SI units as defined by the CGPM at the scientific level and by the ISO for their practical application to industry, technology and trade.

REFERENCES TO CHAPTER 13

1. *World weights and measures* (Statistical Papers: Series M, no 21, rev 1). United Nations, New York, 1966, p. 138.
2. *ISO/R 261-1969 ISO general purpose metric screw threads.* General *plan;* and *ISO/R 263-1962 ISO inch screw threads. General plan and selection for screws, bolts and nuts (diameter range 0.06 to 6 in);* and a series of associated ISO/Rs on screw threads. International Organization for Standardization, Geneva, p. 7, p. 8 etc., respectively.
3. *IEC PUB 72 (Part I and Part II)-1967 Dimensions and output ratings of electrical machines.* International Electrotechnical Commission, Geneva, p. 29 and p. 19 respectively.
4. Berriman, A. E.: *Historical Metrology.* Dent and Sons, London, 1953, p. 224.
5. Vieweg, R.: Kleine Kulturgeschichte der Metrologie (Brief cultural history of metrology). *DIN-Mitteilungen,* 1968, vol 47, pp. 2-11 (*in German*).
6. Vieweg, R.: *Mass und Messen in Kulturgeschichtlicher Sicht* (Measures and measurements from cultural historical viewpoint). Wiesbaden, 1962 (*in German*).
7. International yard and pound. *ISI Bulletin,* 1959, vol 11, p. 165.

8. Cattaneo, A.: The battle of measure systems and the "cocktail series". *Standards Engineering,* 1963, vol 15, no 9, pp. 3-6.
9. Moreau, Henri: Genesis of the metric system and the work of the International Bureau of Weights and Measures. *Journal of Chemical Education,* 1953, vol 30, pp. 3-20.
10. Johnson, J. T.: *The metric system of weights and measures.* Bureau of Publications, Teachers College, Columbia University, New York, 1948, p. 303.
11. Standardization in prehistoric India. *ISO Souvenir.* Indian Standards Institution, New Delhi, 1964, pp. 21-33.
12. Moreau, Henri: Metric Convention and International Bureau of Weights and Measures. *Metric Measures* (India). 1959, vol 2, no 6, pp. 3-12.
13. Terrien, J.: Scientific metrology on the international plane and the Bureau International des Poids et Mesures. *Metrologia,* 1965, vol 1, pp. 15-26.
14. Moreau, Henri:*Le Bureau International des Poids et Mesures et ses activités. 4eme Session: Normalisation.* Association Française de Normalisation, Paris, 1968, p. 13 (*in French*).
15. *Comptes rendus des séances de la onzième conférence générale des poids et mesures* (Proceedings of the eleventh session of CGPM). Bueau International des Poids et Mesures, Sevres, p. 144 (*in French*).
16. Astin, Allen V.: Standards of measurement. *Scientific American,* 1968, vol 218, no 6, pp. 50-62.
17. *International Organization of Legal Metrology (OIML) – field of activities – composition – constitution – operations – budget receipts – budget expenses – entry into force.* OIML, Paris, 1956, pp. 10 + 12 (*in French*).
18. *Premiere conférence générale des états membres de l'Organisation Internationale de Metrologie Légale* (First conference of the member countries of OIML). OIML, Paris, 1956, p. 135 (*in French*).
19. *Report from the Senate Select Committee on the metric system of weights and measures, Part I.* Government of Australia, Canberra, 1968, p. 137 (*see* pp. 8-9 and 117-118).
20. Burton, William K.: *Measuring systems and standards organizations.* American National Standards Institute, New York, 1970, p. 45.
21. Verman, Lal C. and Kaul, Jainath: *Metric change in India.* Indian Standards Institution, New Delhi, 1970, p. 529.
22. Ghosh, A. N.: How India introduced metricization in engineering industry. *The Chemical Engineer,* Sep 1969, pp. 149-152.
23. Metric standards – green light from the government. *BSI News,* July 1965, pp. 4-6.
24. *PD 6245-1967 Going metric, first stages.* British Standards Institution, London, p. 7.
25. *Change to metric system* – Report by the Standing Committee on Metrication. Her Majesty's Stationery Office, London, 1968, p. 10.
26. Ritchie-Calder, Lord: Functions of the Metrication Board. *Going metric.* Her Majesty's Stationery Office, London. July 1969, p. 1.
27. Wynn, A. H. A.: Adoption of the metric system in the United Kingdom. *Chemistry and Industry,* 1968, pp. 1512-1516.
28. Butcher, F. E.: Going metric – Britain's first steps. *Materials Research and Standards,* 1967, vol 7, pp. 357-360.
29. Vickers, J. S.: *Making the most of metrication.* Gower Press, London, 1969, p. 163.
30. *PD 5069-1963 British industry and the metric system.* British Standards Institution, London, p. 7.
31. *White paper on metric conversion in Canada.* Ministry of Industry, Trade and Commerce, Federal Government of Canada, Ottawa, Jan 1970.

32. *Report on the adoption of the metric system in Ceylon.* Bureau of Ceylon Standards, Colombo, 1969, p. 41.
33. *The metric system – report of the working committee of officials.* Department of Industries and Commerce, Wellington (New Zealand), 1968, p. 78.
34. *Report on the metric system of weights and measures.* Council of South African Bureau of Standards, Pretoria, 1965, p. 149.
35. *US Metric study report – international standards* (National Bureau of Standards Special Publication 345-1). US Printing Office, Washington, 1970, p. 145.
36. *Decimal coinage and the metric system – should Britain change?* (A joint committe report appointed by BAAS and ABCC). Butterworth Scientific Publications, London, 1960, p. 107.
37. Gilbert, A. J.: The benefits and implementation of metrication. *Electrical Review,* 1968, vol 182, pp. 246-249.
38. Going metric in the face of difficulties. *Electrical Review,* 1968, vol 183, pp. 822-824.
39. *Report of the 52nd National Conference on weights and measures* (National Bureau of Standards Special Publication 297). 1967, p. 218; and other similar reports. US Printing Office, Washington.
40. Metric study enters data gathering phase. *NBS Technical News Bulletin,* 1970, vol 54, pp. 121-123.
41. *Metric Measures* – a monthly journal issued by the Ministry of International Trade, New Delhi, between 1958 and 1965 in vols 1 to 8.
42. *Introduction of metric system in a developing country – a policy paper* (Annex II to Document AIDC/ASAC (2)/8 – 19 March 1969). Economic Commission for Asia and the Far East, Bangkok, pp. 17-27.
43. *IEC PUB 164-1964 Recommendations in the field of quantities and units used in electricity.* International Electrotechnical Commission, Geneva, p. 63.
44. *Comptes rendus des séances de la dixième Conférence Générale des Poids et Mesures.* Bureau International des Poids et Mesures, Sèvres. 1955, p. 99 (*see* p. 80) (*in French*).
45. *Comptes rendus des séances de la onzième Conférence Générale des Poids et Mesures.* Bureau International des Poids et Mesures, Sevres, 1960, p. 144 (*see* p. 87) (*in French*).
46. Strain, D. C.: Up-to-date SI units for metric measurements. *Instrumentation Technology,* June 1967, pp. 62-64.
47. Mokashi, V. S.: International system (SI) units and their application to engineering. *Journal of the Institution of Engineers (India),* 1970, vol 19, no 7, pp. 2-5.
48. Système international d'unités (SI). *BSI News,* Oct 1968, pp. 6-8.
49. Rao, V. V. L.: SI – a rational metric system of units. *ISI Bulletin,* 1968, vol 20, pp. 399-403 and 438.
50. Pallez, A.: La double structure du système international d'unités SI (Double structure of the international system of SI units). *Courrier de la Normalisation,* 1967, no 195, pp. 384-385 (*in French*).
51. *ISO/R 1000-1969 Rules for the use of units of the international system of units and a selection of the decimal multiples and submultiples of the SI units.* International Organization for Standardization, Geneva, p. 20.
52. Shall we? when? how? *Standards Engineering,* 1962, vol 14, no 12, pp. 4-10.
53. Tamano, M.: *Japan's transition to the metric system* (National Research Laboratory of Metrology Monogram 62-1). Ministry of International Trade and Industry, Tokyo, 1962, p. 17.
54. *Le système international d'unités (SI)* (International system of units). Bureau International des Poids et Mesures, Sèvres. 1970, p. 36 (*in French*).

55. *Comptes rendus des séances de la treizième Conférence Générale des Poids et Mesures*. Bureau International des Poids et Mesures, Sèvres, 1968, p. 120 (*see* pp. 119-120) (*in French*).
56. Van Giesen, Paul: A report to ASTM on metrication in the United Kingdom. *Materials Research and Standards*, 1971, vol 11, pp. 26-28.
57. *A metric America – a decision whose time has come.* Report to the Congress. National Bureau of Standards, Washington, 1971, p. 170.
58. Metric units for pressures and stresses. *New Zealand Standards Bulletin*, 1971, vol 17, no 2, pp. 14-15.

Chapter 14

Implementations of Standards

14.0 The task of securing implementation of standards at all levels has in recent years become one of the important functions of all standardization authorities. It is obvious that the mere writing of standards and publishing them may not always achieve the ends for which they are intended. Though most standards are prepared after due consultation with all those concerned and should, therefore, be automatically found acceptable for immediate use by all interested parties, yet the need for a deliberate effort to ensure their all-round implementation in the various relevant sectors of economy is felt by most standards-making authorities – all the way from company level to the international sphere. The methods adopted and the magnitude of effort involved would vary under different circumstances, depending not only on the nature of standards and their level, but also on the nature of the economic structure of the country or countries involved.

COUNTRIES WITH MANDATORY STANDARDS

14.1 The problem of implementation of standards in countries where the economy is centrally controlled may appear at first sight to be quite simple, firstly because their standards are almost always mandatory to begin with, and secondly because all means of production and distribution of goods and services are subject to one central control. But the very complex nature of such economies, each facet of which touches upon every facet of individual and corporate life of the people, makes the task of implementation rather complicated. Every standard in such countries has a legal status and must necessarily be adhered to by all concerned, in much the same manner as any other law in any other country. But laws, in order to be enforced, require constant vigilance to be exercised by some authority or the other, and due penalties for non-observance have to be imposed through a prescribed process of enforcement. This makes the implementation of standards quite an important concern of the state in those countries where all national standards are intended to serve not only the technical and economic ends but also as legal instruments of state policy.

14.2 Orgizkov[1] has given a comprehensive account of the legal apparatus existing and the procedures followed in this regard in the USSR. He states that there are four categories of laws pertaining to standardization in that country, namely:

(1) *Administrative laws,* to define the competence of standards bodies, to prescribe procedures for the preparation of standards, and to adopt programmes and plans for standardization.

(2) *Civil laws,* to regulate the use of standards in trade and industry, covering guarantees for conformance, delivery conditions and actions to be taken in case of non-conformance.

(3) *Labour laws,* to deal with savings of material and manpower resources and appropriate timing for the introduction of new standards.

(4) *Penal laws,* to prescribe the responsibilities of the officials concerned with the implementation of standards and the penalties that might be imposed in case of non-observance.

14.3 The comprehensive character of the legal apparatus as indicated above amply emphasizes the importance that is attached to standardization in the USSR and, no doubt, in other countries having a similar pattern of economic organization. The USSR has, indeed, given a fresh lead in this matter which, no doubt, is due to a deeper realization of the far-reaching effects that rigid adherence to standards can have on a healthy growth of the economy of the country. It has very recently decided to create a central Ministry of Standardization,[2,3] which has made USSR the first country in the world to elevate the head of its national standards body to the full ministerial rank. This amply reflects the determination of the Soviet government to improve the quality and performance of Russian products, and to enhance industrial efficiency and productivity in a systematic and planned manner under a centralized and authoritative control. The National Standards Body will no longer be dependent on various other organs of the government for securing implementation of standards, but the Ministry of Standards will now exercise direct responsibility for enforcing compliance with them.

14.4 This Soviet decision, which may be considered a big leap forward in the world of standardization, is quite likely to be emulated by other countries, where similar patterns of economy prevail. But in countries where economic patterns differ, the creation of a ministry for standardization may or may not be considered feasible or practical at the present stage of development. Nevertheless, it must be admitted that if a government were conscious of the far-reaching consequences of standardization and were anxious to promote a planned development of its resources at a rapid pace in an organized, rational and least wasteful manner, it would stand to reap handsome dividends by adopting a somewhat similar course of action and assuming direct responsibility for the promulgation and implementation of standards. But it would have to be done in a manner consistent with the peculiar structure of its own economy, which may only lend itself to a partial approach in this direction. In particular, the problem presents an interesting challenge to countries like Brazil and the Philippines where the NSBs issue mandatory standards but where the patterns of economy differ considerably from that of the USSR (*see* Chapter 9, Table 9.1, para 9.6(7)).

COUNTRIES WITH VOLUNTARY STANDARDS

14.5 Countries with economies which are not centrally controlled may, for lack of a better title, be classed as having free or semi-free or mixed economies. These would include most democracies, constitutional monarchies and dictatorships, representing a wide spectrum of patterns ranging from those of the USA and Japan to those of the newly emerging nations of Asia and Africa. The common factor among all these countries, with very few exceptions, would be the prevalence of the practice of voluntary national standards as opposed to the legally enforceable mandatory standards. Roughly 80 percent of the countries having an organized standards movement appear to fall in this category (*see* Chapter 9, Table 9.1, para 9.6(7)). Though some standards may be made compulsory by legislation even in these countries, the basic character of most other standards would remain voluntary. The processes, problems and implications of making some of the originally voluntary standards legally enforceable will be discussed later on in this chapter (*see* paras 14.19–14.26). In the first instance attention will be confined to the implementation of the common run of standards which remain voluntary.

14.6 Countries with voluntary standards may be divided further in two subgroups, namely, the subgroup in which the NSBs are government departments or agencies and the subgroup in which the governments participate along with the private sector interests, together with the purely privately run NSBs (*see* Chapter 9, Table 9.1, para 9.6(2)). The former subgroup is likely to enjoy certain advantages over the latter, because of its proximity to government authority, which might facilitate official actions on matters concerning implementation. Nevertheless, in theory, the latter subgroup of the joint and private NSBs should also be able to receive similar consideration from their respective governments in such matters, though in practice the speed of processing in this case may be comparatively somewhat slow. On the other hand, this latter subgroup of NSBs could be in an advantageous position *vis-à-vis* their relationship with the private sector of industries. But the position of any particular NSB would depend mostly on the peculiar circumstances prevailing in any given country. In view, however, of the identical legal character of voluntary standards issued by both of these subgroups, their problems and their procedures in respect of ensuring implementation of standards would not differ appreciably.

14.7 Furthermore, in their general approach to the initial preparation of standards, both these subgroups have necessarily to ephasize the consensus principle and to insist on the largest possible participation of and agreement among the interests concerned. In this manner, any standards that might emerge would automatically assume an authoritative character, reflecting as they would the views of all those interested in their use. It would thus be in the interest of everyone concerned to make use of such standards to the maximum possible extent. Even the party whose special requirement might have been somewhat compromised for the sake of achieving overall economy of the larger group through a particular standard would be well advised to follow it in the interest of the long-term benefits accruable to the economy of the larger group of which he is a part and with whose progress his own advancement is intimately interlinked. It is, however, necessary to undertake an amount of educational propaganda effort to bring home the benefits of implementing standards to every potential user and keep him fully informed of the variety and the number of standards in various fields. His responsibility in regard to participation in the process of standards preparation is another feature to be kept in view while planning the details of the publicity programmes. Since example is better than precept, it would be most appropriate if a beginning is made in the application of standards by the members of the committee or committees responsible for, or associated, with their formulation.

PUBLICITY

14.8 *Direct Publicity* In the relatively more advanced countries, the National Standards Bodies have not always been so concerned with mounting a specially organized effort at securing widespread implementation of their voluntary standards as has been found necessary in the newly emerging countries where standards movements are of recent origin. This is perhaps so chiefly because in the developed countries adequate consciousness of the manifold advantages of standards prevails among the various sectors of economy. Even so, many advanced countries have felt the need for maintaining a constant flow of information between the standards-making authority on the one hand and the standards-using interests on the other. Practically every NSB of an advanced country, as of course many others, regularly issue periodicals and some of them issue two or three. According to the statistics given earlier (*see* Chapter 9, Table 9.1, para 9.6(8)), it will be seen that 41 national standards bodies, constituting some 60 percent of the total number of NSBs of the world, issue 59 periodicals between them, most of which are monthly, but a few are published more and others less frequently. The main objective of such periodicals is to keep the actual and potential users of standards informed of the up-to-date developments in the field of standards not only within the country itself but also elsewhere in the world, and including of course those at the international level. In addition, these periodicals serve as forums for the exchange of views on standardization matters between the different interests concerned with the use and development of standards. Besides, they also function as media for reporting advances in technology which might have a direct or indirect bearing on the preparation of standards or their implementation in practice.

14.9 *General Publicity* The direct publicity carried out by NSBs through their own periodicals is, however, hardly adequate to reach all sections of the community. At best it would reach most of the organized users of standards such as the members of the NSB itself and the interested subscribers, corporations and production units, government agencies and departments, and so on. The general public which constitutes the major bulk of the unorganized consumers, the small-scale producers and entrepreneurs would, by and large, be left out. Constituting as these groups do the backbone of the community, it is important that they are not lost sight of and are equally made conscious of the benefits that standards could bring them and the manner in which they could avail themselves of such benefits. A valuable survey of information activities of National Standards Bodies has recently been conducted under ISO auspices, which might serve as a valuable guide to others for planning their own publicity campaigns.[4]

14.10 In certain countries the NSBs find it useful to undertake publicity work on a much wider scale and make use of all types of mass media for this purpose, such as the following:

(1) Daily, weekly and monthly press, as also the technical periodicals and trade journals. They are kept informed of the progress of the NSB's work by the issue of material for publication which may include:

(a) press notes briefly describing the contents of draft standards circulated for eliciting public opinion and announcing the publication of new standards, indicating the importance of their subject matter to the economy of the country and to the consuming public;

(b) descriptive articles dealing with different aspects of standardization and their importance to the consumer, the producer and the technologist, as also to the economic growth of the country;

(c) advertisements pointing out the functions of the NSB and the advantages of standards to the country and the people. These advertisements may be paid for by the NSB members or other clients using its services, or by the NSB itself;

(d) a collection of material including informative articles written by specialists and advertisements from producers of products covered by standards may be made available by the NSB to the publishers of newspapers and journals for the issue of special numbers or special supplements at certain occasions of importance, such as conferences, conventions, seminars, and the like; and

(e) press interviews with standards executives and other authorities concerned with standardization.

(2) Radio and television[5] publicity and advertisements along lines similar to those indicated under (1) above. These media also lend themselves excellently to the presentation of discussion between a number of knowledgeable people on subjects of topical interest as also special interviews with important standards personalities and other authorities who may occasionally be visiting the locality.

(3) Films, film strips, slides and commentaries[6] are prepared by some

NSBs for presenting standards philosophy in a popular manner, which could be used for public exhibition in general, for special audiences at conferences and other similar occasions, for educational institutions, or could even be loaned out for exhibition to other groups.

(4) At certain important expositions, exhibitions and trade fairs, both national and international, some NSBs have been known to arrange specially designed pavilions, stalls or exhibits to convey to the visiting public the message of standardization and its relevance to the theme of the exhibition or the fair. This activity also includes permanent exposition halls established for public view at the headquarters of some NSBs (*viz.* India,[37] Israel) and mobile exhibitions (*viz.* Germany).[7]

(5) Publicity pamphlets in popular language, handbills and handouts may be issued for distribution on suitable occasions, when they can help promote the cause of and disseminate information about standardization.

(6) Posters may also be displayed at strategic locations or on hoardings situated at key positions.

In other words, all means of mass communications may be marshalled, depending on the need of the occasion and the financial means available to the NSB.

14.11 *Specialized Publicity* In addition to the publicity addressed to the public in general, it is also quite important, perhaps more so, to make a special effort to reach the various sections of the community, which are quite directly concerned with standards and standardization, such as:

(1) producers of goods and providers of services, both in the private and public sector enterprises, together with their organizations comprising associations and chambers of commerce and industry;

(2) consumers and users at various levels, particularly the organized consumers of goods and users of equipment and services both in the public and private sectors enterprises: these cover a wide cross-section of economy, including:

(a) government departments and services, comprising, among others, the defence, the communications, the construction and public works, the railways, and the electric supply;

(b) state governments, municipal corporations, local authorities and other official agencies;

(c) industrial and commercial houses, contractors, exporters and the like;

(3) procurement agencies, official and others serving the needs of the groups covered under (1) and (2) above;

(4) professional institutes of scientists and technologists, particularly those of engineers and architects, who are responsible for design and construction and who are concerned with rendering expert advice for the utilization of goods and services in general;

(5) inspection agencies and authorities, and industrial testing and development laboratories concerned with the application of standards in their daily work: these are naturally placed in an advantageous position to suggest improvements in current standards and the development of new standards for the future needs of advancing and developing economies; and

(6) educational and technical institutions concerned with the teaching and training of future engineers and technologists.

14.12 All these and similar bodies and institutions and the individuals concerned with them may be considered to be alert enough and presumed to be quite conscious of their responsibilities towards standardization. Yet, being fully preoccupied with their normal functions and duties, they are sometimes quite likely to be unaware of the latest developments taking place in the standards world and may continue to follow or even insist on following some of the traditional practices and using some of the outdated standards. There is no single device which could be used to reach this very widespread audience of varied interests and character. Though the direct and general publicity measures described in the foregoing paragraphs would reach this class of audience as well and would have some influence, it must be recognized that the generalized approach involved in those measures cannot be expected to meet the specific needs of these rather specialized groups.

14.13 It is important, therefore, that these groups may be reached either individually or collectively, but always in a specialized manner, so as to be able to cater to their specific fields of interest and to meet their special needs. Some of the measures adopted for this purpose may include:

(1) national conventions and conferences organized annually or otherwise at convenient intervals, where a number of these groups could be represented to present and discuss their problems and evolve possible solutions and recommendations for action to be taken by the appropriate bodies;[8-10]

(2) regional or state-wise conferences in which participants from public or private sector groups or both could participate for similar purpose;[11]

(3) industry-wise conferences, where problems of a given sector of industry or trade could be highlighted, the relevant existing standards

critically examined, and future lines of action proposed to be taken at various levels of standardization could be discussed;[12]

(4) seminars, symposia and discussion groups for detailed examination of specifically important subjects of common interest may be organized, where specialists from all the concerned groups could participate;[13,14]

(5) special commentaries on certain important groups of standards or codes of practice[15,16] and collected presentation of reference data from well established standards dealing with certain sectors of industry,[17,18] may be published to facilitate their understanding by students and to promote their use by professional engineers, designers, and technologists;

(6) special lectures delivered and informative papers may be presented by knowledgeable persons to learned and professional societies for salutary effect in propagating the knowledge and practice of standardization among certain classes of professional people who may not otherwise be quite in close contact with standardization movement.[19–22]

EDUCATIONAL INSTITUTIONS

14.14 The key personnel for implementing standards are the engineers and technologists engaged in their professional activities, who are directly in charge of development, production and testing of goods, and are responsible for the designing and construction of machines, structures and other installations. Yet during the course of their basic education at the universities, technical colleges and institutions, they receive little, if any, instruction about the fundamental philosophy and the economic implications of standardization or about the manner in which standards are produced and used. In some institutions an odd code of practice or a basic standard may be used for reference purposes, but the background of standardization as a discipline and its main objectives are seldom explained to the students of engineering and technology as a part of their curriculum. It is obvious that the more a young engineer is made familiar with these matters, the better equipped he would be in later years during his active professional life to implement standards in a well-informed manner and may even be able to make valuable contribution to their preparation and improvement.[23] Teachers of engineering have a special responsibility in this matter. However, the problems of education and training will be dealt with later, in a little more detail, in Chapter 25.

GOVERNMENT'S ROLE

14.15 Governments can play an important part in securing implementation of standards[3] which are initially voluntary – both by virtue of their capacity as regulating agencies and as organized consumers of a sizable portion of the economic production of any country. Furthermore, in certain countries governments control an appreciable proportion of the industrial production capacity as well, especially in respect of certain basic industries, such as petroleum, steel and atomic energy and certain public utility services, such as railways, posts, telecommunications and electrical power generation and supply. Their role as a standards-implementing authority in such cases, therefore, becomes correspondingly enhanced. When one talks of governments in the context of implementation of voluntary standards, one includes all government authorities – states, provinces, county councils, district boards, city corporations, municipalities and others, by whatever name they may be known.

14.16 Some governments, like that of the USA, have well organized departmental activity for the preparation of their own standards for defence services and the like, but others like those in India and the UK depend primarily on national standards issued by their respective NSBs. That is not to say that no government agency need issue any standards when national standards are available. The policy, particularly in developing countries, should be that items of particular interest to a government agency or department are preferably covered by the agency or departmental standards, but items of common interest to the public and the nation as a whole are dealt with in national standards. For example, a standard for a letter box could quite appropriately be issued by the postal authorities, on the same lines as any company standard may be issued by a company, but the standards for the materials of construction, workmanship and finish of the letter box should better be those having a national status. Only in this manner could the requirements of the official agencies and those of the general public be inter-woven most economically and expeditiously. This feature becomes particularly vital in times of national emergency when defence forces, for example, have to draw heavily on all the available resources and the productive capacity in every sector of the economy.

14.17 In respect of procurement of stores for use by various services, it has been found useful by most governments to adopt, as a matter of policy, the procedure to place orders and contracts on the basis of departmental standards whenever applicable and on the basis of national standards whenever items of general utility are involved, to the exclusion of all other standards. In developing countries it is particularly necessary to deliberately enunciate and follow such a policy, because in most of these countries the officials concerned with procurement of stores and services are likely to have got used to basing their operations on the standards of the metropolitan powers which originally controlled the country. It is quite possible that national standards for every item of interest may not all be available at a given time to cover all the needs, in which case it would be necessary to continue the old practice for an interim period. In any case, the general policy directive should cover such contingencies and insist on the authorities concerned to bring their needs for new standards to the attention of the NSBs and collaborate with them actively to help expedite their preparation. It also sometimes happens that the current standards of an NSB may not always suit some specific departmental requirements. But this can only come about when effective collaboration between the NSB and the department concerned happens to be lacking at the preparation stage of the standard in question. Such a situation can readily be remedied by bringing the matter to the notice of the NSB for an early amendment or revision and by exercising proper vigilance when future drafts are circulated for study and comment.

14.18 Wherever governments are directly concerned with the control of industrial production or distribution, they are in an advantageous position to play an even more effective role in the implementation of national standards by simply directing that all public sector production and distribution units should base their production programmes and sales and purchase policies exclusively on the basis of sound standards wherever they are available. Insistence on organizing company standards departments within each unit would be quite a reliable means of ensuring all-round implementation of standards from whatever level they might have originated – company, industry, national or international. In fact, these questions of implementing national standards and organizing the company standards effort should be borne in mind at the earliest stage of planning all new production units and appropriate provisions should be made in the original project framework. Furthermore, wherever a government happens to be in a position to be able to influence the private sector industry, it is important that its first concern should be to take similar steps in relation to the design and operation of its own production units.

LEGAL ASPECTS OF IMPLEMENTATION

14.19 In countries where national standards are issued as mandatory documents, the governments may be said to have recognized that all these standards are important enough for compulsory implementation in the public interest, in the same manner as any other law of the land. The legal aspects of such situations have already been referred to in earlier paragraphs of this chapter (*see* 14.1 to 14.4). In what follows, attention will be confined to the legal aspects of implementation in those countries where national standards are initially issued as voluntary instruments. It has been recognized that even in these countries it becomes necessary in the public interest to enforce legally some of the standards in certain sectors of economy. Such standards may be concerned with a wide variety of subjects and items, depending on the needs of a particular economy. The decision to make certain standards mandatory would also depend upon the government's own inclination and capacity to create the requisite machinery to ensure compliance with these standards. Some of the subjects and items which have been brought under control in this manner, with advantage, include the following:

(1) safety of persons against accidental injury and of those working in dangerous surroundings, through factory byelaws, mine regulations, electrical wiring codes, etc.;

(2) safety of machines and other equipment against dangers to human life, such as standards for domestic electrical appliances, automobiles, grinding wheels and centrifugal machines;

(3) safety of structures against collapse, through design codes for buildings, bridges, cranes, boilers, etc.;

(4) safety of property against fire, explosions, and earthquakes, for example, specifications for determining the flammability of materials and codes for earthquake-proof design and construction;

(5) ensuring healthy surroundings, through regulations concerning water and air pollution, codes for the design and installation of sanitary piping, etc.;

(6) purity and potency of drugs, for example, through various pharmacopoeias and labeling regulations;

(7) purity of foodstuffs and their nutritive value, through rules governing sanitation of kitchens, food processing factories and abattoirs; standards for food preservatives, colouring materials and other food additives; standards for tests for bacteriological or other harmful contamination, and so on;

(8) important matters vitally affecting the economic development of a country and its public welfare, such as the regulation of weights and measures; control of quality and pre-shipment inspection of exports

designed to promote the expansion of the country's international trade; conservation of certain scarce materials and resources,* and the like.

14.20 Some of these matters may be considered so vital that a government may have to enact detailed legislation in the absence of the existence or possible future preparation of a voluntary national standard in the field. Indeed, in countries where there is no National Standards Body, it becomes necessary for governments to undertake such legislation in many of these fields. For example, long before the Indian Standards Institution (ISI) came into being, there existed a Boilers Act,[24] a Mines Act[25] and other legislation of a similar type in the country. These acts empowered the government to make rules and regulations[26,27] amounting to codes of practice and standards that could be enforced directly through the appropriate departmental machinery. After the creation of ISI, the basic acts and the rules and regulations made thereunder continue to be enforced as usual, but the process of their being kept up-to-date and their actual operation in practice have been considerably facilitated, because of the ready availability of the numerous related standards that the NSB has been able to bring out. On the basis of these related standards several indigenous sources of supply of materials and services have been developed. For example, the availability of national standards for boiler plate and for mines hauling ropes, codes for welding and training of welders, the various test methods and many other standards have facilitated the work of the regulating authorities by eliminating the need for the latter to have to undertake the detailed and specialized work of standards-making in several related fields or to have to adopt overseas standards for the purpose, which may not always be applicable to indigenous production. The Indian example cited here to illustrate the point would apply equally, though with some variations, to most other countries, developed or developing, wherever the practice of voluntary standards might prevail or where national standardization is still to be organized.

14.21 In the process of adopting ancillary and related standards for the operation of an existing legislation, it would become necessary to make the standards a part of the regulations concerned with the particular

* Silver was officially required to be used for electrical bus bars and transformers in the United States during the second World War for strategic reasons to conserve copper supplies. Similarly aluminium has been specified for the manufacture of a large variety of electrical conductors in India during recent years to the complete exclusion of copper, because of its shortage and abundance of aluminium among the available mineral resources.

legislation. Besides, there would be many other situations, as indicated under para 14.19, which would require direct legal enforcement of certain standards or codes of practice such as those which are in the interest of public safety, health or economic development. It should also be borne in mind that such legislative action may be required to be taken at the national or state level or at the very bottom level of administration such as the municipality or the village council. In the case of building byelaws, in particular, a central or state legislation is seldom considered adequate, not only due to the varying conditions, natural and man-made, prevailing in different regions and localities, but largely because of the juridical privileges of the different local governments. In such cases the adoption of model byelaws and other standards issued by a central authority like an NSB is generally considered advisable, appropriate alterations to suit the local conditions being introduced as considered fit. However, in any such legislative adoption of a standard or code issued by a body other than the legislative body itself, several legal considerations arise, which have become the subject matter of special studies carried out in several countries, for example, in New Zealand and the USA.[28,29]

14.22 Briefly stated, the legal problems that arise are concerned with the method to be adopted for the incorporation of a standard in the text of the legislation. In the first place it is to be appreciated that standards are always subject to amendment and revision so that they may be kept up-to-date with the advancement of technology and time. A legislative authority having once legally recognized a standard as a legal document for enforcement must, therefore, always be prepared to bring its legislation up-to-date everytime the standard in question is amended or revised; it cannot delegate its legislative authority for such subsequent alterations to another body, such an NSB, even though the latter be an official organ. This amounts to having to undertake an extensive operation on a technological front by the legislative body, which will need to create a special apparatus of its own to keep it constantly advised of the changes being made or desired to be made in the standard. Indeed, it is exactly this sort of situation that has to be faced by a legislative authority, when it enacts a law of its own for the enforcement of certain technological requirements in the absence of a requisite national standard, as, for example, in the case of boilers codes in India[24] or the food, drugs and cosmetics legislations in the USA.[30] Thus the very advantage of adopting a suitable national standard as a legal instrument is lost, namely, the ready availability of services of an expert body specially and constantly occupied with standardization work, which can be relied upon to keep technological decisions in line with the latest developments of the economy and whose conclusions can be safely used as a guide for legislation. Of course, it sometimes does happen that a government authority does not consider it necessary or advisable to depend on the standards of an outside body and prefers to enact its own legislation incorporating all the technical requirements. But this approach would perhaps involve avoidable duplication of expenditure, particularly in countries where a strong NSB already exists. In the newly developing countries, it would definitely be more desirable to expend extra funds on the strengthening of an NSB than having to create a parallel standards apparatus for legal purposes.

14.23 The legal studies referred to above and others have revealed that three methods of approach for the incorporation of national standards and codes in the law have been found to be effective, namely:

(1) reproduction method, in which an available and acceptable standard or code is reproduced in its entirety in the body of the legislation (which incidentally is identical in the legal sense to enacting a complete legislation in the absence of a standard);

(2) reference method, in which reference is made by number, title and the date of publication of a standard or code in the statute, thus incorporating the document in the law without reproducing it *in extenso*;

(3) means of compliance method, in which the law stipulates certain requirements to be fulfilled in very general terms, followed by a statement that such requirements would be considered to have been satisfactorily fulfilled, if the relevant standard or code issued by such and such authority has been complied with.

14.24 The first two of these methods are almost identical in principle, both requiring the statute to be revised as and when the standard in question is revised. But the former of the two, the reproduction method, has the added disadvantage of requiring an expensive reproduction of the full text of an already published standard. The cost of such reproduction would become quite prohibitive in certain cases,[29] for example, in the case of bulky national building code documents which are usually adopted by numerous municipal authorities in every country. Nor would it be satisfactory to reproducc only a part of a standard or an extract of the essential requirements, thus leaving the provisions of the law incomplete. Furthermore, the cost of reproduction does not extend only to the main document, but it is also necessary to reproduce all the other related standards that may have been quoted within the body of the main document, which, in turn, may have referred to others, and so on. The second so-called reference method overcomes this difficulty, but does continue to involve the necessity of having to make publicly available all the standards material which may be quoted in the statute, as well as the related documents referred to above. But this is relatively a simpler affair and can be taken care of by depositing a stipulated number of copies of the material in question at a place where the public could have free access to it at all reasonable times and notice of which fact is given in the normal manner.[29] In any case, standards would be freely available on sale.

14.25 However, the third alternative, the so-called means of compliance method, is considered by far the most preferred method by a number of legal authorities. While it may require public availability of the documents involved, as in the case of the reference method, it has the great advantage of not requiring either their reproduction or the need for keeping the statute up-to-date. This it accomplishes without in any way delegating legislative authority. The conditions of the law are so stated as to ensure the satisfactory fulfilment of the stipulated requirements, and compliance with certain standards quoted in the legsilation is taken to be *prima facie* evidence of such fulfilment. In this way the most up-to-date version of the standard or code becomes applicable and the onus of proof is passed on to the user of the standard or code. An added advantage of this method is that a general statement of the requirements to be fulfilled is found to be quite adequate, since the precision of the requirement together with the method and proof of satisfying them would be covered precisely enough in the standard or the code with which compliance is required. The following example[29] from an American municipal ordinance would illustrate this point:

> All installations of electrical equipment shall be reasonably safe to persons and property and in conformity with the provisions of this ordinance and the applicable statutes of the State of . . . and all orders, rules and regulations issued by authority thereof.
>
> Conformity of installations of electrical equipment with applicable regulations set forth in the National Electrical Code, the National Electrical Safety Code, or electrical provisions of other safety codes, which have been approved by the American Standards Association, shall be *prima facie* evidence that such installations are reasonably safe to persons and property.

14.26 Another important legal aspect of standards is that concerning their use as bases of contracts for supply or construction, when they are cited in order to secure compliance with certain technical provisions of the contract.[31] This is perhaps by far the most common legal use of standards and is governed in most countries by the normal laws concerning contracts. More recently, another legal aspect of standards has been attracting some attention, particularly in the United States,[32] and that is in relation to the possibility of some standards running foul of the anti-trust laws by, for example, retarding innovation and progress. But these are quite special matters perhaps of local concern and need not be discussed here in any detail.

IMPLEMENTATION DEPARTMENT

14.27 In certain countries, it has been found extremely useful for an NSB to create a special cell or department within its directorate to look after the task of securing implementation of standards. Such a department has proved particularly useful in India, where the Indian Standards Institution (ISI) was perhaps the first NSB to initiate this activity in a cell as early as 1954, and to create a formal Implementation Department in 1960. The functions of an implementation department would include the pursuit and encouragement of the various fields of activity described in the previous paragraphs of this chapter. In addition to organizing nation-wide publicity and public relations work, it would help organize and promote the various conferences, symposia, seminars, lectures and the like and offer consultancy service to those faced with technical and other problems arising out of the application of standards in practice. In India, the ISI Implementation Department, in addition, organizes personal visits to industrial units, carries out continuous surveys of various sectors of industry through the issue of questionnaires to determine the extent to which each unit of the industry may be utilizing Indian standards in relation to various activities of their own, for example, purchasing, planning, designing, production, quality control, packaging, sales, etc.[33,34] The information so gathered is compiled and indexed, and kept up-to-date so that it may serve as the basis for replying to the numerous enquiries that are continuously received in connection with the availability of materials, products and equipment conforming to national standards. Such a compilation also serves as a guide for planning the activities of the NSB itself, for it furnishes an index to the relative demand for the various types of standards and their respective popularity among the various interests within different sectors of industry. A word of caution may be useful here. It must be recognized that a claim by a given firm that its products comply with certain standards should be taken at its face value and not a proof of standards being factually complied with. Experience, on the whole, appears to indicate that for a developing economy the existence of an active implementation department within the structure of an NSB can go a long way in promoting the widespread use of national standards at all levels.

14.28 The importance of implementation even at the international level has been recognized. It has, thus, been found necessary to make special surveys to determine the depth to which international recommendations for standardization have penetrated into the fabric of standards in different countries. ISO has so far conducted two such surveys and has come up with very interesting results which have been compiled in a series of reports.[35, 36] Apart from the factual data that these surveys have brought to light, they have proved invaluable in giving a new orientation to the ISO programme of work. They also furnish a valuable guide to different countries as regards the nature and content of standards of other countries on various subjects, indicating the extent to which they may be in line with ISO Recommendations. This is an important factor in the promotion of international trade.

ALLIED ACTIVITIES

14.29 Almost any activity of which standardization forms an integral part would have the potential of promoting the implementation of standards. The encouragement of such activities by NSBs and other agencies would amount to propagating standardization. Some of the more important activities of this nature, with which NSBs are often connected directly or indirectly, include:

(1) certification marking
(2) informative labeling
(3) pre-shipment inspection of exports
(4) statistical quality control
(5) company standardization and
(6) organizations of professional standards engineers.

The first four of these will be discussed in subsequent chapters (Chapters 15, 16, 17 and 23 respectively), but the last two have already been covered in chapter 7 (*see particularly* para 7.3).

REFERENCES TO CHAPTER 14

1. Orgizkov, V. M.: Standardization, quality and law. *Standarty i Kachestvo,* 1970, no 5, pp. 35-37.
2. Soviet Union elevates standardization to ministerial level. *ISO Bulletin,* Dec 1970, p. 1.

3. Standardization, government and industry (Editorial). *ISI Bulletin,* 1971, vol 23, pp. 47-48.
4. *Survey of the present situation on information in the ISO countries* (Document ISO/INFCO). International Organization for Standardization, Geneva, 1970.
5. *Catalogue of scripts for television programmes on standardization* (Document ISO/INFCO (USSR-2) 19E). International Organization for Standardization, Geneva, 1970.
6. *Catalogue of films on standardization* (Document ISO/INFCO (USSR-1) 18E). International Organization for Standardization, Geneva, 1970.
7. *Catalogue of exhibitions on standardization* (Document ISO/INFCO (USSR-3) 20E). International Organization for Standardization, Geneva, 1970.
8. Indian Standards Conventions (ISC) held annually from 1955 to 1968, and now held biennially, whose full proceedings may be obtained from the Indian Standards Institution, New Delhi, but summaries of which have been published as follows:
   First ISC, Calacutta. *ISI Bulletin,* 1955, vol 7, pp. 1-7.
   Second ISC, Bombay. *ISI Bulletin,* 1956, vol 8, pp. 37-58.
   Third ISC, Madras. *ISI Bulletin,* 1958, vol 10, pp. 39-79.
   Fourth ISC, New Delhi. *ISI Bulletin,* 1959, vol 11, pp. 1-33.
   Fifth ISC, Hyderabad. *ISI Bulletin,* 1960, vol 12, pp. 47-89.
   Sixth ISC, Kanpur. *ISI Bulletin,* 1962, vol 14, pp. 61-69 and pp. 129-157.
   Seventh ISC, Calcutta. *ISI Bulletin,* 1963, vol 15, pp. 117-160.
   Eighth ISC, Ahmedabad. *ISI Bulletin,* 1964, vol 16, pp. 58-59 and pp. 103-135.
   Ninth ISC, Bangalore. *ISI Bulletin,* 1966, vol 18, pp. 45-101.
   Tenth ISC, Ernakulam. *ISI Bulletin,* 1967, vol 19, pp. 93-136.
   Eleventh ISC, Chandigarh. *ISI Bulletin,* 1967, vol 19, pp. 521-573.
   Twelfth ISC, Bhubaneswar. *ISI Bulletin,* 1969, vol 21, pp. 51-98.
   Thirteenth ISC, Bombay. *ISI Bulletin,* 1971, vol 23, pp. 49-101.
9. The American National Standards Institute (ANSI), formerly known as the American Standards Association (ASA), has been holding a National Conference on Standards (NCS) annually for many years. Brief reports of these appear in print. For example, *see*:
   Eighth NCS, San Francisco. *Standards – key to progress and profits,* 1957, p. 160.
   Ninth NCS, New York. *Standardization – what's in it for me?* 1958, p. 128.
   Tenth NCS, Detroit. *Standardization – keystone of industrial progress,* 1959, p. 124.
   Eleventh NCS, New York, *Standards for a dynamic decade,* 1960, p. 116.
   Twelfth NCS, Houston, *Philosophy and benefits of standardization,* 1961, p. 96.
   Thirteenth NCS, New York. *Voluntary standards – the American way,* 1963, p. 100.
   Fourteenth NCS, Washington. *Market standards and profits,* 1964, p. 96.
   Fifteenth NCS, Chicago. *Standards and the consumer; industry, family and government,* 1965, p. 48.
   Sixteenth NCS, San Francisco. *Standards – roadblocks or building blocks?* 1966, p. 72.
   Seventeenth NCS, Cleveland. *Safety – can it be standardized?* 1967, p. 63.
10. The British Standards Institution regularly holds two sets of conferences on standards; one, the BSI Standards Associates Conferences (SAC) and the other, Women's Advisory Committee Conferences (WACC), whose proceedings are available from BSI. For example, *see*:

SAC, London. Better management through standards. *BSI News,* Aug 1968, pp. 11-17.

SAC, London. Face to face with metrication. *BSI News,* Nov 1969, pp. 14-18.

WACC, Wales. *BSI News,* Oct 1969, pp. 14-15.

WACC, London. *BSI News,* Oct 1970, pp. 15-16.

11. The following regional conferences held in India may be of interest:

Zonal conference of ISI members of southern region. *ISI Bulletin,* 1959, vol 11, pp. 251-252 and 257.

Indian standards and organized purchasers: State conferences in Orissa, Kerala, West Bengal and Punjab. *ISI Bulletin,* 1960, vol 12, pp. 167-169 and 178.

Himachal Pradesh Conference on implementation of Indian standards. *ISI Bulletin,* 1961, vol 13, p. 21.

Conference on implementation of Indian standards in Bihar. *ISI Bulletin,* 1961, vol 13, pp. 124-125.

Conferences on implementation of Indian standards in Maharashtra and Mysore. *ISI Bulletin,* 1961, vol 13, pp. 230-233.

Conference on implementation of Indian standards in Gujarat. *ISI Bulletin,* 1962, vol 14, pp. 83-85.

Conference in Madhya Pradesh on implementation of Indian standards. *ISI Bulletin,* 1963, vol 15, pp. 22-24.

Conference in Madras State on widespread implementation of Indian standards. *ISI Bulletin,* 1965, vol 17, pp. 319-322.

Conference in Andhra Pradesh on implementation of Indian standards. *ISI Bulletin,* 1968, vol 20, pp. 103-104.

12. The following brief reports about some industry-wise conferences in India may be of interest:

Orthopaedic surgeons: Summer conference discusses standardization's role. *ISI Bulletin,* 1969, vol 21, pp. 395-396.

Standardization in refractory industry: Jamshedpur Conference reviews industry's problems. *ISI Bulletin,* 1969, vol 21, pp. 513-529.

Quality control of animal feeds and feedingstuffs: Conference stresses standardization and certification marking. *ISI Bulletin,* 1970, vol 22, pp. 3-8.

Textile accessories – need for implementation of Indian standards (seminar). *ISI Bulletin,* 1970, vol 22, pp. 65-67.

Design, fabrication and erection of steel structures: Calcutta conference reviews problems. *ISI Bulletin,* 1970, vol 22, pp. 91-100.

Pump industry: Conference reviews impact of standards. *ISI Bulletin,* 1970, vol 22, pp. 148-160.

13. AIMO (All Indian Manufacturers' Organization) discusses implementation of Indian standards. *ISI Bulletin,* 1962, vol 14, p. 349.
14. Implementation of Indian Standards. *ISI Bulletin,* 1960, vol 12, pp. 56-58.
15. *BS 971: 1950 Commentary on British standard wrought steels – En series.* British Standards Institution, London, p. 91.
16. *IS: 1871–1965 Commentary on Indian standard wrought steels for general engineering purposes.* Indian Standards Institution, New Delhi, p. 70.
17. *The mechanical and physical properties of the British Standard En steels* (3 vols (1964, 1966 and 1969)). British Iron and Steel Research Association. Pergamon Press, London, pp. 442, 488 and 602 respectively.
18. *IS: 1870–1965 Comparison of Indian and overseas standards for wrought steels for general engineering purposes.* Indian Standards Institution, New Delhi, p. 131.

19. Good, Percy: *The history and philosophy of standardization* (Presidential address to the Institution of Electrical Engineers). CK(OC) 2772. British Standards Institution, London, Oct 1947, p. 10.
20. Mehta, V. N.: Establishment and implementation of a company standards programme. *ISI Bulletin,* 1966, vol 18, pp. 447-449.
21. Verman, Lal C.: *Standardization – a triple point discipline* (Presidential address to 57th Indian Science Congress, 1970, Calcutta). Indian Science Congress Association, p. 9. Also printed in adapted form in *ISI Bulletin,* 1970, vol 22, pp. 47-50.
22. *Catalogue of lectures on standards subjects (1960–1969)* International Organization for Standardization, Geneva, p. 112.
23. Technical Education. *ISI Bulletin,* 1969, vol 21, p. 57.
24. *Indian Boilers Act, 1923 (modified up to 1 July 1962).* Government of India Press, Delhi, 1963, p. 21.
25. *The Mines Act, 1923 (revised in 1952).* Government of India Press, Delhi, 1953, p. 36.
26. *Indian Boiler Regulations, 1950 (amended up to 1969).* Central Boilers Board, New Delhi, 1970, pp. 570 + 63.
27. *Mines Rules, 1955 (amended up to 1960).* Ministry of Labour, Government of India, New Delhi, 1961, p. 37.
28. Hitchcock, E. H.: Standards, technology, and law. *New Zealand Engineering,* 1968, vol 15, pp. 271-279.
29. *Nationally recognized standards in state laws and local ordinances* – A presentation of the problem and possible solutions. American National Standards Institute, New York, 1949, p. 43.
30. *Federal Food, Drug and Cosmetic Act as Amended.* US Department of Health, Education and Welfare, Washington, 1969, p. 78.
31. Zemlin, Hans and Budde, Eckart: Is quality guaranteed in purchasing, when there are DIN standards or by reference to DIN standards? *DIN-Mitteilungen,* 1970, vol 49, pp. 360-361 (*in German*).
32. Standards and the law. *ANSI Reporter,* 20 Nov 1970, vol 4, pp. 1-2.
33. Implementation of Indian Standards (Editorial). *ISI Bulletin,* 1954, vol 6, pp. 75-77.
34. Implementation of Indian Standards. *Sixth Indian Standards Convention (Kanpur) Brochure.* Indian Standards Institution, New Delhi, 1961, p. 29 (*see* pp. 6-8).
35. *Degree of application of ISO Recommendations (Report). Document ISO/RAP (GS-1) 1-1962 and related 3 annexures bearing the same document number.* International Organization for Standardization, Geneva, 1962.
36. *Degree of application of ISO Recommendations (Report). Document ISO/RAP (GS-2) 2-1964 and related supplement.* International Organization for Standardization, Geneva, 1964.
37. Visvesvaraya, H. C.: Standards Museum at Manak Bhavan. *ISI Bulletin,* 1962, vol 14, pp. 13-15 and 19.

Chapter 15

Certification Marks

15.0 The certification mark, sometimes referred to as the certification trade mark, has long been recognized in the legislations of many countries as a special type of trade mark. Both the trade mark and the certification trade mark have the status of industrial property registered and owned by interested parties and enjoy protection as may be prescribed under the law of the land. The chief difference between the two is that while a trade mark is used by its owner to identify his goods purporting to indicate their origin and quality, during the course of trade in which he is himself directly involved, a certification mark is used or licensed out by its owner for being applied to goods by others to indicate their compliance with a set of regulations including the specifications prescribed by the owner of the certification mark. The owner of the certification mark, incidentally, is not himself interested in either the sale or purchase of the goods so marked. In fact, he acts as a third party, standing guarantee that the goods, on which his certification mark is affixed, comply with the declared specifications and can thus be bought and sold with a degree of confidence not conveyed by goods not so marked. He thus assumes the responsibility for exercising due check and control over the production and distribution of such goods to ensure that the guarantee given by him is, in fact, meaningful.

15.1 Certification trade marks were originally intended for being administered by trade and industry associations, chambers of commerce, technical institutions, professional bodies and the like. But very few indeed were registered and even fewer actually used in practice to any large extent, that is in comparison with the number of ordinary trade marks. Perhaps the chief factor in this relative lack of popularity was the inherent requirements imposed on their use, firstly, by way of having to adopt a set of specifications for each item, which would be generally acceptable to those concerned, and secondly, to have to frame regulations for administering the marks and to organize an adequate system of controls for ensuring compliance. Both these functions had to be carried out by the type of organizations which, with a few exceptions perhaps, were otherwise engaged in tasks closer to their major objectives, which did not include certification as such but were somewhat allied to it. With the advent, however, of the National Standards Bodies, the first of these hurdles was automatically removed because the chief concern of any NSB was to issue nationally acceptable standard specifications for all items of trade interest. All that remained for an NSB to do was to provide the requisite regulations and a control machinery to be able to enter the field of certification marks operations.

15.2 Gradually, in time, an increasing number of NSBs in advanced countries began to recognize this rather closely associated avenue of service to the community and to act as certifying agencies. As a result, the movement today has spread all over the world. Practically all NSBs are concerned with the work of certification marking. In most cases where this work has not so far been included among the activities of an NSB, it is being planned to be taken up in due course. This happens to be the present state of affairs not only among the developed but also in the developing countries. All countries, regardless of whether their national standards are mandatory or voluntary, appear to be equally interested in a national system of certification marks. From Table 9.1 para 9.6(9) of Chapter 9, it will be seen that 48 of the 67 NSBs listed, that is some 72 percent of the total, have certification marks already in operation, while others are preparing to adopt them.

NEED

15.3 In the competitive world we live in, the place that the trade marks have come to assume is quite clear – in short, they are of great help to their owners in selling their goods. All kinds of claims are made on their behalf for the excellence of their goods,[1, 2] on the labels and in their

Fig. 15.1 Designs of some national certification marks indicating conformity with standards.

advertisements. A good many of them are genuine, but others are exaggerated and still others have been proven false as a result of experience.[3] Trade marks by themselves do not, therefore, always help the consumer in making a wise choice from among the multitude of goods offered to him, when a series of claims is made in favour of each trade mark for the buyer's attention. He is neither technically qualified to adjudge the value of the claims on a theoretical basis nor is he so placed as to be able experimentally to prove or disprove the claims within the limits of the resources available to him. His purchases are generally too small in value to justify a laboratory test to be made; nor has he such facilities available at his doorstep. Furthermore, it takes time to have the test conducted and to evaluate its results. The consumer guidance literature[3] is indeed helpful to him (*see also* Chapter 18), but to make use of it, he has either to search for the recommendation in respect of a given item of immediate interest or to have to draw on the store of information carried in his memory, or in a comprehensive index he may be maintaining. All this is too involved for an ordinary consumer who is almost daily concerned with making many small purchases.

15.4 With a view to helping such a consumer in making a wiser choice, several seals of approval have come into use. Some of these seals have been introduced to serve a special class of clientele only, such as members of an automobile club[2] which may issue seals of approval to the approved list of garages and hotels. Others may be licensed out for use by their owners to manufacturers of commodities without requiring a strict regime of control on quality or suitability for use.[4] Still others may be simply self-certification seals – some administered by the producers of goods themselves,[5] some in cooperation with an official agency[6] – but all of them depending, by and large, on the integrity of the producer. A common feature of a majority of these seals is that they mostly remain unregistered as certification trade marks, though the facilities for their registration under the trade marks acts are available[7-10] in most countries. Nevertheless, some seals have attained a stature of eminence in the fields to which they pertain. For example, the seal of the Underwriters Laboratories founded in 1894 in the USA has become a recognized symbol of safety against fire, electrical and other hazards. Unfortunately, in the industrially advanced countries there are far too few reliable seals of this nature and quite a few of those not-so-reliable. But in the newly developing countries, the field is practically wide open for creating institutions and procedures based on the lessons learnt from the experience of more advanced countries.[12]

15.5 It is in this context that the value of certification marks may be appreciated. A product bearing this mark carries a third party guarantee, which implies that

(1) the product has been produced according to an accepted standard, which is publicly available for everyone's inspection and study;

(2) its production has been carried out under continued supervision;

(3) it has been appropriately inspected and tested, with a view to determining its conformity to an authoritative and agreed standard; and

(4) if, for some reason known or unknown, it does prove to be otherwise than as claimed, the owner of the certification mark can readily be reached to help redress the grievance; and he being not a party to the sale or purchase of the article and being anxious to retain his reputation as a third party guarantor could be relied upon to do the right thing.

In the case of trade marks, the consumer has no such assurances or after-purchase redress facilities. He is entirely dependent on the will of the manufacturer or the seller who is himself an interested party. In case of highly reputed manufacturers, however, the consumer may often be on safer grounds. But the reputed manufacturers usually price their goods at a level much higher than what prevails normally, even though equally reliable goods of comparatively less known makes may be available at a reasonably lower price.

15.6 In addition to their value to the everyday small consumer, the certification marks render valuable service to the industrial purchaser especially in the small-scale sector. Large-scale purchasing units, industrial or otherwise, can well afford to have their own inspection and testing facilities to help determine the quality of goods they buy, either at the time they are offered for sale or when they are actually delivered. In fact, it is sometimes necessary to have a check at both stages. The quantities involved are usually so large that even if the purchaser does not have his own testing and inspection facilities, he can well afford to get an outside agency to undertake the task on his behalf. However, there is the very large number of industrial consumers and users operating on a comparatively small scale which could hardly justify the establishment or the use of independent testing and inspection facilities. Furthermore, such cases not only involve the small and medium-scale industrial buyers and contractors, but also a large majority of municipalities, town councils, and other similar public and private bodies. For this class of organizations certification marks can prove as advantageous as they may be to the common consumer, or perhaps even more so, because of the relatively large quantities of goods involved and the complexity of their nature requiring rather expensive types of tests, as in the case of building and construction materials like portland cement, electrical equipment like motors and transformers, electronic components and the like.

REQUIREMENTS

15.7 In order to fulfil all these needs, it is essential to keep the certification marking operation absolutely impartial and beyond all reproach. It is, therefore, extremely important that each scheme should be carefully designed and meticulously administered. Many good examples are available for close study and emulation.[13-19] The ISO has also given some consideration to this matter and published two sets of Recommendations dealing with the basic principles to be observed in organizing national standards marks and their significance.[20,21] In addition, the Commonwealth Standards Conference has made several recommendations from time to time for the creation and administration of certification marks schemes within the framework of the National Standards Bodies.[22] Based on these internationally agreed recommendations and in the light of experience gained in both the advanced as well as some of the developing countries, it is considered essential that while preparing a certification marking scheme of national importance certain pertinent points, such as the following, may be kept closely in view:

(1) status of the agency controlling the certification marks;

(2) character of the standards to be used with which the marks would indicate conformity;

(3) laboratory facilities required for an effective control of the use of the marks;

(4) production process control essential for ensuring conformity;

(5) inspection force to be trained and organized;

(6) vigilance required to be exercised before issuing a certification marks license and during its operation;

(7) procedure for dealing with possible complaints and essential follow-up action;

(8) desirability of keeping costs at the barest minimum consistent with the integrity of the marks;

(9) features to be incorporated in the design of the marks;

(10) methods to be employed for applying the marks to the certified goods;

(11) provision for the feedback of field experience to the standards-making process and to the operation of the scheme as a whole; and

(12) legal protection available in the existing statutes or to be specially provided.

STATUS OF CERTIFICATION AGENCY

15.8 That the certification marks administration should be entrusted to a recognized authority of unquestionable independence has been indisputably established. In this connection, reference is invited to the report of the high level UK Committee – the Molony Committee on Consumer Protection.[4] During the course of its investigations the Committee has repeatedly emphasized that the agency administering the marks should be "at arm's length" from the manufacturers, which, in effect, means that it should really be free from any undue influence of any group, be it manufacturers, consumers or even technologists. It should indeed enjoy the confidence of all such interested groups. Only then can it make a really effective contribution to the cause of the conflicting interests within the economic community – producers and manufacturers on the one hand, consumers and users on the other. Judged in this light, the National Standards Bodies would meet all the criteria mentioned above. This does not mean to say that there is no room for other organizations. Several specialized agencies like the Underwriters Laboratory in the USA[11] and the Gas Council of the UK,[23] for example, have amply justified the position of prestige they presently enjoy in their respective countries. But in the newly developing countries it may well be considered most advisable to encourage the development of certification activity under the central control and authority of the National Standards Bodies, which could undertake the operation on a national scale to meet the needs of the common consumer as well as the industrial and corporate user of goods and services in most fields of interest for which national standards may have been issued or could be issued.

CHARACTER OF STANDARDS

15.9 Obviously, national standards are most suited for the purpose of certification marking, not only because of their general acceptability and authoritative character, but also because of their public availability. It must, however, be recognized that not all national standards would lend themselves for certification marking, as for example, those dealing with terminology or other basic matters, like preferred numbers or limits and fits. On the other hand, those dealing with products and materials are excellently suited for this operation. Standards concerning certain types of processes and operations may also be amenable to certification. For example, a given structure might be certified as having been designed and/or constructed in accordance with a standard code of practice, or a house or a factory could be certified to have been wired according to a standard code for electrical wiring, and so on. Though organizations like the Underwriters Laboratories[11] undertake a great deal of such work, only very few of the National Standards Bodies appear to have evolved procedures to certify processes and installations of this type.

15.10 Specifications type of standards dealing with materials and products, meant for the common consumer and the industrial and corporate user, evolved in accordance with the procedure outlined in Chapter 10, have been found to be most suitable for the purpose of certification marking. But special attention should be drawn to the need for clearly specifying in such standards the characteristics which are considered essential from the viewpoint of the user or consumer, together with the quantitative limits within which these characteristics are to be found or incorporated in the article. In doing so, emphasis should be placed more on such characteristics as make the article functionally serviceable for the intended purpose and adequately safe and reliable in use, rather than on characteristics which may have a theoretical importance, or constitute manufacturing details or directives. Again, the test methods should be precisely stated in an unambiguous manner, for it so happens that the test results quite often depend on the nature of the method employed and on the amount of care taken to control the prevailing conditions during the test. Furthermore, statistical sampling procedures for controlling the quality during production and/or for batch sampling and testing for determining conformity to the relevant standard must be carefully worked out and specified, together with the criteria for acceptance and rejection. These statistical matters are further referred to in some detail in a later chapter (*see* Chapter 23).

15.11 In designing the statistical sampling and related clauses of a specification, it must be borne in mind that the cost of sampling and subsequent testing of the sample is maintained at a reasonably low level, so that the article in question does not become uneconomic. Some tests are destructive in nature, in which the sample is expended to obtain the results, but even in non-destructive testing the number of samples tested has to be kept reasonably small, consistent with cost and level of quality assurance desired. In practice, the statistical criterion commonly found adequate for designing the sampling clauses is that they be based on 5 percent consumer risk and 5 percent producer risk. This means that in the long run there would be a 5 percent chance of a defective article being passed as satisfactory and a 5 percent chance of a satisfactory article being rejected. If a much greater degree of statistical assurance than this is sought, the certification operation is likely to become unduly expensive. But certain circumstances in which lower risks are considered justified are discussed in the following paragraph.

15.12 The UK Committee on Consumer Protection has considered this issue from a purely legal point of view[24] in relation to a specific case law in which the well-known BSI kite mark was found to have been applied to a particular pillow whose fillings turned out to be unclean in terms of the relevant British Standard. This failure of strict compliance with the specification was ascribed to the consumer risk of sampling procedure referred to above. After thoroughly analysing the legal pros and cons, the Committee made the recommendation that those provisions of a standard which belong more properly to the certification mark scheme operation may not form a part of the standard but may be contained in an annexure to the main body of the standard. This is considered to be quite a sound recommendation. In order, however, to protect the consumer further, some

NSBs like the Institute of Standards and Industrial Research of Iran undertakes to reimburse the consumer in case he is found to be justifiably dissatisfied with an article certified by ISIRI. The Indian Standards Institution, on its part, has made it a condition of the scheme that in case a consumer does happen to find that a particular certified article purchased by him happens to be defective and thus falls within the limits of the consumer risk of defectives, his money may be refunded by, or the article replaced at the cost of the certification license holder. This liability for replacement or for cost refunding may be quite adequate in most cases of articles of ordinary use, but it does not meet the need of the cases where safety of health or property is involved, in which case the discovery of the defect may be made after the damage has been done. It is most essential, therefore, that in such cases where important safety considerations are involved, the 5 percent consumer risk may be further decreased to a safer lower limit and other safeguards may also be provided, such as tightening the limits of specification values to a narrow enough safety margin at which the chance of risk will be minimal. This indeed is another suggestion which has been made by the UK Committee on Consumer Protection.[24]

15.13 Reverting to the question of the essential elements of a specification suitable for certification marking, it might be added that wherever a product is expected to be made in more than one grade or of more than one quality, each grade or quality should be assigned an appropriate designation which is clearly defined in the standard. This designation should preferably be so brief that it could readily be incorporated in the design of the certification mark as applied to the product without making the mark too complex, as for example Gr A. Long designations such as "prime quality" or "super grade" do not readily lend themselves to be used in a certification mark design, as sometimes the mark has to be applied to a very small size product, where the reduction in size of designation would make it illegible (*see also* para 15.33).

LABORATORY FACILITIES

15.14 Facilities of an adequately equipped laboratory are most important for a satisfactory operation of any certification marks scheme; indeed several laboratories may be required, because different sectors of industry have to be covered, such as chemicals, textiles, electrical, structural and metallurgical. Equipment and personnel for testing every item which may be under certification marking operation must be available and provision should be made for those which might be expected to be brought under this operation in the near future. Though it is clearly desirable to have these facilities organized under the direct control of the certifying agencies themselves, some facilities may conveniently be available outside the agency's own organization. Under these circumstances it may be more economical to make use of these outside facilities instead of duplicating them by creating new ones within the agency. Such an arrangement would specially be considered suitable in smaller countries and in situations where the overall testing load is comparatively light. In other situations it may be more desirable to create internal facilities for certain types of work which is more frequent and to utilize outside facilities for less frequent demands. Under certain circumstances, it may become necessary to utilize the laboratory facilities of a licensed manufacturer himself, but appropriate precautions should be taken in such cases; firstly, the calibration of the instruments and reliability of the equipment should be carefully checked, and secondly, the staff conducting the tests should be that of the certifying agency itself.

15.15 It will be seen from Chapter 9 (*see* Table 9.1, paras 9.6(9) and (10)) that 48 of the 67 NSBs listed administer certification marks of their own but as many as 29 of these 48 have no testing facilities within their own organization and depend on outside laboratory organizations for the purpose of testing samples of the certified goods and others. This means that only about 40 percent of the NSBs operating certification mark schemes have their own testing facilities. But even this does not imply that those having their own facilities do not depend for some of their testing work on outside agencies. One of the chief hurdles encountered in utilizing outside facilities has been the difficulty of securing adequate priorities for the testing work farmed out to them. It is obvious that a testing organization would undertake the work in the order that it is received, without necessarily giving any higher priority to the work of an NSB. For an NSB, however, it is highly important that the test results be available soon after the sample has been taken, because the very objective of any such testing is to provide a check on the internal system of quality control and to utilize the test results for correcting the production line, if found necessary. If the results are not available within a reasonable period of time for judging whether the production line is satisfactorily under control or that it needs some attention, no timely action could be taken for correcting the production process. Thus, under the circumstances, a proportion of the product may have been packaged for despatch or even passed into distribution channels, before a possible defect could be pinpointed and eliminated. It is for this reason that certain certifying agencies provide for the essential part of testing work to be carried out right at the site of production by their own inspectors during their periodical inspection visits. Furthermore, a good deal of reliance has to be placed on the continuous quality control maintained by the licensee during production in accordance with the scheme prescribed.

15.16 In any case, extreme care has to be exercised in selecting outside laboratories for carrying out sample testing on behalf of a certifying agency. The points to be borne in mind are the adequacy and suitability of the available facilities, satisfactory training and experience of the operating staff, accuracy of day-to-day work, provision for occasional cross-checking the work of one operator against another, the possibility of securing adequate priorities for certification testing, integrity of the personnel of the organization, and so on. It is advisable that after satisfying itself on these points, the certifying agency should prepare a list of approved laboratories whose facilities might be drawn upon as required. It would also be advisable that the work of one laboratory be occasionally checked against another by running the same test on a common sample and the list of approved laboratories be revised from time to time.

PRODUCTION CONTROL

15.17 In any scheme of certification marking, it is not only important to determine in the first instance that the product to be certified meets all the criteria laid down in the relevant standards, but what is equally important is to ensure that it would continue to do so throughout the validity of the certification marking license. In this respect, making occasional checks of the production plant and testing the samples taken at random is essential but not adequate for all purposes. In addition, it is most helpful to create and instal a regular system of quality control to be exercised during production within the plant. This not only assists in the maintenance of consistent quality but also helps in reducing the frequency of periodical inspection and the overall cost of operation. In some modern plants effective quality control systems may already exist which might be found quite satisfactory for certification purpose. In others it may be necessary to introduce them anew or to modify the existing ones.

15.18 In most well-organized systems of certification marking, it has become quite customary to incorporate in the original license a detailed scheme of quality control to be installed by the licensee within his plant, which would include, among other things, the following essential features:

(1) designating the points of production from where samples would be

drawn for tests and/or inspection: this will start with raw materials, through semi-finished products (or sub-assemblies), to the finished product, including product packaged and made ready for shipment or stocking, as well as that at various distribution and sales outlet points;

(2) specifying the frequency, the manner and the size of such samples to be drawn;

(3) tests and inspection to which the samples are to be subjected;

(4) the manner in which the test and inspection results are to be recorded by the licensee, which would be made available for periodic examination of the inspectors of the certifying agency;

(5) the criteria for judgement of the test results to determine whether the production should be considered under control or otherwise; and

(6) the various possible actions which may have to be taken by the licensee in case the product is not found quite in conformity with the standard and the form in which the report is to be made to the certifying authority in this regard.

15.19 Such schemes of control are generally based on statistical methods of quality control (*see* Chapter 23) and designed to ensure 95 percent confidence limit. A 100 percent assurance is seldom sought for reasons already discussed (*see* paras 15.10 to 15.12).

INSPECTORS

15.20 Specially trained and appropriately briefed staff of qualified inspectors are most essential for the satisfactory operation of a certifying agency. The inspectors should not only be familiar with the production processes in the concerned industries but should also be knowledgeable on the statistical theory of quality control, sampling techniques, acceptance criteria, etc. Their services would be required in all phases of certification work; for example:

(1) to inspect the premises of a prospective licensee who might have applied for the use of a certification mark, with a view to determining whether the production process used is satisfactory and whether an adequate system of quality control for ensuring conformity of the product to the standards exists;

(2) to suggest the requisite modification or augmentation of the applicant's quality control system wherever such an action may be called for;

(3) to draw for independent testing and to adjudge test results before a decision to issue a licence is taken;

(4) in the case of operative licenses, to maintain a continued check on the production process and on the associated quality control system, and to draw occasional samples, extending from the raw material stage to the finished product stage, from factories, warehouses and other premises under the control of the licensee;

(5) to purchase samples of certified products in the open market for the purpose of check testing; and

(6) to investigate complaints from purchasers and consumers of certified goods or from competitors of licensees.

15.21 In organizing such a force of inspectors, it is important to bear in mind that their professional integrity must be safeguarded and that they must receive adequate remuneration in order to ensure that they are free from ordinary temptations. While placing full confidence in their integrity it would be well to provide for a system of checks and counter-checks. For example, in allotting the various tasks to the different inspectors, it has been found useful frequently to change their assignments and transfer them from one locality to another at convenient intervals. This also helps to broaden their experience by exposing them to varying situations. In addition, this approach contributes to the building of public confidence in the integrity of the operation, besides avoiding the possibility of any malpractice. Furthermore, a perfectly impartial check is provided, if different inspectors are assigned the task of taking samples and testing the product of a given licensee at other different stages of operation, namely, at the application stage, during the operation of the license, from the open market, and for investigating complaints, if any.

VIGILANCE DURING DISTRIBUTION

15.22 It is important not only to check the quality of raw materials before they enter production, and to control the production process itself, but also to keep an eye on the distribution and marketing channels to ensure that during the handling of shipments, and the inevitable storage involved at various points subsequent to manufacture, no undue variations would occur in the quality of the certified goods. Such variations can often arise either as a result of unsuitable storage conditions or exposure to adverse weather conditions. For example, portland cement tends to begin to set and partially harden prematurely on exposure to high humidity. In other cases, defective packaging, such as the inadequately moisture-proof packets used for biscuits, or simply the expiry of shelf-life of the product, as in the case of dry cells, may be responsible for deterioration. Obviously, in order to ensure that the consumer receives the certified product in as sound a condition as intended by the requirements of the standard, it is the certifying agency's duty to maintain a close vigilance on the various distribution channels, in which the product is likely to suffer deterioration during exposure, weathering, storage, handling and/or shipment. It is, of course, primarily in the interest of the producer to exercise his own control and provide for the recall of such batches of his product as might have lost their claim to conformity with the standard before they reach the hands of the ultimate consumer. Serious consideration should be given to the incorporation of such conditions in the license and in certain cases perhaps in the standard as well.

15.23 Sometimes a producer may not be in a position to control each and every point of distribution. For example, an interesting case arose in relation to lead acid storage batteries which are usually shipped to the distributors and through them to small retailers in a dry condition, without the electrolyte. Just before their sale, the electrolyte made from battery grade distilled water and battery grade acid has to be added; the quality of both, being rather critical, is usually covered by the standard on batteries. Certain manufacturers of batteries using the certification mark had claimed that they were powerless in ensuring that the small retailers handling their product would use only the correct grade of water and acid and for that reason and to that extent they could not undertake the assurance of conformity of their batteries to the standard. It was found necessary in this case to discontinue the licenses, for neither the national standard nor the terms of the license could be amended to accommodate the manufacturers' difficulty. In cases like this there is hardly any simple solution, except to consider that, if the user's interest is to be properly protected and the reputation and the integrity of the mark is to be maintained, the only way open would be to insist on making the necessary improvements in the distribution system.

HANDLING OF COMPLAINTS

15.24 In view of the fact that each and every item of a product cannot be tested for conformity to standards, because of technical as well as economic reasons and that under certain circumstances the product is likely to suffer deterioration or damage subsequent to manufacture and marking, it is not unlikely that under the best of conditions of control and with all the good intentions in the world, a particular item of product may pass into the hands of the consumer, which does not give him full satisfaction (*see also* paras 15.10 to 15.12 and 15.19). In a case like this, which is likely to arise in any well organized system of certification marking, though seldom indeed, the consumer should have the means for redressing his grievance. The certifying agency inspectorate, as suggested above, should be assigned a specific duty in this regard and should thoroughly and expeditiously investigate any complaint that might arise. Some NSBs operating the certification marks schemes, as mentioned earlier (*see* para 15.12), go even further and incorporate in their license a specific condition that in case a complaint is proved valid, the licensee would be responsible for replacing that defective item or refunding the cost, and that in this respect the decision of the NSB shall be considered final.

15.25 As it happens in practice, most consumers do not always realize the importance of giving all the pertinent information in regard to their complaints. Some of them have even filed complaints about a product which has carried no certification mark whatever. The complainant has merely assumed that a product like an electric bulb, for example should have been certified and, therefore, the NSB must take up his case. Sometimes a batch of products carrying a clandestine mark has been sold to an unsuspecting buyer by a fly-by-night outfit which cannot subsequently be traced. In such a case, it is already too late when the unwary buyer becomes alert and files a complaint. There is little that a certifying agency can do in such circumstances. But in case a complaint is found to be genuine, it is up to the agency concerned to initiate immediate follow-up action to plug the loop-hole in the quality control process and ensure that no cause for future complaint would arise. Lessons learnt from one case should be used to review all other possible cases where a similar situation might arise.

COST OF CERTIFICATION

15.26 As the cost of operating a certification marks scheme has necessarily to be added to the production cost of the certified product, some producers have complained that their competitive position in the open market would suffer seriously if the overall cost of production and marketing goes up unduly as a result of certification. But this fear is seldom based on a close examination of the factual position. First of all, it must be recognized that, in order to ensure a consistent and well-defined quality level, a reputed manufacturer has necessarily to incorporate, in his production line and in his distribution system, a procedure for controlling and maintaining the quality of his goods, and this has to extend up to the stage of their reaching the hands of the consumers. This, in itself, should be considered an integral part of the production and marketing costs, irrespective of whether the goods carry a certification mark or not. It is true that some manufacturers do not pay adequate attention to this obligation, but then these are the ones who are more likely than not to fail anyhow in establishing a reputation for their product. In order that this group also is able to compete successfully, it will necessarily have to adopt similar quality control measures as those of the more successful competitors. It is only after this pre-condition is satisfied that they would become eligible for a certification mark license.

15.27 Furthermore, it has been found by experience that any cost incurred on maintaining a systematic control of quality is usually more than paid back with bonus, through the savings that it brings about on several accounts including, for example, the reduction in the frequency of rejections and rework, more economical use of material and manpower resources, better assurance of continued long production runs without undue shutdowns[25] and, above all, by improving the marketability of the product through improved consumer preference.

15.28 The overall cost of quality control varies a great deal from product to product. Its magnitude depends largely on the complexity of the product itself and that of the tests involved in determining its characteristics. Usually, for articles of normal industrial and consumer use, it varies from less than 1 up to 3 percent of the total cost chargeable to production. Surely this is not a great deal to pay for ensuring consistency of output and quality of the product which guarantees consumer satisfaction. As such, this cost has necessarily to be considered a legitimate part of the cost of production and is indeed so treated by most manufacturers. The extra expenditure involved in obtaining and operating a certification marks license is usually even more nominal. Given the existence of a satisfactory system of quality control, the cost of operation of a license may seldom exceed the level of 0.01 percent of the product cost, which amounts only to one part in 10,000. In view of the increased marketability that the certification mark brings to the product, this extra bit need hardly be considered as a real increment of production cost.

15.29 Some large-scale and reputed manufacturers claim that since their own trade marks and brand names furnish an adequate guarantee of the quality of their product, they have no need for a certification mark. But then there are others of this class who strongly feel that a certification mark gives them the competitive advantage they need over those who, because of their lack of quality control, can afford to and usually do undersell them. Then, there is always the consideration for building up consumer confidence in the far-off overseas markets, which is facilitated through a national certification marks scheme applicable to all types of goods and recognizable as the mark of the country of origin. This is of particular importance to the newly emerging countries for whom export promotion has a special economic significance, in which all reputed manufacturers as well as others should be vitally interested.

15.30 It is sometimes argued on behalf of very small-scale manufacturers that the cost of quality control as well as that of operating the license itself cannot readily be absorbed in the production cost, because of the scale of their operation. The need for control of quality cannot, however, be denied even in this case, for after all, unless a product meets the minimum requirements of the consumer, it can scarcely find a market and would, therefore, be hardly worth producing. There is, nevertheless, a real difficulty in providing each small-scale unit with adequate testing facilities, mainly because of the utilization factor being extremely small. But this hurdle may be overcome either by setting up a cooperative testing centre collectively by several units themselves, or through resources made available by official or non-official organizations which have positive interest in assisting the development of such industries. It must also be recognized that in many cases the small-scale and cottage-scale industries function as feeder industries to the medium and large-scale sector. In such cases, the medium and large-scale sectors should be able to absorb the cost involved in organizing the quality control operations in their capacity as consumers and users of the products of a large number of small production units.

15.31 In conclusion, it may be stated that the whole subject of quality control is too complex and specialized to be dealt with here under certification marks costs. Suffice it to say that the costs involved are not only nominal but legitimate and every effort should be made to keep them low and means should be found to meet them in the interest of both the producers and consumers. In this task both the statisticians and technical specialists must cooperate and coordinate their effort to arrive at satisfactory solutions.

DESIGN OF THE MARK

15.32 It is well worthwhile to give some thought to the details of the design of the certification mark itself. The simpler it is in concept and more sparing of lines, the more distinctive it will be. Furthermore, it would be desirable to adopt one basic design for a national standards mark. This practice has indeed become quite common, but some authorities still consider it desirable that several different designs should be adopted for use with different classes of goods. For economy in early stages of development, having an eye on export markets, it would perhaps be preferable to adopt one central design of the national mark, around which minor variations could be introduced, if necessary. This would help build up an image of the country as a whole in overseas markets rather than that of its individual producers. Even for use within the country itself, it is desirable to have one basic design, since the consumer is likely to be confused by the existence of several marks purporting to convey the same significance of a national symbol of certification issued by one and the same central standards authority.

15.33 Simplicity of the design is also important for rendering it clearly legible even when it is applied to a very small article, as, for example, a writing pen or drawing instrument, a spark plug or a piece of jewellery. Flowery and intricate designs do not lend themselves readily to this kind of reduction. Sometimes it is considered necessary to incorporate in the mark the designation number of the standard with which it signifies conformity. On other occasions it is considered important to indicate in the mark the particular grade or type of the article which has been certified from among the several grades or types specified in the standard. Thus it will be seen that the extra information required to be added to the basic design of the mark would tend to make it complex enough and aggravate the problem of size reduction. The accompanying illustrations would amply illustrate these points.

15.34 In certain exceptional cases demand may arise for an entirely distinctive mark for a class of products requiring pointed attention of the buyer to be drawn to its extraordinary character, such as the safety features of fire-proof equipment for use in mines, or the purity of gold and platinum jewellery. In such cases, it may well be justified to create a certification mark which is quite distinctive in design and different from the normal national mark. But even here an appropriate variation of the normal design could be evolved which would at the same time be distinctive enough to serve the purpose.

15.35 There are some other pertinent pieces of information, which it is useful for the consumer to know about the article he purchases; for example, the instructions for its use, its care and maintenance, date of its manufacture, its shelf-life expiry date, name of the manufacturer and his license number. Some of this important information, like the instructions and other data, would be required to be supplied along with the certified article as a condition of the license, but the date of expiry of shelf life is important enough to be indicated on the label or container very close to

Fig. 15.2 Specimen of some certification marks of Indian Standards Institution containing significant extra information.

Top row IS:10 is the number of the standard specification for plywood tea chest, covering all its components. The three marks are designed to certify respectively the three components indicated.

Middle row "Type 1" is one of the two types of fire fighting hoses covered by the specification IS:636 and "Pure" is one of the two grades of carbon tetrachloride.

Bottom row The two marks are used to certify only the containers and not the contents – an 18-litre square tin and a glass milk bottle respectively.

the certification mark, if it is not made a part of the mark itself (*see also* Chapter 16, para 16.1).

METHODS FOR APPLICATION OF THE MARK

15.36 This is perhaps the most intricate feature of the whole operation of certification marking and requires very careful and detailed attention. The mark may be applied by stamping or printing on the article itself, or on a label or tag attached to it, or on its container or package; every case

FOR ARTICLES OF A GENERAL NATURE

FOR BLOCKS IN CONCRETE

FOR ARTICLES MADE OF STERLING SILVER

Fig. 15.3 Examples of national certification marks of France (left two) and South Africa (right two) for articles of general nature (top row) and for special purposes (bottom row).

presents its own problems. Take, for instance, bulk material such as portland cement or a fertilizer mix, which by itself cannot be marked – only its package or container can be marked or perhaps a tag attached to the package. Sometimes the packaging material is a bag of coarse gunny, or of paper or plastic. If such a container is marked with a certification mark, there is always the possibility of its being re-used for packing similar material for sale or in some cases even some other material, which is not certified. A tag is also liable to be put to a similar unauthorized re-use. Careful consideration has to be given to the marking of such packages. The problem arises more often in developing countries because sometimes there is shortage of essential materials, which is usually conducive to such malpractices. In industrially advanced countries, the small-scale operations of the type involved in such malpractices would hardly be considered worthwhile or profitable.

15.37 A similar situation arises in the case of liquid bulk materials such as acids, solvents and beverages. In this case, however, it is not so difficult to devise simple methods to avoid re-use of containers because the marking can be applied to the stoppers and other closures in such a way as would make them unsuited for re-use. In applying the mark at the point of the closure to the container itself, be it of glass, tin plate, plastic or other material, methods may be employed which would require a seal to be broken when the container is opened. Such a seal need not be a complex one; even a simple strip of paper carrying the mark applied over the cork would sometimes suffice. Another interesting situation arises when a re-usable container, such as a barrel or a cannister alone has to be certified without in any way referring to the contents. In such cases, it has been found absolutely necessary to incorporate some qualifying words in the mark itself; for example, "TIN ONLY". A similar case may also arise when only one or a few of the several characteristics specified in a standard are certified; for example, the safety of an appliance or the interchangeability of a component.[26] Similar precautions will have to be taken in such cases.

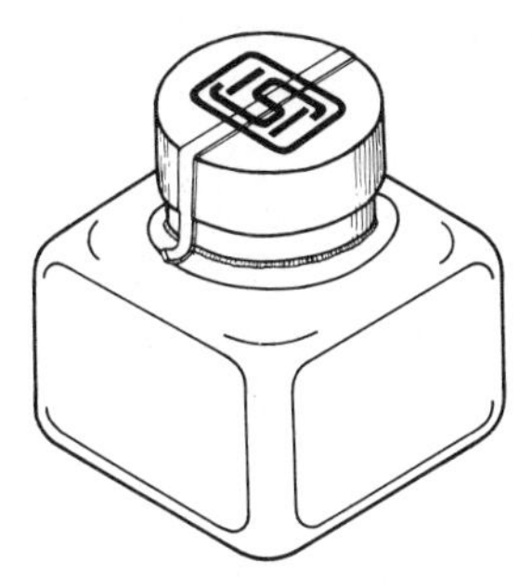

Fig. 15.4 Stopper of ink bottle illustrating the use of tear-ribbon which destroys the certification mark in the process of opening.

15.38 Sometimes an article requires the application of the mark at an early stage of production before it goes through other finishing operations involving considerable expense. Typical examples of this kind are to be found in the ceramics and clay products industries, such as china-ware, roofing tiles, bricks and the like. It is obvious that after being finally fired, when a product like this is subjected to inspection for defects and tested for other characteristics, a number of pieces may be found to be unsatisfactory which would have to be rejected. In normal trade practice such rejects could be passed on as seconds, but in the case of the certified goods they cannot be so readily disposed of, because they would carry the certification mark intended to indicate full conformity with the standards, which they actually cannot claim. In certain cases such as in mass produced glassware, the product could be broken up and re-used as cull material, but in the case of burnt clay products very little could be done to salvage the rejects.

15.39 Three solutions appear feasible; namely:

(1) to obliterate the mark or somehow deface it from the defective lot of the product;

(2) to consider the possibility of applying the mark after the product has gone through all the stages of manufacture and is ready for packing or despatch after due inspection so that the rejects could be disposed of as seconds;

(3) to control the production processes so precisely in all respects and at all stages that the incidence of rejections is reduced to the barest minimum, and within the economically tolerable limits.

It is obvious that the last of these solutions is decidedly the best, because it has the potential of leading to the most economical production. But it may not be found feasible in many cases. The first two methods, on the other hand, might involve extra expenditure, because they entail an additional operation for obliterating or applying the mark. Ordinarily, in the case of clay products, the mark is inscribed in the mould itself and involves no separate operation for application, unlike obliteration, removal or subsequent application. It is thus open for study that when method (3) is not found to be feasible, which of the other two methods would be more economical for adoption. Similar considerations would apply to stamped, pressed, forged, cast and moulded products of metals, plastics and other similar materials. But in these cases controls are relatively simple to apply both at the processing as well as the subsequent finishing stages. Nor is it always so difficult in these cases to salvage some of the rejected material, as it is in the case of ceramics.

15.40 It would be clear from the above examples that every class of product presents its own problems of marking and that each case has to be dealt with on its own merits. In most of the difficult cases that may arise, it is always possible to find a satisfactory solution through mutual discussion and cooperation between the prospective licensee and the certifying authority. Whatever the solution agreed upon, it should be such that, without unduly increasing the cost of production, the mark would convey the requisite message to the purchaser and the consumer, and that it would not in any case increase the chances of fraudulant use or re-use.

FEEDBACK

15.41 The NSBs which have a certification marks scheme in operation enjoy one great advantage over others which do not have such schemes. This is in respect of the abundance of feedback of field experience they receive, which is gathered during the actual implementation of standards in practice and which is so essential for their continuous improvement. Certain standards, when implemented under a quality control system used during production, may be found wanting in some respect or the other. Furthermore, the users of certified articles are naturally bound to be more alert to their shortcomings than the users of ordinary uncertified goods. The former are, therefore, more likely to bring such shortcomings to the attention of the producer or the certifying agency. Every complaint so received, every flaw thus detected during inspection and testing constitutes valuable feedback information for the NSB. It can be of great national and economic importance to pin-point the lines along which national standards might usefully be improved. Every effort should be made to encourage and even seek this feedback through various channels. An amusing case of such feedback is cited, which led to the correction of an obvious oversight in a New Zealand standard,[27] when it it was brought to light that there was a need for developing a new breed of bulls 50 percent larger than the largest size available in the famous cattle breeding country, if the New Zealand standard in question were to be complied with.

LEGAL PROTECTION

15.42 A well-run certification marks scheme depends a great deal for its success on the integrity of each licensed manufacturer as well as on the confidence which the certifying agency places in him. This mutual understanding and trust is considered most essential for achieving progressive development and all-round acceptance of the scheme in different sections of the business community. Nevertheless, the possibility of existence of unscrupulous elements or a badly organized factory or careless administration in any society cannot be ruled out, which would require constant vigilance and helpful guidance. It is also essential to protect the legitimate rights of a licensee against a competitor who may choose to use the mark without securing a license. Safeguards will have to be provided against the misuse of the mark and its use without the requisite authority; as also against licensed parties who wilfully, negligently or otherwise do not conform to the conditions of the license. In addition, certain rights and protection will have to be given to the personnel of the certifying agency. All these conditions require certain legal sanctions to be provided. However, the form in which these have been provided in different countries varies a great deal.

15.43 Generally speaking, most countries would fall in three broad classes, namely:

(1) those in which the standard certification mark is registered as a certification trade mark under the normal trade marks act by the NSB;

(2) those where the law establishing the NSB includes provisions for the creation of national certification marks under its authority; and

(3) those countries where an independent legislation for certification marking exists authorizing the NSB to administer the mark.

It is obvious that from the legal point of view, the last two categories namely (2) and (3) could be considered together, for the legislation in question would deal specifically with a national standard mark operated by the NSB, while in the case of (1), the legal provisions would be more of a general character, applicable to a certification trade marks registerable by any competent agency.

15.44 Leading examples of a standard certification mark being registered and operated by the NSB of a country are those of the BSI[28] in the UK and the ANSI[18] in the USA. In such cases, although the normal trade mark legislation provides adequate legal protection against the misuse of marks by unauthorized parties, certain other safeguards referred to above are not usually provided directly and have to be taken care of by way of mutual agreement between the certifying agency and the licensee. Nor could the penalties for misuse be as severe as might be considered advisable in certain countries in the case of a mark of national status. Withdrawal or cancellation of the license is perhaps the greatest penalty that could be levied against a licensee, but the remedy for unauthorized use of the mark by others would lie in the lengthy civil suit proceedings under the trade marks law. Furthermore, it is required under trade mark legislations to register along with the mark a set of regulations pertaining to its operation. Usually, there are several classes of goods provided for in these legislations, each one of which has to be covered by a separate registration and a separate set of regulations including the relevant standards. In the UK Act, for example, there are 50 such classes specified which would need separate registration of the mark in each class. A similar situation originally existed in India where the legal authorities felt that it would be necessary to include among the regulations for every class the particular standards pertaining to each item covered by that class together with details of the scheme for controlling the quality of the article which would form a part of the license. The number of standards available at that time was quite limited but the pace of their development was on the increase. This situation would have entailed a considerable amount of delay and paper work, if trade marks legislation had to be relied upon. It was found more expedient to adopt a self-contained comprehensive legislation which now prevails in the country. In the UK, however, it is understood that the trade marks law has had to be suitably modified to simplify this procedure of registration, at the request of the BSI.[29] Nevertheless, in certain countries the situation may still be complicated in regard to the registration of a national certification mark and its control under the normal trade marks act.

15.45 Examples of countries in class (2) include Iran,[30] Ireland,[31] Japan,[32] the Philippines, Malaysia, Thailand and some others. Countries with separate independent legislations in class (3) are exemplified by France[33] and India.[19] Provisions in these legislations are very similar and provide for most of the safeguards required for an effective control and administration of the marks scheme. Briefly, they stipulate:

(1) that the NSB is authorized to prepare national standards and to prescribe and administer certification marks;

(2) that the marks are protected against being copied by unauthorized persons;

(3) that the authorized persons are required not to use them in an unauthorized manner;

(4) that the NSB is authorized to prescribe the conditions of license and charge the requisite fees from the licensee, and enjoys powers of entry into and inspection of his premises;

(5) that the personnel of the NSB is legally safeguarded for all acts carried out in good faith, but is required to treat as confidential all information obtained during the course of work;

(6) the powers of the NSB to suspend and cancel a license or take other appropriate action, as described, in cases of malpractices or inability of the licensee to continue to satisfy the conditions of the license; and

(7) the penalties for various offences and procedures for legal actions together with the rights of appeal in different situations.

15.46 Evidently, the chief merit of the two systems adopted in the two classes of countries (*see* (2) and (3) in para 15.43 above) is that they leave the NSB independent of the trade mark registration authorities which are more concerned with trade mark affairs rather than certification questions of a more specialized character. Under such legislations, the NSB acquires a national status and the standard mark a position of national prestige. For the developing countries particularly, the systems prevalent in either of the two classes of countries (2) or (3) may preferably be adopted, depending on the background of the existing legislations. But the system in (2) would perhaps appear to be more generally suited, because the law for establishing an NSB, which has quite often to be promulgated anyhow, could readily be extended to provide for certification marking. Only in a few developing countries particularly, where the NSB happens to be organized as a joint body or a private institution under the societies or companies act or some other similar law, there may be a need for an independent legislation, as in a country of class (3) like India.

15.47 In the Indian legislation[19] on certification marking, a special provision has been made for the creation of "competent authorities," which does not appear to exist in others. This means that on the recommendation of the NSB, the government may authorize certain organizations other than the NSB to exercise some of the powers of inspection and overall supervision of marking of the certified products in specified areas of industry or in specified geographical regions. Such competent authorities are intended to assist the NSB in extending the scope of its certification marking operation beyond the limits imposed by its own resources but under the overall control and supervision of the NSB itself. In large countries there may exist certain agencies fully qualified and equipped to undertake some of the responsibilities connected with the administration of marks in a given area or an industry, and it would be in the interest of the nation to make full use of such facilities. It is under such provisions of the Indian Standards Institution (Certification Marks) Act, that some of the industry organizations, states as well as central government departments have been recognized as competent authorities, which, while working in harmony with the NSB, form an integral part of the nation-wide network of the certification marking operation. In this manner the need for registering separate certification trade marks under the trade marks act is eliminated and a unified pattern of national marking emerges under a single authority.

15.48 Another feature in the ISI Certification Marks Act,[19] which is rather uncommon and which is of interest to the developing nations, is that concerning the possibility of ISI being able to recognize temporarily as an Indian standard a standard issued by an authority other than itself, purely for the purpose of certification marking. Such a standard may emanate either from an Indian source such as a private industry association or a government department, or even an overseas institution. However, it is required to be replaced by a duly processed full-fledged Indian Standard within a stipulated period of time. This provision enables certain industries to start the use of the national certification mark for goods which may not have so far been covered by an Indian Standard, but which need to be covered. By drawing pointed attention to the demand for certain national standards, it helps to accelerate their preparation. In the developing countries such a provision can be of great help in advancing the cause of standardization generally and may be well worthwhile keeping in view when framing the legislation on certification marking.

INTERNATIONAL CERTIFICATION MARK

15.49 The possibility of creating an international system of certification marking has been attracting serious attention of most standardization authorities in the world for the past couple of decades or more, but so far very little headway seems to have been made in this direction. ISO considered this question in the early 1950s and came to the conclusion that it would be difficult to secure legal protection for any international mark among the participating countries. Even the question of bilateral recognition of the certification marks of one country by another involves the problem of one country being able to give the same legal protection to the mark of another country as it may enjoy at home or as does the mark of one's own country. The international conventions on trade marks are not very helpful in promoting the cause of international certification trade marks, and national standards marks created under independent legislations in different countries are not covered by any international conventions so far.

15.50 Keeping all these difficulties in view, the International Commission on Rules for the Approval of Electrical Equipment (CEE) has successfully launched a programme of mutual recognition of test certificates for approval of electrical equipment.[34] This is not claimed to be a complete solution, but only a step towards the introduction of an international approvals mark. The scheme is not open to all countries of the world but only to member nations of the CEE, and even among them to those members who accept its rules. It is administered by a committee of representatives of the participants. As far as is known the following 16 countries of the CEE participate in this scheme.

Austria	Finland	Norway
Belgium	France	Poland
Czechoslovakia	Greece	Sweden
Denmark	Hungary	Switzerland
Germany (Federal Republic)	Italy	UK
	Netherlands	

15.51 Another scheme, which is neither international nor regional but more national in character and covers electrical equipment only, like the CEE scheme, is that organized by the Canadian Standards Association (CSA).[16,17,35] It constitutes a system encompassing several nations including Australia, Japan, UK, USA and certain western European countries of the CEE group, which export electrical goods to Canada. The success of the CSA operation depends largely on two factors – firstly, the Canadian electrical code issued by the CSA, which forms the basis of the scheme, has been legally adopted and made mandatory by every provincial government of Canada, with a view to ensuring the safety of electrical equipment. The second factor is that, though the scheme was originally intended for Canada itself, it has now spread over many other countries because Canada constitutes a large market for the household and other electrical goods for the various participating countries. Under this scheme, the CSA maintains its own staff of inspectors in the recognized laboratories of the cooperating countries like those of the BSI in the UK and the CEE in the Netherlands, who, together with the local staff, test and approve equipment on behalf of the CSA and qualify it for the CSA monograph indicating certification. It will be seen that this operation can hardly satisfy the requirements of the international mark because, though it involves several nations, it is in fact intended to serve the needs of only one nation and there is no question of mutual exchange of goods on the basis of this mark, the flow being mainly unidirectional, that is, into Canada. Of course, all Canadian goods approved by CSA scheme may also be accepted in other countries, but this is not the same thing as legal recognition of the Canadian certification system in the cooperating countries.

15.52 The European Standards Coordinating Committee (CEN) has quite recently begun to emphasize that it is not enough just to issue agreed standards to help overcome the various non-tariff barriers that exist within a common market, but it is also essential to have mutually recognized standard certification marks to indicate conformity with the common standards. CEN is therefore presently investigating the possibility of creating such a system.[36,37] This system as conceived would be based generally on ISO Recommendations on the subject of certification marking[20,21] but would be made suitable for adoption by all nations within CEN, which would have the option to accept or not the mutual obligations involved. A set of articles of association together with regulations covering the use of the CEN mark have been drafted and are presently under the active consideration of the various member countries involved.

15.53 Simultaneously, the ISO has begun once more to consider the question of evolving a truly international mark[38, 39] which may be somewhat on lines similar to those of the CEN system. As indicated by the Chairman of the ISO special committee appointed to study this subject: "There is increasing public demand for assurance that machines and fabrics and domestic appliances of all descriptions are in conformity with high standards of utility and safety. It is our task to harmonize all these certification schemes and thereby eliminate some serious obstacles to international trade." The IEC has also begun to interest itself in this move and to consider the possibility of a system of its own for certifying internationally the reliability of electronic components.[37] In view of the legal implications involved, both these efforts pose a big problem. But given the goodwill and cooperation among the various NSBs and the governments involved, it may not be difficult to overcome the hurdles and evolve a successful system, which might take the form of an international convention. Standardization discipline appears to have recently come of age and with the growing influence of the NSBs within their respective national economies, the solution should be much more within the range of possibility today than it was perhaps a decade ago. But in any case ultimate success of such a system would depend largely on the harmonization of the various national standards in the first place, which appears by and large to be well on its way; and in the second place on the mutual recognition of the common interest of all the nations concerned.

REFERENCES TO CHAPTER 15

1. Coles, Jessie V.: *Standards and labels for consumers' goods.* Ronald Press, New York, 1949, p. 556.
2. Wilhelms, Fred T. and Heimerl, Ramon P.: *Consumer economics.* McGraw-Hill, New York, 1959, p. 534.
3. See files of the consumers' magazines, such as the following:
   - International: International Consumer (International Organization of Consumers' Union, The Hague, Netherlands).
   - Canada: *Canadian Consumer* (Consumers' Association of Canada, Ottawa).
   - UK: *Which?* (Consumers' Association, London).
     Consumer Affairs Bulletin (International Cooperative Alliance, London).
   - USA: *Consumer Bulletin* (Consumers' Research, Washington, NJ).
     Consumer Reports (Consumers Union, Mount Vernon, NY).
4. *Final report of the Committee on Consumer Protection.* Her Majesty's Stationery Office, London, 1962, p. 331 (*see* pp. 111-119).

5. Gardner, Annesta R.: Seals of approval – what they do for your product. *Product Engineering,* 1966, vol 37, no 24, pp. 137-139.
6. Jennings, Charles R.: Self-certification – its birth and growth. *FDA Papers* (Food and Drug Administration, USA). Nov 1969, vol 3, pp. 16-18.
7. *UK Trade Marks Act, 1938.* Her Majesty's Stationery Office, London, p. 56.
8. *UK Merchandise Marks Act, 1887, as amended in 1953.* Her Majesty's Stationery Office, London, p. 5.
9. *Trade Marks Laws, January 1959.* United States Government Printing Office, Washington, p. 38.
10. *The Trade Marks Act, 1940 as modified up to April 1952.* Manager of Publications, Delhi, p. 37.
11. Farrell, William H.: What's behind the U & L? *ASHRAE Journal* (American Society of Heating, Refrigerating and Air-conditioning Engineers). July 1970, pp. 46-48.
12. Verman, Lal C.: Certification marks for standard products. *ISI Bulletin,* 1951, vol 3, pp. 3-5.
13. See p. 100 of reference 4 above.
14. Sherman, W. A.: Certification – Tower of Babel. *Industrial Quality Control,* 1966, vol 23, pp. 72-75.
15. Tebo, Gordon B.: Product certification – putting standards to work. *Magazine of Standards,* 1964, vol 35, pp. 131-135.
16. Hopkins, Stuart K.: Safety equipment certification in Canada. *Magazine of Standards,* 1966, vol 37, pp. 134-136.
17. *Approvals manual, respecting inspection, test, and approval of electrical equipment.* Canadian Engineering Standards Association (now CSA), Ottawa, 1940, p. 20.
18. Rockwell, William H.: The whys and hows of USASI's certification programme. *Magazine of Standards,* 1969, vol 40, pp. 112-113.
19. *The Indian Standards Institution Certification Marks Act, 1952, with its Rules and Regulations amended up to 1968.* Indian Standards Institution, New Delhi, p. 13.
20. *ISO/R 189–1961 Principles of operation of standards marks.* International Organization for Standardization, Geneva, p. 4.
21. *ISO/R 526–1966 Significance to purchasers of marks indicating conformity with standards.* International Organization for Standardization, Geneva, p. 3.
22. See the references listed under reference 22 of Chapter 12.
23. See p. 118 of reference 4 above.
24. See pp. 107-110 of reference 4 above.
25. Bicking, Charles A.: Cost and value aspects of quality control. *Industrial Quality Control,* 1967, vol 24, pp. 306-308.
26. See para 1.1 of reference 21 above.
27. Ward, T. S.: New Zealand standard certification mark scheme. *New Zealand Standards Bulletin,* July 1970, vol 16, pp. 9-12.
28. *Notes on the certification marks of the institution.* British Standards Institution, London, 1958, p. 27.
29. See p. 3 of reference 28 above.
30. *Constitution and Law.* Institute of Standards and Industrial Research of Iran (ISIRI), Tehran, 1965.
31. *Industrial research and standards act, 1961.* Stationery Office, Dublin, 1962, p. 51.
32. *Industrial standardization law, JIS.* Standards Division, Agency of Industrial Science and Technology, Ministry of International Trade and Industry, Tokyo, 1959, p. 17.

33. *National mark of conformity to the standards, NF – Legislation texts and regulations.* Association Française de Normalisation, Paris, p. 7.
34. *CEE PUB 21 Certification scheme.* International Commission on Rules for the Approval of Electrical Equipment, Arnhem (Netherlands), 1961, p. 15.
35. Tebo, Gordon B.: Product certification in Canada – some national and international aspects. *BSI News,* July 1966, pp. 10-13.
36. CEN system of certification and marking. *BSI News,* April 1970, pp. 15-16.
37. Rockwell, William H.: International trade – the impact of certification. *Magazine of Standards,* 1970, vol 41, pp. 77-79.
38. New ISO committee seeks to cut red tape. *ISO Bulletin,* Feb 1971, vol 2, p. 1.
39. Certification on global basis. *ISI Bulletin,* 1971, vol 23, p. 103.

Chapter 16

Informative Labeling

16.0 Labeling of goods and their packages has been practised from time immemorial to facilitate stocking and marketing. Labels are designed to inform those who handle, purchase or utilize them, about the nature of goods and weight or volume of contents in packages. Until the advent of standardization, and to a large extent even now, the practice of labeling has varied from one individual marketing agency to another. Everyone is, by and large, free to choose the pattern of labeling and the manner in which the consumer or user might be informed. This has sometimes led to malpractices.[1] With the growth of standardization activity, as the goods became more and more standardized, their labeling also began to become more and more related to standards.[2] Consumer product labeling particularly attracted considerable attention and became more and more useful both for the producer in selling the goods as well as for the consumer in guiding his buying.[3] The labeling of dangerous goods and equipment in which certain hazards were inherent came specially to be emphasized for the protection of handlers and users, specially when safety requirements began to be specified in the laws of the land and in certain standards of mandatory character.[4, 5, 6] In spite of these evolutionary changes that have gone to influence the labeling practices of today, it is hardly possible to claim that any degree of uniformity or coordination has been achieved either among countries or among industries, or even within a given industry in a given country. In the absence of such uniformity, most consumers in the world are by and large left to their own devices to compare the relative merits of different makes of products offered to them for sale.

CONSUMER NEEDS

16.1 Apart from the information regarding the place of origin or the name of the producer, a label, to be of service to the consumer, should give information about several aspects of an article, which would vary with the nature of the article. For example, a textile product would require, among other things, information on:

- size – length, breadth, fit, etc.;
- weave;
- weight per unit area;
- fastness of colour to various agents, such as washing, sunlight, perspiration, etc.;
- washing or dry cleaning instructions, including permissible solvents and detergents, temperature of wash, method of drying, temperature of ironing, etc.;
- shrinkage on washing;
- finish – water-proof, crease-proof, shrink-proof, etc.; and
- flammable or otherwise.

In case of food products, the consumer would like to know the:

- quantity of contents;
- mode of preservation;
- strength of syrup, if any;
- seasoning used;
- grade or quality; and
- date of expiry of usefulness.

Household electrical appliances, on the other hand, would require information, among other things, on:

- voltage;
- frequency;
- power consumption;
- performance level, *viz*., air delivery of fan, time for kettle to boil, etc.;

- special instructions for use, if any;
- safety precautions;
- instructions for maintenance;

and so on for other classes of goods.

16.2 It will be seen from these examples that each class of goods would require information on several points peculiar to itself and that there is a large number of points in each case, on which information could be usefully given. In order, however, to compare one brand of a product with another brand it would be essential that the information given is in similar terms and that the data for this information have been obtained by uniform standard methods of test. For example, there are several methods available for determining the shrinkage of fabrics, each of which would give a result slightly different from those obtained from the others. The grade or ripeness of a fruit may be determined by several different methods and expressed in several different ways. The consumer, in order to be appropriately guided, would require a uniform practice to be followed by the labeling agencies, which would require a degree of discipline.

SYSTEMATIC LABELING

16.3 It is only in recent years that the question of bringing about such uniformity has received serious attention and that too only in a few advanced countries where a standardization movement has existed for some time. It was in 1951, in Sweden, that the first step was taken to organize an Institute for Informative Labeling which has since been carrying on pioneering work in this field.[7, 8] The Swedish effort has attracted a good deal of attention in other countries, particularly those of Western Europe and the USA. International discussions[9, 10, 11] have led to a general appreciation of the usefulness of the system of informative labeling as originated in Sweden. An International Labeling Centre has been established[17] with the object of promoting and unifying the practice of informative labeling. Its membership comprises organizations actively engaged in the operation in Denmark, Finland, France, Netherlands, Norway, Spain, Sweden and Switzerland. Application of the system to the developing countries has not yet attracted much attention.

16.4 The basic motivation of the system is that, while a certification mark ensures the product quality to be in conformity with an agreed authoritative standard (*see* Chapter 15), a well informed consumer might prefer to act as his own judge of the level of quality acceptable to him. Informative labels enable him to exercise his own discretion in this manner, because the label informs him about the characteristics of the product in a standardized manner and style and assures him that this information has been obtained by the use of specified methods of test. The information thus given is, therefore, directly comparable from label to label. In addition, the informative labeling system would include instructions for use, care and maintenance of the article, which in the case of certified articles has necessarily to be supplied separately from and additional to the certification mark particulars (*see also* para 15.35 of Chapter 15).

INFORMATIVE LABELING AND CERTIFICATION MARKING

16.5 So far as the producer is concerned, the informative label does not bind him down to any specified level of quality or performance except as implied in the test results given on the label. On the other hand, in the case of certification marking, he is bound by the specification of the standard in question. This makes informative labeling attractive to certain categories of producers – those who produce goods below the standard level as also those who produce them significantly above the standardized level of quality and performance. The goods below the standard level can sell at a price preference which may be welcome to a certain class of consumers with modest means, but if such goods carry an informative label, the buyer can judge whether the price differential and the difference between the quality level of such goods and others with better characteristics are acceptable to him. It is in the making of such a judgement that a consumer needs to exercise more than ordinary level of discerning ability which would often involve the possession of a quantum of technical knowledge. Such an ability may be presumed to exist among well-informed consumers in countries where informative labeling has been in vogue for some years. In any case the producer of the goods is not to blame if the consumer later becomes dissatisfied with his own decision. On the other hand, in the case of certified goods the responsibility of the producers extends much further and he remains involved throughout the useful life of his product.

16.6 At the other end of the quality spectrum are the goods which are claimed to be much superior in quality and performance to those covered by standards. Such goods have to be significantly superior to be classed above standard, because in order to qualify for compliance to a standard, the quality of a product during production has necessarily to be maintained at a level somewhat higher than the standard level. If this is not done, there is always the danger of some individual items occasionally failing to pass all the requisite tests. Thus the manufacturers who claim to produce goods above the standard level and find certification marks schemes unattractive would welcome the informative labeling schemes, which might give them the opportunity to withstand successfully the competition from cheaper goods, on the basis of substantial quality differential. But all this demands, on the part of the manufacturer, an acceptance of the same discipline entailing constant vigilance and testing required for the maintenance of quality characteristics, whether it be for the purpose of a label or for achieving compliance with a standard.

16.7 Does this mean that wherever an informative labeling scheme is in force there is no place for certification marking, or that the two schemes are mutually exclusive? The experience of a pioneering country like Sweden in this regard indicates that both systems may prevail and prosper side by side. Some of the reasons contributing to this could perhaps be listed as follows:

(1) Not many members of the consuming public feel competent enough to exercise a choice based on the technical data given on informative labels and would prefer to depend on the judgement of a standards authority as implicit in the certification mark.

(2) There are several rather complex types of goods and equipment which a consumer buys; the judgement of the quality or performance of such goods often creates serious enough problems even for the well-informed technical expert. It is only logical that in such cases the ordinary consumer should depend on a certification mark rather than on an informative label.

(3) Certain goods are inherently more suited to certification while others lend themselves more readily to informative labeling. For example, to certify a motor car would not only involve quite a comprehensive set of standards including many complex methods of test but would also have to cater to many different tastes and varying capacities of the users to pay. On the other hand, an informative label could easily give the buyer valuable information about its performance and upkeep in which he is mainly interested. In contrast to this situation, a step-down transformer for a household appliance or a TV receiver could preferably carry a certification mark.

(4) The process of preparation of national standards on which certification marks are dependent is inherently slow (*see* paras 10.36 to 10.47, Chapter 10), while the preparation of standards on which informative labels are based is comparatively faster, since the latter do not

have to incorporate agreements on quality levels but only on methods of test and details of label designs. It is natural, therefore, that labeling could be introduced for certain products much more speedily than certification marking. Ultimately, however, both may come to exist side by side.

In reality there is nothing against a given product carrying both a national certification mark as well as a formally licensed informative label. This approach should be particularly welcome to manufacturers who produce goods significantly above the standard level. They will thus be able to cater both to the less informed public as well as the well-informed consumer.

16.8 From the point of view of the developing countries, it would appear that the system of certification marking would be more readily acceptable to the consuming public in preference to informative labeling, largely because of the comparatively lower level of technical information that a common consumer in these countries generally possesses. On the other hand, labeling standards being more simply prepared, it might be more expeditious to introduce informative labeling in such countries, than to have to wait for the lengthy time-consuming process involved in the preparation of national standards. In practice, however, it is not so simple as that, because to set up a successful centralized labeling authority, it is essential to have the active cooperation of the National Standards Body, which must pre-exist. Conditions in various countries being as variable as they are, no generalization can be attempted, except to say that the prevailing conditions will dictate the course of action.

PROCEDURES

16.9 On the pattern of Sweden, considerable amount of work has been done on informative labeling in several other countries such as Denmark, Germany, Norway and the UK.[12-16] Basically, it is a voluntary scheme on the usual model of standardization and certification marks schemes. Of necessity it has to be worked with active participation of the National Standards Body of the country. Just as in the case of standardization, its success depends upon the cooperation of several interests including manufacturers, wholesalers, retailers, technologists and consumers. Consumers and their organizations have a special role to play in this scheme, for it is they who should determine the kind of information they would wish to be given on the labels and the manner in which it might be given so as to be easily comprehensible and useful to a majority of the buying public. At times it becomes necessary to conduct a survey on such matters to determine the views of a representative set of consumers.

16.10 Thus, first of all special standards for informative labeling have to be drawn up by committees or otherwise, ensuring the concurrence of all the interests concerned. In preparing such standards the first concern is to determine the particular characteristics which are to be covered by the label. In deciding this, only those characteristics are considered which can be measured or objectively determined. Subjective elements such as appearances, styles, colour shades, patterns, taste, workmanship, etc., are not covered. Then, methods of tests have to be delineated to measure the characteristics decided upon to be included. In this respect the national standards come in very handy. But it is not always possible to find a suitable method of test among those covered by the national standards, because it often happens that national standards for many consumer goods are not always available. These methods are often referred to as SMMP – standard methods for measuring performance.[15]

16.11 As a general rule, no attempt is made while preparing these standards to specify any particular level of quality or performance, which is expected to be left to the judgement of the buyer, based on the characteristics data given on the label. But in certain cases, in Sweden for example, a minimum value of characteristics measured is expected to be achieved for a product to qualify for an informative label. This, however, is quite different from the quality and performance level specified in an ordinary national standard which is not necessarily minimum, but one that has to be achieved for a product to claim compliance with a given standard on the basis of satisfactory service. Ordinarily, such levels are those which are considered generally acceptable to the majority of users and consumers on the one hand and economically feasible for the manufacturers and producers on the other. In informative labeling practice only minimal levels need be adopted for qualifying a product for a label, leaving the exact levels achieved to be given on the labels so that a buyer may be in a position to make his choice by comparison with similar data on competing products similarly labeled. There is, however, one exception where the labeling practice must adopt strict standard levels of performance and that is where safety or health hazards are involved.

16.12 In order, however, to enable the buyer to make such a choice intelligently without precaution or analysis, a great deal of emphasis is placed on the manner in which the data are given and the language in which they are expressed. The manner of transcribing the data has to be uniform and the language easily comprehensible. Great care is, therefore, taken in producing standard designs for labels applicable to each class of goods.[8, 14, 15]

16.13 Having prepared the informative label standard for a given product, the institution responsible for the operation gets busy in selling the label to the manufacturers of the product.[15] Applications for the grant of licenses for the use of labels are processed in much the same manner as those for standards certification marks (*see* Chapter 15),

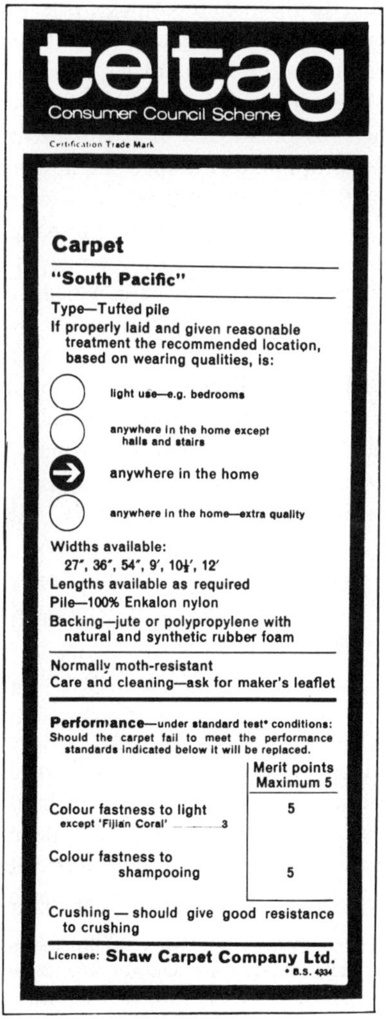

teltag
Consumer Council Scheme
Certification Trade Mark

Carpet

"South Pacific"

Type—Tufted pile
If properly laid and given reasonable treatment the recommended location, based on wearing qualities, is:

light use—e.g. bedrooms
anywhere in the home except halls and stairs
anywhere in the home
anywhere in the home—extra quality

Widths available:
27", 36", 54", 9', 10½', 12'
Lengths available as required
Pile—100% Enkalon nylon
Backing—jute or polypropylene with natural and synthetic rubber foam

Normally moth-resistant
Care and cleaning—ask for maker's leaflet

Performance—under standard test* conditions:
Should the carpet fail to meet the performance standards indicated below it will be replaced.

	Merit points Maximum 5
Colour fastness to light except 'Fijian Coral' 3	5
Colour fastness to shampooing	5

Crushing — should give good resistance to crushing

Licensee: **Shaw Carpet Company Ltd.**
* B.S. 4334

Fig. 16.1 Teltag – the informative label issued by the Consumer Council of UK.

namely, by testing the product, through inspection of the production line, by examination of the associated quality control system which exists or the evolution of a new one that has to be introduced, the grant of license for the use of the label, and the subsequent sampling and testing of components and the product during manufacture and of the labeled product samples purchased from the open market. The main object to be kept in view is to ensure the continued correctness of the information given on the label. In case the information on the label is found not to be in conformity with check test results, the licensee becomes liable to corrective action stipulated in his license, which may extend to its cancellation and breach of contract proceedings. In certain countries, as in the case of the Consumer Council in the UK,* the labels are printed by the licensing authority itself, while in others, as in Sweden, it is left to the licensee to print and use his own labels according to the specified design. But in the latter case it is necessary for the licensing authority to keep a careful check on the contents of the labels to ensure that neither more nor less information is included on the label than is required by the relevant standard design. Of course, extra information can always be included by the manufacturer outside the area of the formal label, presumably by concurrence of the licensing authority.

16.14 Usually the scheme is operated under a license controlled by a contract entered into between the licensing authority, which is usually a national informative labeling institute, and the manufacturer or distributor or even importer. In these cases the operation is subject to the contract obligations under the laws governing such contracts. But, in the UK, the Consumer Council has gone a step further and registered its "TELTAG" labels as certification trade marks. Teltags, therefore, are additionally subject to the protection of the Trade Marks Act of the country and come under the general supervision of the Board of Trade.

CONCLUSION

16.15 As indicated earlier, informative labeling schemes, to be successful and useful, require the existence of a body of well-informed consumers who are in a position to exercise correct judgement on the basis of the technical and other information given on competing labels. While in advanced countries such labeling has made some headway, the practice of certification marking appears to be much further advanced. In developing

* Recent information indicates that the Consumer Council of UK has been dissolved and its Teltag scheme may become gradually phased out, unless some other consumers' organization takes it over.

countries, on the other hand, the value of informative labels appears to be quite limited and so far as is known, they have hardly been introduced anywhere. In fact, the first requirement in these countries is to bring into being consumers organizations which can undertake the study of consumer problems and consumer education before thinking of introducing informative labels. In the meantime, certification marking should be adhered to as far as possible to assist the consumer in the newly emerging countries, where a standardization movement is beginning to be organized.

REFERENCES TO CHAPTER 16

1. *Read the label on food, drugs, devices, cosmetics.* Miscellaneous Publication No 3. US Department of Health, Education and Welfare – Food and Drug Administration, Washington DC, 1953, p. 35.
2. Coles, Jessie V.: *Standards and labels for consumers' goods.* Ronald Press, New York, 1949, p. 556.
3. Wilhelms, Fred T. and Heimerl, Ramon P.: *Consumer economics.* 2nd ed, McGraw-Hill, New York, 1959, p. 534.
4. Mellan, Ibert and Mellan, Eleanor: *Encyclopedia of chemical labeling.* Chemical Publishing Co, New York, 1961, p. 111.
5. *ISO/R 780–1968 Pictorial markings for handling of goods (general symbols).* International Organization for Standardization, Geneva, p. 15.
6. *ISO/R 884–1968 Pictorial marking of transit packages containing photographic materials sensitive to radiant energy.* International Organization for Standardization, Geneva, p. 3.
7. *Informative labeling in Sweden – coordinated system introduced by national labeling institute.* Swedish Institute of Informative Labeling, Stockholm, p. 24.
8. *Whose Choice?* Swedish Institute of Informative Labeling, Stockholm, 1963, p. 13.
9. *Consumers on the march* – 3rd biennial conference of the International Organization of Consumers Unions, Oslo. International Organization of Consumers Unions, The Hague, 1964, p. 143 (*see* pp. 20-39).
10. The role of consumer organizations in institutes of informative labeling. *International Consumer*, 1966, no 4/5, pp. 78-80.
11. *ISO/R 436–1965 Informative labeling.* International Organization for Standardization, Geneva, p. 3.
12. Forbrukerradet, Eoj: Informative labeling – what information and which authority? *International Consumer,* 1964, no 29, pp. 16-18.
13. Bergne, Villiers: British plan for informative labels. *International Consumer,* 1964, no 29, pp. 12-15.
14. Informative labeling – new scheme tried by Consumer Council, UK. *International Consumer,* 1966, no 1, pp. 10-11.

15. Hanworth, Lord: The British "TELTAG". *International Consumer,* 1969, no 3, pp. 4-5.
16. Carpet labels. *Which?* Sep 1970, pp. 264-268.
17. International Labeling Centre has two offices:
    (a) Secretariat, Bezuidenhoutsweg 60, The Hague, Netherlands.
    (b) Technical Section, Taptogaten 6, S-11528, Stockholm, Sweden.
18. *Progress Reports* (issued annually). VDN Swedish Institute for Informative Labeling. Tegnérgaten 11, Stockholm.
19. Lee, Stewart Munro: Consumer information. *International Consumer,* 1965, no 2, pp. 14-16.
20. Publications of the National Labeling Committee of USA. 910 Seventh Street, NW, Washington DC, 20006.

Chapter 17

Pre-shipment Inspection

17.0 Even before the voluntary standardization movement had gathered worldwide momentum, legislations in several countries had begun to be promulgated, with the object of introducing standard practices in certain important fields of their respective economies, particularly in the spheres of health and safety. Examples may be cited of the *Boilers Explosion Acts of 1882 and 1890* in the UK, the *Food and Drugs Act of 1906* in the USA, and many more in these and several other countries. Another type of standards legislation is of more recent origin; this had for its object the promotion and protection of export trade in certain commodities in which the country concerned happened to have a special interest. Thus India covered its coffee in 1942, tea in 1953[1] and textiles in 1963,[2] Malaysia its rubber in 1949 and timber in 1968,[3] New Zealand its dairy products in 1938 and meat in 1940, and so on. The list could be extended to a large number of commodities exported from several countries in different parts of the world.[2]

17.1 The existence of this legislation proved greatly helpful in regulating the trade in such commodities and checking malpractices which had crept in to hamper progress and in certain cases had even brought the exporting countries a bad name. But independent legislation to deal separately with each and every commodity could hardly be considered a satisfactory method of dealing with the general problem of regulating the export trade as such, because a wide range of products is usually involved, a large number of which may be of vital interest to the export trade. This fact was perhaps first realized in post-war Japan soon after the reconstruction activity had started. Having lost its pre-war markets, it had become essential for Japan to rebuild extensive export markets in overseas countries where political influence no longer existed. In pre-war days cheap Japanese goods had flooded the world markets and were generally reputed for their lack of quality. Such goods, it was now recognized, could hardly stand open competition. Post-war Japan, therefore, decided to capture the sophisticated markets of the world with high quality goods. But this required a considerable amount of discipline on the part of the manufacturers, and proper guidance in the form of standards and quality control techniques. Some sort of legislation was also considered essential to enforce the requisite discipline.

17.2 It was thus that in 1948 Japan enacted its comprehensive export inspection law which was perhaps the first of its kind in the world.[4] It dealt with export trade in general, and its provisions were made applicable to any export commodity as may be designated by the government for being brought under control. This legislation was later revised in 1959[5,6] to be brought up-to-date in the light of the experience of the previous decade. India and Iran followed suit in 1963. A cabinet decree authorized the Institute of Standards and Industrial Research of Iran[7] to control the quality of export of a large number of designated products. In India, a special *ad hoc* committee of the government under the author's chairmanship was appointed in 1960 to study the situation in detail and make suitable recommendations in regard to legislative and administrative measures required and to suggest the various commodities which might be brought under control as a first step. The Committee submitted a detailed report[1] early in 1961 and a comprehensive legislation based on its recommendations[8] was enacted in 1963, which came into force early in 1964.

THE NEED

17.3 All international trade largely depends on good faith and mutual confidence of the trading partners. But, as in all human affairs, failures do take place and complaints do sometimes arise about short weight or short measure, unsatisfactory quality, delayed payment or delivery, damaged or unsatisfactory packaging, or for some other reason. To handle such complaints no world court exists, nor is one practicable due to the distances involved, the differences in jurisdictions concerned and the disproportionate expense that would have to be incurred by the litigants. But, during the recent decades, the International Chamber of Commerce (ICC) has been abe to set up machinery for arbitration and conciliation[10] between the aggrieved parties in accordance with certain rules.[11] In addition to their having been evolved through a consensus of views among the member nations, these rules are actually implemented by the direct involvement and participation of the National Committees of the ICC.[10] The existence of these rules and the associated machinery for arbitration is undoubtedly a great help to the encouragement of smooth flow of international trade, but it must be admitted that they can be brought into action only after the occasion for complaint has arisen. Any redress obtained by these means necessarily takes time and involves some expense. But the greatest drawback of the system is that the trade nevertheless suffers because doubts about the genuineness of the trading partners are generated which may persist even after the complaint has been redressed. In order, therefore, to build a clear image of the trade of a country, it is desirable to take all such measures as would help prevent any occasion arising for reconciliation or arbitration proceedings. One such measure happens to be the legislative enforcement of pre-shipment inspection which can take care not only of the quality and standard of exported goods but also of their proper packaging. In addition, it can provide for checking and guaranteeing the weight and quantity of the goods shipped.

17.4 Apart from one developed country, Japan, and two developing countries, India and Iran, there appears to be no other country where such comprehensive legislation for quality control and pre-shipment inspection of exports appears to have been enacted. In many other countries, as mentioned earlier, there exists specialized legislation covering particular commodities of special interest to their respective export trades. The object of all such legislations, whether general or particular, is to provide positive means to prevent any occasion for complaints arising and thus to promote good will among trading partners. This is achieved by implementing standards which have either been previously agreed to between the trading partners or declared by the exporting country itself as being applicable. In many countries this has been found to be most useful in protecting the existing export trade, promoting it further and expanding it to newer markets.

17.5 For maintaining its balance of payments, every country has to export at least as much as it imports. Import requirements of a country are determined by its economic needs which are dependent on many factors, but the export quantum required to balance the imports has to be achieved by a deliberate and well planned effort. In most, but by no means all, developed countries, this effort need hardly be made in a concerted manner, because of the very nature of their economy which is so oriented and which is so technically advanced that their products find ready markets all over the world. A good proportion of such markets resides in the developing countries which in today's context consitute large-scale buyers of industrial goods, including capital goods, industrial raw and semi-finished materials, machine components and parts, as well as consumer goods.

17.6 In contrast, in the developing countries, the problem of maintaining exports at the requisite level is much more complex. Most of them have acquired independence only recently and are anxious rapidly to convert their colonial pattern of economy to a self-sustaining type. Under colonial regimes, they had been content to export mainly the agricultural and mineral raw materials, in exchange for which they obtained a quantum of consumer goods. Today, to their demand for consumer goods must be added that for expertise and industrial goods. The aggregate of the demand is, therefore, rising so fast that the amount of agricultural and mineral surpluses available for export are hardly adequate to balance the total import bill.

17.7 In trying to meet this situation the developing countries are pushing their programmes for industrialization and for building up the infrastructure that goes with it. In addition increased agricultural productivity has become a must. This means even heavier import needs for industrial equipment and other sophisticated products. In order, therefore, to maintain an adequate measure of trade balance, these countries need to make a special effort in promoting their exports – maintaining the existing markets and creating new ones. This promotional effort may consist of several measures – legislative, administrative and managerial. Among them are many forms of export incentives such as subsidies, foreign exchange vouchers, open import licenses, establishment of special supporting organs such as export promotion councils and training centres, sending of trade delegations abroad, and so on. But the most far-reaching and rewarding among these measures are perhaps the ones which are aimed at building up the reputation and image of a country's products.

17.8 Japan's example referred to earlier, which has been most successful, is being emulated in several developing countries. Its success must be attributed to the permanent and lasting effect it has had on the establishment of healthy practices of widespread and scrupulous quality control in industry at home and on producing a sustained and growing demand abroad. Most developing nations can learn a great deal from what has been accomplished by Japan in establishing a high reputation for her products all over the world. For example, today, if one says that such and such a thing is made in Japan, it is automatically taken as a certificate of its excellence. Under these circumstances, sales promote themselves without much promotional effort, and the demand swells.

17.9 The temporary imbalances of trade experienced occasionally by the UK during the past couple of decades have been remedied largely by the advances made in technology and the otherwise excellence of quality of its products. No legislative measures of the Japanese pattern have been found necessary. Similarly, the present imbalance being experienced by the USA may soon be overcome by a concerted effort at producing quality goods at the price level that Japan is able to do. Of course the question of USA and Japan is quite complicated by the differences that prevail in wage levels and standards of living. But as far as the developing countries are concerned, legal pre-shipment inspection can go a long way to help develop quality control systems in their newly developing enterprises and thus help promote their export trade on a lasting basis. In addition to legal steps, intensive promotion of quality control techniques within the country and wide publicity abroad of the measures taken would pay handsome dividends.

ORGANIZATION AND OPERATION

17.10 For organizing any successful operation of a pre-shipment inspection system, it is most helpful that a National Standards Body exists which can be depended upon to perform several essential functions, such as:

(1) preparation and promotion of standards to guide production within the country and those required to satisfy the consumer and user demand in overseas markets;

(2) to promote and establish quality control procedures among the industries;

(3) to institute national certification mark schemes which, besides serving the local needs, could be used as basis for export worthiness of goods in general; and

(4) to assist in several other ways; for example, by creating an atmosphere of standards consciousness in the country, imparting training in testing and inspection methods, quality control techniques and standardization at company level.

17.11 In the absence of a National Standards Body, some of these functions could well be performed by the pre-shipment inspection authority, but then it would amount to the authority itself duplicating the role of the standards body. On the other hand, it is desirable that the two functions be separately entrusted to two different organizations, for the simple reason that while a standards institution has to work in close co-operation with producers and consumers to prepare and promote standards largely on a voluntary and always on a collective basis, the pre-shipment inspection organization would chiefly be concerned with what may be termed a policing function. It would be required to screen the exports in such a manner that only those goods are permitted to enter overseas trade which would satisfy the customer abroad. This may in certain cases conflict with the interest of the exporter who may also be a producer. In fact, both in India and in Japan, it is the government function to control and supervise export inspection, while the national standards bodies in these countries enjoy a separate independent existence. In other countries, where specialized commodity-wise legislation prevails, there too some government department or the other looks after this work. In Iran, on the other hand, it is the NSB which carries the responsibility of pre-shipment inspection. But here the NSB itself happens to be a governmental organ. The combination of the two functions has so far been operating quite well. It is largely a matter of socio-economic background whether such a combination would be successful in any given country.

17.12 As regards the nature and origin of standards to be used for export inspection, practices vary in different countries. In respect of specialized legislations for specific commodities, it is almost invariably the rule that the government or the inspection agency concerned would prepare and enforce the standards, or it would have the option to adopt a national standard. But where generalized legislations prevail, national standards may play quite an important part. In Iran, for example, all the standards used for export inspection are those issued by the Institute of Standards and Industrial Research (ISIRI), the NSB of Iran. In Japan, on the other hand, most standards employed for the purpose are specially issued by the pre-shipment inspection authority itself, but what part the national standards play in this process is somewhat uncertain. In India, the practice is to examine the national standards to determine their suitability before they are adopted for export inspection. If found unsuitable they are either modified or new ones issued to take their place.

17.13 The main consideration that prevails in the selection of standards is that they should be adequate to serve the purpose of the importing countries without in any way damaging the reputation of the exporting country in respect of quality or adequacy to meet certain requirements. A national standard is often devised with an eye to the future and aims at improving the existing quality levels. On the other hand, export inspection authorities are sometimes satisfied with a standard that would meet the demand of the day, without taking any chances of a possible effect that a stiff standard may have on the level of this demand. In certain cases, like the optical and electronic equipment exported from Japan, the export quality is specified at a level equal to, if not higher than, that for goods intended for the home market. In Iran, the export inspection operation has so far dealt with dried fruit and a few other agricultural products and the national standards have been prepared largely with the export point of view in mind. In India while a large number of national Indian Standards have been found suitable and adopted, a fairly large number has had to be amended in respect of certain requirements to suit the export needs.

17.14 The direct use of national standards for export inspection has the added advantage that certification marking machinery of the NSB concerned can be pressed into service and duly certified goods allowed to be exported without further inspection. This not only saves inspection time and permits prompt shipments to be effected but also gives a much higher degree of assurance of compliance with standards than would be possible if inspection were to be carried out at the point of export or at a stage after the goods had been manufactured. It is an integral part of any certification marking operation (*see* Chapter 15) that the control of quality is initiated at the raw material level and continued through all the processing stages up to the point of packaging and delivery. No export inspection authority could afford to go into such extensive details of quality control and yet be adequately efficient to act promptly enough without in any way delaying shipments. In this connection it is also important to bear in mind that most goods are manufactured simultaneously for the home as well as the overseas markets in the same factories, and that promptness of inspection and despatch of shipments without delay are the basic requirements of all effective export promotion measures.

17.15 In certain countries there exist agencies other than the NSBs, which operate certification schemes of national importance. These too must receive due consideration for recognition as being adequate for export worthiness on similar footings in the NSB schemes.

17.16 Irrespective of the choice of the standard for export inspection, whether it is a national one or an official document, specially issued for the purpose, it often happens that a purchaser abroad may wish to impose his own and quite different specifications. To take care of such cases, it is desirable that export legislation and rules made thereunder should be flexible enough to permit exemptions to be made from the recognized standards and allow production and inspection of goods to be carried out in accordance with the specifications of the particular purchaser as may be contained in a particular export contract.

17.17 One word of caution in regard to accepting the purchaser's own specifications for export inspection may, however, be sounded here. It must be recognized that very few importers abroad are users of the goods they import. Most of the goods are sold through trade channels to users and consumers and it is with this latter class of people that the exporting country has to establish its reputation. If for some reason the importer abroad, who is just a middle man, happens to specify his own quality which is appreciably below the recognized level of the relevant standard of the exporting country, it becomes necessary for the export inspection authority to weigh the pros and cons carefully before agreeing to such a proposal. Against the desirability of accepting such an export order and adding to the export statistics, must be balanced the interest of the ultimate consumer abroad whose goodwill is so essential for the maintenance of export trade and its further promotion. As a rule it is suggested that the export of low quality goods against a specific demand of a particular purchaser should generally be discouraged as far as possible. Such a procedure would be most advisable especially when the purchaser does not happen to be the direct user or consumer of the goods concerned.

17.18 On the other hand, there can be no question about accepting a special request for higher quality goods than those specified in a normally applicable standard. Such a demand would help stimulate improvement of quality of indigenous production. A similar situation would exist in the case of goods in respect of which the importing country may insist on satisfying its own official or legal requirements of standards. Several western countries, for example, prohibit the import of meat and meat products unless the animals have gone through a specified ante-mortem inspection before slaughter. Similar strict provisions apply to the purity of foodstuffs, efficacy of drugs, fire hazards of explosives, safety of electrical appliances and other items affecting health and safety. In all such cases the legislation of the exporting country must provide for accepting the stipulations of the importing country. The best way to deal with these problems would perhaps be to find a common denominator of the requirements of the customer countries in general and incorporate it in the export inspection standard. But this may not always be possible, in which case while a majority of cases might be covered by an export standard, the exceptional few will have to be dealt with individually by accepting the regulations of the importing countries.

17.19 There is another category of goods for which it may be difficult or inadvisable to formulate standard specifications, such as textile fabrics or wearing apparel, where fashions, styles and consumer preferences constantly undergo quick changes. It would be useful in such cases to introduce factual inspection and marking of the goods with factual data before export so that the overseas importer is made fully aware of the nature of the consignment. Normally this procedure may not be required to be made mandatory, but where the pattern of trade is such that there is a possibility of collusion between the importer and exporter at the expense of the ultimate consumer, it may be found useful to require compulsory labeling of goods with factual data. At the same time it is important to discourage factual inspection and labeling where standard specifications are available or feasible and normal pre-shipment inspection and certification procedures are applicable.

LEGISLATIVE PROVISIONS

17.20 The provisions of law designed to promote export trade through quality control and pre-shipment inspection would have to be made to suit not only the legal system of the country concerned but also the general pattern of its trade and business practices. The institutional background of the country would also play its part. Being as it is a comprehensive operation, it would be most desirable, in designing the legislation, to provide for the utilization of all the existing institutional facilities and services, such as those concerned with standardization, testing, inspection and certification. In addition, the control of quality should extend to ensuring the soundness of packaging, both for retailing and for transporting, and also the correctness of weight and/or measurement of goods contained in a consignment. As regards ensuring the correctness of weight and measurements, certain trade agencies and chambers in several countries provide facilities to look after this aspect. It would simplify matters, however, if this point were also covered in the comprehensive legislation for pre-shipment inspection. The actual task of ensuring correctness of weight, etc., could be delegated under legislative provisions to the existing trade agencies, if any. Otherwise, it would be entrusted to the same export inspection agency as may be in charge of quality control operations.

17.21 Keeping these and other important points in mind, the legislation should be framed to provide for certain basic requirements which would include:

(1) the creation of an agency to undertake the operation, to which certain powers might be delegated by the legislation, while other powers may flow from higher authorities such as the cabinet or the minister-in-charge;

(2) power to "declare" certain commodities for being controlled under pre-shipment inspection, which could be exported only after due process of inspection and/or quality control;

(3) power to establish, promulgate, recognize or otherwise adopt certain standards to which the "declared" commodities should conform;

(4) power to recognize the existing institutions, private and public, and to create new ones, for the purpose of carrying out inspection and quality control work and to lay down terms and conditions for their oganization and operation;

(5) power to supervise the administration of a private agency or institution which may be so recognized;

(6) power to recognize existing certification marks or to establish new ones, which when applied to a "declared" commodity would not require any further inspection to establish its export-worthiness;

(7) directions to customs authorities to prevent the shipment of consignments of "declared" commodities which have not been appropriately inspected and passed for export or suitably certified for the purpose; and

(8) penalties, including appropriate levels of fines and imprisonment, for offences committed in contravention of the provisions of the law, which should be severe enough to discourage wilful infringement, yet not too severe to discourage legitimate export activity.

7.22 In addition, it is desirable that a high level council may be created under the law to advise government on various matters connected with export inspection policies and operations. Such a council should include representatives both from public and private sector enterprises and from the institutions which are concerned with standardization, inspection, testing, certification and export promotion activities in general. Such councils exist both in India and Japan under the title of Export Inspection Councils.[6, 8, 9] They have proved most useful not only in promoting the export inspection work but also in projecting the correct image of this operation both at home and abroad. A great deal can be learnt from a close study of the working of these two councils and the laws governing them.

REDRESS OF COMPLAINTS

17.23 In any system of inspection, no matter how perfect, there is always a chance of an individual item or even an occasional batch of goods passing the inspection when it should not have been passed. Statistical considerations and economical implications entering the design of an inspection system have been briefly referred to earlier (*see* Chapter 15, paras 15.11–15.12) and will again be discussed in Chapter 23. Suffice it to say here that in view of the small probability of a defective item or batch reaching the hands of a buyer abroad, the pre-shipment inspection system should provide a means of redressing the complaint. Under the Iranian practice such a provision exists, which should be of great interest to any new country that may be contemplating the adoption of the system of pre-shipment inspection. When a complaint arises about a particular consignment shipped after due inspection and certification, the Iranian inspection authority, ISIRI, undertakes a detailed enquiry which may include on-the-spot inspection of the consignment in question. If after such an enquiry, it is established that the complaint is justified and that the importer abroad has indeed suffered a financial loss, the latter is duly compensated for by ISIRI.

17.24 A similar but slightly different type of safeguard is provided for in the Indian pre-shipment inspection system, whenever the ISI certification mark for a declared product is recognized as a certificate of export worthiness under the pre-shipment inspection procedure. This automatically protects the buyer against an occasional substandard item passing the inspection inadvertently. As mentioned in Chapter 15 (para 15.12), if ISI is satisfied about the correctness of the complaint, the certification mark licensee becomes liable to compensate the buyer or to replace the article at his own cost. This protection, however, does not extend to all types of goods shipped under the Indian pre-shipment inspection system, but only to those carrying the ISI mark. In all other cases, the appeal lies to the Export Inspection Council, who would undertake due investigations and give a decision in each case. Similar procedure presumably would apply to the Japanese system of export inspection.

17.25 In any case it is important that the buyers abroad be made aware of these matters – not only about the existence of a rare chance of an article inadvertently escaping detection during inspection but also about the possibility of their being able to have their grievances redressed in a relatively simple manner by bringing it to the attention of the inspection authority. No protracted or expensive arbitration proceedings would be involved in such cases, because the inspection authorities should be more concerned with the preservation of the good name of the country and not so much with the legal niceties that carry important weight in all arbitrations.

REFERENCES TO CHAPTER 17

1. *Report of the ad hoc committee on quality control and pre-shipment inspection.* Ministry of Commerce and Industry, Government of India, New Delhi, 1961, p. 100 (*see* pp. 30-31).
2. *Quality control and pre-shipment inspection for export in ECAFE countries – a status paper.* Economic Commission for Asia and the Far East, Bangkok, 1970, p. 64 (*see* p. 11).
3. See pp. 45-48 of reference 2 above.
4. See pp. 38-39 of reference 1 above.
5. See pp. 39-43 of reference 2 above.
6. *Inspection system of Japan's exports and the governing law.* Japan Export Trade Promotion Agency, Tokyo, 1959, p. 70.
7. See pp. 36-38 of reference 1 above.
8. *The Export (Quality Control and Inspection) Act, 1963.* Government of India Press, New Delhi, p. 9.
9. See pp. 11-35 of reference 2 above.
10. *Guide to ICC arbitration.* International Chamber of Commerce, Paris, p. 40 (*see also* Practical Hints on International Arbitration).
11. *Rules for conciliation and arbitration – in force on 1 June 1955.* International Chamber of Commerce, Paris, p. 20.
12. Modawal, C. N.: Export promotion, quality control and pre-shipment inspection. *11th Indian Standards Convention Souvenir.* Northern India Chamber of Commerce and Industry, Chandigarh, 1967, pp. 34-38.
13. Quality control and pre-shipment inspection – Marked progress achieved. *Journal of Industry and Trade.* Directorate of Commercial Publicity, Ministry of Commerce, New Delhi, March 1967, vol 17, pp. 262-268.
14. *EIC News Letter* (monthly journal issued by the Export Inspection Council of India, Calcutta), since 1967.
15. *Compulsory quality control and pre-shipment inspection in India.* Directorate of Commercial Publicity, Government of India, New Delhi, p. 36.
16. *India – quality control and pre-shipment inspection.* Directorate of Commercial Publicity, Government of India, New Delhi, p. 36.
17. *Checks on quality of products* (Report of the Committee on Market Research of the International Chamber of Commerce, Paris), 1964, p. 40.
18. *A scheme for inspection of textiles meant for export.* Cotton Textile Fund Committee, Bombay, 1962.
19. *AGMARK in India's life and trade.* Agmark Packers and Fruit Products Manufacturers Association, Delhi, 1969, p. 116.
20. Saraf, D. N.: Quality control for export. *The Seminar no 62 – The Consumer.* New Delhi, Oct 1964, pp. 26-29.

Chapter 18

The Consumer and Standards

18.0 As already stated in earlier chapters, consumer interest is always sought to be safeguarded in the preparation of every standard. In order, however, to protect this interest to the full satisfaction of different classes of consumers, it is important that the standards finally established should reflect all their varying needs. It is obvious that the needs of consumers would differ from class to class. For example, there are consumers of industrial goods who buy materials and components in bulk quantities for direct use in their plants. At the other extreme, there is the individual consumer, a housewife, for instance. She buys in small quantities to meet her daily or weekly needs for direct consumption of her family and family guests, as against the industrial consumer who buys goods and services with the object of producing other goods and services for further sale for profit. In between these two extremes there are many other classes of consumers. These include office establishments, hotels and catering houses, small-scale industries, artisans and craftsmen engaged in cottage-scale enterprises and so on. Most of these classes of consumers may readily be included among the industrial consumers, as distinct from the common consumer.

18.1 Obviously, the needs and outlooks of the various classes of consumers vary according to the nature of their enterprises and vocations. It must also be understood that the word "consumer" includes user as well. An article may be consumed either as food or drink which brings its usefulness to an end, or it may be used as raw material for making or finishing other goods. In the former case, the article is finally or ultimately consumed and leads to the production of no saleable goods or service. But in the latter case, a new or better article emerges, which has an economic value, such as coal used for steam raising or cotton for making cloth. This kind of consumption is generally known as "use," but the concept of "use" would also include the use of a piece of equipment or machinery for performing specific operations, which may be in a home, a factory or an office; for example, cooking utensils in the kitchen, a punch press in a factory or a typewriter in an office. Again, a given article may be suitable not only for use by a common consumer but also by one who requires it for further production; for example, lubricating oil for a private motor car engine or for industrial machinery, hand tools for home repairs or for an assembly plant.

18.2 In addition, all classes of consumers are interested in the use of "services," which vary over quite as wide a range as goods, from shoe shining to TV maintenance, from newspaper delivery to electric power supply, from credit card systems to hire-purchase agreements, from tree-trimming (to clear telephone lines) to computer programming, and so on. A cursory reflection would show that services, like goods, may be useful for ultimate and final consumption or for producing further goods or other saleable services. Thus, as far as their utility is concerned, both goods and services represent economic needs of the community which have necessarily to be covered by standards, if their proper functioning and regulation is to be secured in the interest of their users and consumers of whatever class.

18.3 In recent years the word "consumer" has come to acquire a narrower meaning than implied in the above discussion. It has become more and more limited in its scope and now denotes a person who purchases goods and services for his own private use or for that of his family or a group closely associated with him. Furthermore, the new concept of "consumerism,"[1] which has recently been evolved, tries to envelope all the affairs and problems of just such a consumer. The industrial or commercial consumer is no longer considered a part of this new consumer class as it has now come to mean, nor has the former any concern with "consumerism" except in his capacity as a producer or supplier of consumer goods or services, if he happens to be one such. For the purpose of this chapter, it will be in this restricted sense of the consumer that his inter-relationship with standards will be discussed. For a better understanding of the concept and terms discussed in the paragraphs above, a schematic presentation is given in Fig. 18.1.

18.4 It has become necessary to clarify this distinction between the various classes of consumers because of a number of new developments that have taken place during recent years. Every day the common consumer is becoming more and more vocal in asserting himself. He has organized himself in numerous cooperatives for meeting his immediate needs. For collective action he has set up regional, national and international organizations. He is acting on his own for adjudging the quality and worth of goods and services he buys and making the results of his findings widely available to all concerned. He is demanding attention and getting it from standards-making authorities as well as from governments and regulatory bodies. He is demanding and securing legislative protection in most of the advanced countries as also in some of the developing ones. While touching briefly on the various aspects of consumer activity and citing some references from the extensive literature that has come into existence, it will be the concern of this chapter to concentrate attention mainly on the question of standards as they relate to consumer needs and how standardization could assist and further promote the cause of the consumer.

EARLIER DISCUSSION

18.5 In some of the previous chapters, several aspects of standardization have already been discussed, which are directly or indirectly related to the cause of promoting consumer interest. Without repeating the material, it may suffice to cite references to the following:

(1) *Chapter 6 – Individual Level Standards.* Based on national, industrial and company level standards, individual level standards can help the consumer in satisfying some of his needs most economically; for example, in designing and building a house for himself and family.

(2) *Chapter 13 – Units of Measurements.* It is on the basis of standardization of units of measurements that any weights and measures service can be organized, which is so essential for regulating the retail distribution trade, with which the consumer is in daily contact.

(3) *Chapter 14 – Implementation of Standards.* One of the chief concerns of organized and active pursuit of implementation effort is to secure for the consumer the various economic benefits, safety protection, serviceability and reliability advantages that flow from standards.

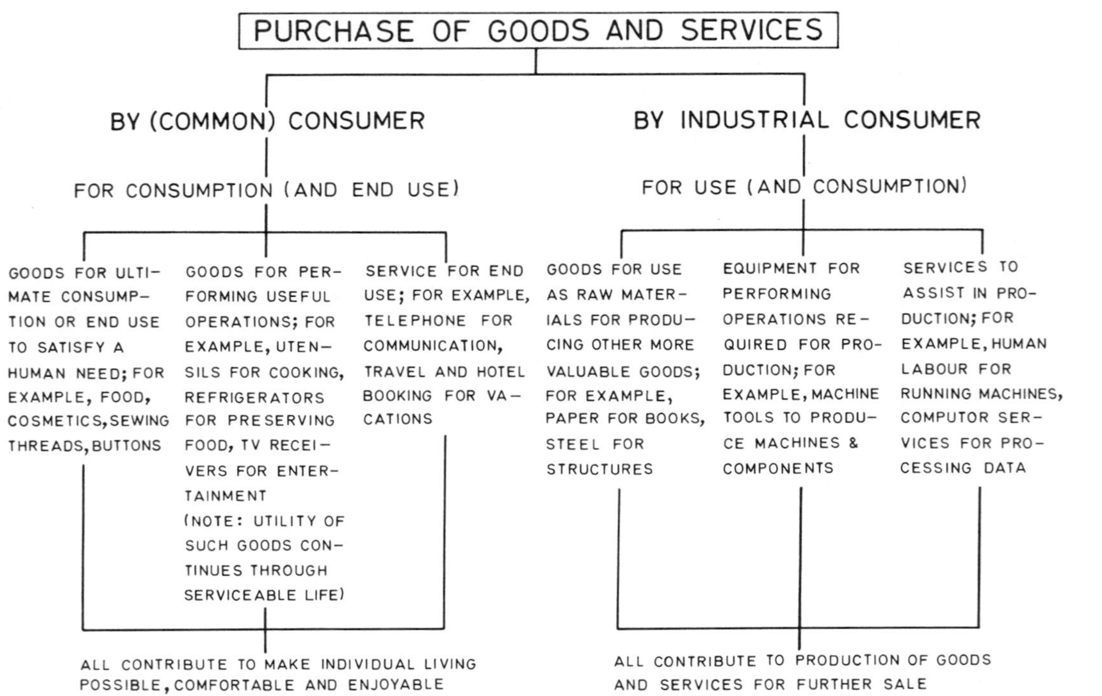

Fig. 18.1 Schematic presentation of the concepts and terminology of consumption and use, as employed in the text:
Note: Although bracketed terms ( ), are included in the concept, it is the unbracketed main concept terms which are used in the discussion.

(4) *Chapter 15 – Certification Marking.* The *raison d'etre* of certification marking schemes based on national or other standards is to provide a third party guarantee of quality of the goods that the consumer purchases, without in any way himself being concerned with having either to incur the extra expenditure for testing the goods or to expend time in acquiring the know-how for the purpose.

(5) *Chapter 16 – Informative Labeling.* Similarly, informative labeling is designed for the specific service of the consumer. This scheme is especially useful where certification marks are either inapplicable or non-existent, either for lack of standards or because of the difficulty in preparing them, or for some other reason.

(6) *Chapter 17 – Pre-shipment Inspection.* Designed primarily to promote export trade, pre-shipment inspection procedures have of necessity to be geared to the protection of consumer interest in overseas countries. In doing so they automatically further the cause of the indigenous consumer as well, because of the commonality of the producers who supply both the markets, at home and abroad.

CONSUMER ORGANIZATIONS

18.6 The earliest form that consumer action took was perhaps to organize retail cooperatives[2] which attempted a direct solution of the problem of supply of reasonably priced goods of acceptable quality directly to consumers, whether members of the group or otherwise. Such societies became so successful, in certain countries at least, such as Sweden,[3] for example, that they began to experience a serious difficulty in procuring all the necessary supplies from normal trade channels. To overcome this difficulty, which was partly an expression of resistance of the established trade, the consumer cooperatives opened up their own production units. And today there is full harmony between the cooperatives and the normal trade channels in most of the Scandinavian countries. Cooperatives have also become extensively popular in the field of agriculture, but here they are concerned chiefly with the collectivized marketing of farm produce on the one hand and distribution of farm inputs on the other. In all these operations, whether of the consumer or farm cooperatives, it is obvious that standards can play an important part at each level of their activity.

18.7 Another quite active form of consumers' organizations that have recently acquired prominence are the so-called consumers' unions or consumers' associations.[4-6] Their main line of activity is centered around comparative testing of consumer goods and reporting to their members and others on the results of such tests. They regularly issue periodical journals on the results and comparative evaluation of the products tested. Some of these journals have become quite popular and acquired widespread circulation.[7-13] These comparative test reports on branded and trade marked products are generally considered to be highly objective, and as such, appreciated both by the consumer and the producer. Some of the producers have indeed actually made use of the results and, on their basis, gone on to improve the products. In fact, there has been no known legal action taken against a consumers' union for finding fault in a manufacturer's product in the test reports. This may, in part, perhaps be ascribed to the general practice of limiting the circulation of the test reports among the members of the union and subscribers of the periodical publishing them. Furthermore, no advertisements are carried in these periodicals, which tends to keep them free of bias in favour of advertisers.

18.8 In conducting these comparative tests, the methods of tests generally used are those devised *ad hoc* by the laboratory personnel for the purpose. This practice leads to non-compatability of tests carried out by different consumer unions. This situation is quite often unavoidable because recognized standards, national or otherwise, are not always available. The products tested and reported upon are generally quite complex and very few national standards have been published to cover them in a comprehensive manner. Besides, the consumers' unions may take into account certain features, such as shape and design of products and convenience of their use, which cannot always be covered in a national standard in an objective manner. Nevertheless, since the resurgence of consumer activity, many NSBs are beginning to give high priority to standards for consumer goods and particularly to standard methods of measuring performance (SMMP) which are really in great demand. By and large, however, active consumers' unions in advanced countries have come to rely more or less on their own devices. To some extent this is perhaps inevitable, but the responsibility of meeting their needs expeditiously lies squarely on the shoulders of the National Standards Bodies.

18.9 With a view to pooling together their resources of know-how and exchanging their experiences, at the same time securing a degree of coordination and uniformity in their operations, the consumers' unions in several countries have organized themselves into an international body – the International Organization of Consumers Unions (IOCU),[4, 14, 50] of which there are more than thirty constituent bodies. IOCU meets biennially in different centres of consumer activity and discusses and plans collective lines of action. It has established close liaison with ISO and

IEC[15] which have agreed to assist it in providing international standards for test methods applicable to consumer products. The bimonthly journal of IOCU – *The International Consumer* – published from its Hague secretariat carries information on consumer activity all over the world.

CONSUMER STANDARDS

18.10 Standards required for the consumer differ in no special manner from those required by any other interest. But for adequately serving the well-developed activity of comparative testing, consumer organizations stand particularly in need of standardized test methods more so than for other types of standards. This is not to say that they can do without specifications type of standards or codes of practice. Particularly, if standards could be evolved to measure performance and reliability accurately and objectively, and their levels could appropriately be defined for each class of service, they would constitute an ideal type of standards for consumer purposes. Such standards would eliminate to an extent even the need for comparative testing, for all testing could then be reduced to testing in terms of the standard methods prescribed and adjudging the quality in terms of the performance and determining the reliability levels. A consumer is after all not so much interested in knowing about the materials or the processes used in manufacturing, or the extent to which the specification requirements have been met or exceeded, as he is interested in knowing the quality of service he may expect from what he buys and how far and how long can he rely on continuing to get this service.

18.11 Perhaps an example will illustrate this point more fully. Take standards on paints for example. They are usually confined to specifying features such as composition, consistency, drying time, covering power, and so on. Some types of paints specifications also include an artificial weathering test. But technology has not yet advanced far enough to be able to devise a test for determining the durability of paint in relation to a given set of environmental conditions, except of course, the long-term actual weathering trial which is indeed not a test at all. What is needed in this case is for a reasonably accurate estimate to be made of paint durability within a reasonable test period which will make compositional specifications redundant.

18.12 General realization of the importance of performance and reliability standards has been comparatively recent. It must also be recognized that there are limitations placed on the preparation of such standards, not the least being the relative inadequacy of technological advancement. Consumer testing, or for that matter most other kinds of testing, has, therefore, necessarily to depend on empirical procedures standardized or adopted after due trial. This at least has been the general practice so far. But wherever possible performance and reliability standards are being introduced. Some authorities feel that in this fast changing world, the task of keeping up with the ever-increasing volume of demand for consumer standards would itself be quite unmanageable. To quote one view: Malcolm Jenson[16] of the National Bureau of Standards "pointed out that the specification for a refrigerator, for example, would need definition of some twenty-seven specific factors in order to assure an adequate product. He felt that the enormous number of variables would make a specification programme virtually unmanageable. Despite the very dark picture painted of the consumer voluntary standards programme, it is evident the critical problem is one of size rather than procedure. Should the public demand specifications for every one of the millions of products on the market, there would be no hope of fulfilling their demands." Under the circumstances the consumer approach to comparative testing of certain class of goods based on *ad hoc* procedures may be the only answer, at least for the time being.

18.13 Nevertheless, it must be appreciated how heavy the responsibility is that has to be carried by the National Standards Bodies of the world and their international counterparts, if the consumer is to be effectively served. Some consumer enthusiasts have been outspoken enough to have hinted that most standards programmes have been oriented to cater to producers' needs,[17] implying that this is so because it is the vested interests who subsidize the standards organizations and support their operations. While this contention may or may not be true in all cases, there is ample evidence to indicate that many NSBs go out of their way to meet the consumer demand for standards as well as the associated services, namely, certification marking, informative labeling and so on. Many NSBs, particularly those in the developing countries, have a stipulation in their constitutions to ensure that consumer representation predominates in their sectional and other committees responsible for the preparation of standards.[18]

18.14 In advanced countries where the common consumer is well enough organized, there can be no serious difficulty in securing adequate consumer participation in the standards-writing work of the NSBs. In such cases consumer representatives are well enough informed, have definite views on the subject and can make useful contributions. While doing so they are acting on behalf of the organized consumer and would be assisting in taking care of the technological as well as the economic aspects of standards from the consumer's viewpoint. But in the developing countries where consumers are not generally so well organized, it is not so simple a matter even to satisfy the basic conditions of the NSB's constitution, such as those of the Indian Standards Institution cited above.[18] It was expected at one stage that under these circumstances representation may be given to the organized trade unions. This did not always prove useful, perhaps because the trade unions are normally geared for action towards their employers and not so much towards protection of workers' interests as consumers, which would require among other things a great deal of technological skill.

18.15 The next best thing that an NSB may be able to do to satisfy the condition for securing predominance of consumer representation might be to rely on government departments and public institutions to look after consumer interest in standards-writing committees. Of course, in certain cases, the interest of the organized industrial consumers may also coincide with that of the common consumer, in which case an amount of relief might be expected. Certain NSBs have gone further and have actually inspired and promoted the organization of consumers' associations,[19, 20] which, if successful, could be of some assistance. But by and large it has been observed that, if such organizational effort does not emanate from the initiative of the consumer class itself and from the compulsions which they feel, no amount of outside prodding can go very far.

18.16 In certain countries, NSBs have organized within their own frame-work consumer advisory groups such as the Women's Advisory Committee (WAC) in the British Standards Institution[21] which with the help of BSI resources attempts to spread consumer education among the housewives. The WAC of BSI also issues periodical consumer reports for the guidance of its constituent organizations through which it attempts to carry out a regular programme of study of British Standards drafts by consumer groups and individuals. The object is to crystalize consumer viewpoint on the contents and scopes of the drafts, which might help shape the British Standards more in line with consumer needs. At one time, BSI also had set up an Advisory Council on Standards for Consumer Goods, briefly known as the Consumer Advisory Council, which was disbanded later when the officially sponsored Consumer Council was created[22] in that country. Other NSBs, like ISI, have also experimented with creating women's advisory committees,[23] with varying amount of success. All these efforts can at best be poor substitutes for a really viable consumer movement having its roots in well-organized consumer societies, who can take independent action and have their voice heard by governments as well as the NSBs.

GOVERNMENT ACTION

18.17 In addition to all that has been done by consumer unions and national standards bodies, there remains a wide field for government action for the protection of the consumer. It must be recognized that every member of a country's population is a consumer, and governments have a duty towards the welfare of their peoples. No matter how well and how efficiently the consumers may have organized themselves, they have necessarily to depend for many things on their governments. It becomes natural, that besides providing police protection and judicial remedies to redress grievances, governments should be concerned with protecting the consumer against dangers from unsafe appliances, adulterated foods, inefficient drugs, fraudulent trade practices, misleading information contained in advertisements, unfair pricing practices and so on. President John F. Kennedy in his message to the US Congress in 1962[24] very clearly brought out the obligations of a government towards the consumer and outlined an extensive programme of administrative and legislative measures, which is worth a close study by all government authorities in the world.

18.18 Among the several measures taken by the US government was the appointment of a National Commission of Product Safety.[25] The Commission made an exhaustive enquiry and submitted many recommendations, which resulted in a spate of legislations.[26] In the United Kingdom also the government has become seriously concerned about consumers' affairs. It appointed a Committee on Consumer Protection in 1959, whose interim report led to the passing of the 1961 Consumer Protection Act of the UK,[27] which was aimed chiefly at the safety aspects of consumer goods and services. The final report of this Committee[21] presented quite and exhaustive study of consumer affairs in the UK and recommended, among other things, the creation of an officially sponsored Consumers' Council. Such a Council was indeed brought into being and did some very useful work including the introduction of TELTAG label (*see* Chapter 16). The Council was ultimately disbanded in 1971, ostensibly due to the reason that other consumer organizations had developed far enough to be able to undertake its functions. In other countries like Canada, France, Germany, Japan and Sweden similar actions have been taken.[49]

18.19 Whatever the case may be for or against an officially sponsored consumers' council or a similar organ in a given country, it must be conceded that the consumer, as President Kennedy said in his message,[24] has a

- right to safety,
- right to be correctly informed,
- right to free choice and
- right to be heard.

In order to safeguard these fundamental rights of the consumer, a variety of laws are necessary[28, 29, 30] and their scrupulous enforcement most essential. In the process of providing legal prescriptions, what role standards can play, has already been discussed in Chapter 14 (*see* paras 14.9 to 14.26).

DEVELOPING COUNTRIES' PROBLEMS

18.20 In most of the advanced countries, whatever has been described above is being done to the extent dependent on the consumer pressure and government's own willingness and capacity. But in developing countries, a great deal still remains to be achieved. Up to a point, the problems of the developing countries are the same or similar to those of the advanced countries, namely those of organization and education.[31] As far as standards are concerned, the situation in various countries ranges from an adequate supply to a serious inadequacy and even to an entire lack of them. Technological skills available for any programme of comparative testing and reporting are sadly limited. What is even more serious is the level of standards of living of people in some of the countries and the limited demand for consumer goods, which is mostly centred around unmanufactured products of agricultural and forest origin. The Research Institute for Consumer Affairs (RICA) of the UK has recently published quite a dismal picture[32] of consumer needs of some of the developing countries, which seems to indicate that, before any consumer protection measures are taken, many other things must receive priority. First of all, standards of living of the people must be improved through a planned programme of economic development. Side by side, standardization and quality control movements must be initiated and all other steps taken to educate and inform the public about a better way of life.

18.21 But all developing countries are not in such early stages of development as the RICA report describes. There are many others, as discussed in previous chapters, having their own progressive standards institutions and even well-developed certification marks schemes. In several countries such as those, time is quite ripe to initiate consumer movement to advantage. This has indeed been done, for example, in India and Malaysia. The Selangor Consumer Association of Kuala Lumpur in Malaysia[13] has already quite an impressive record of work to its credit. And most of this work is proceeding side by side with the emergence of a strong National Standards Body. The main prerequisite for organizing a healthy consumer movement anywhere would appear to be the availability of a band of dedicated workers who can devote their time and energy freely to the cause which is the cause of every citizen. This applies equally to developed countries, but only more so to the developing countries. The example of the emergence of one-man consumer organization in Ralph Nader[33] is an outstanding example of what a single dedicated man can achieve even in as complex a society as that of the USA. The Malaysian example is another typical one from among the developing countries, where just a handful of devoted workers are carrying most of the burden.

18.22 There is of course the problem of unearthing such dedicated people and getting them involved in the consumer movement. In developing countries there are so many social, educational, health and other problems to attract the attention of dedicated people, that the consumer's cause gets easily overlooked. It is essential, therefore, that having got a small core of selfless workers, every attempt be made to enlist the active support of a large number of consumers and build a broad base for the movement to sustain itself. In all this, governments can play an important part with both financial as well as legislative and administrative assistance.

REFERENCES TO CHAPTER 18

1. Gulati, G. L.: Consumerism and standardization. *Business Times,* Madras, Jan 1970, pp. 9-10.
2. Coops as consumer organizations. *Circular letter.* International Organization of Consumers' Unions, The Hague, Dec 1969, pp. 2-7.
3. International Cooperative Alliance. *Cooperation in European market economies.* Asia Publishing House, Bombay, 1967, p. 185 (*see also* reference 44).
4. Consumer's representation in Europe. *International Consumer,* The Hague, July-Aug 1965, no 4, pp. 3-9.
5. Lee, Stewart Munro: Consumer information. *International Consumer,* The Hague, Mar-Apr 1965, no 2, pp. 14-16.
6. The demands of the Soviet consumer. *International Consumer,* The Hague, no 5-6, 1965, p. 13.
7. *Which?* (monthly). Consumers' Associaion, London.

8. *Consumers on the march* – Proceedings of 3rd Biennial Conference of International Organization of Consumers' Unions held at Oslo. IOCU, The Hague, 1964, p. 143 (*see* pp. 64-74).
9. *Shoppers' guide.* Issued by the British Standards Institution during 1957 to 1963. Now closed. (*See also BSI News,* July 1963, p. 9.)
10. *Consumer Reports* (monthly). Consumers' Union, Mount Vernon, NY.
11. *Consumer Bulletin* (monthly). Consumers' Research, Washington, NJ.
12. *Canadian Consumer* (bimonthly). Consumers' Association of Canada, Ottawa.
13. *Consumer News* (monthly). Selangor Consumers' Association, Kuala Lumpur (Malaysia).
14. Bhagwan, Hari: The problem. *Seminar*, New Delhi, 1964, no 62, pp. 10-13.
15. Labeling and standardization. *International Consumer,* The Hague, 1966, nos. 4 & 5, pp. 76-85.
16. Officers' conference held at ASTM headquarters. *Material Research and Standards,* 1970, vol 10, no 12, p. 28.
17. Boggis, F. D.: International standardization. *International Consumer,* The Hague, 1966, nos. 4 & 5, pp. 80-82.
18. *Indian Standards Institution Constitution.* ISI, New Delhi, 1966, p. 21 (*see* Article 26, p. 17).
19. The consumer – a symposium on the measures necessary to safeguard consumer interest. *Seminar,* New Delhi, Oct 1964, no 62, pp. 9-43 (includes a bibliography of 100 references).
20. *All-India Seminar on consumer problems: report.* Consumers' Association of India, Delhi, 1961, pp. vi + 78.
21. *Final Report of the Committee on Consumer Protection.* Her Majesty's Stationery Office, London, 1962, p. 331 (*see* clauses 265-266, p. 84).
22. See pp. 288-289 of reference 21 above.
23. Women's Advisory Committee. *ISI Bulletin,* 1959, vol 11, p. 214.
24. Kennedy, John F.: Presidential Message of 15 March 1962. *Congressional Quarterly Almanac,* 1962, vol 18, pp. 890-893.
25. *Final Report of the National Commission on Product Safety.* US Government Printing Press, Washington, June 1970, pp. 167 + 32.
26. Strong product safety bills make Senate debut. *ANSI Reporter,* New York, June 18, 1971, vol 5, no 13, p. 1.
27. See p. 71 of reference 21 above.
28. Cole, Harvey R. and Diamond, Aubrey L.: *The consumer and the law.* Co-operative Union, Loughborough, 1960, p. 132.
29. *The law for consumers.* Consumers' Association, London, 1962, p. 120.
30. *A consumers guide to USDA Services.* US Department of Agriculture, Washington, 1966, p. 49.
31. Seminar on consumer education. *International Consumer,* The Hague, 1966, no 3, pp. 1-24.
32. *New nations – problems for consumers.* Research Institute for Consumer Affairs, London, 1964, p. 47.
33. The US's toughest customer. *Time,* Dec 12, 1969, pp. 41-47.
34. Haglund, Elsa: *Note on consumer education – an important aspect of home economics.* Food and Agriculture Organization, Rome, 1959, pp. 22 + iv (includes a bibliography).
35. *Annual reports of the Advisory Council on Standards for Consumer Goods.* British Standards Institution, London, 1955 to 1962.
36. *Outline for discussion groups on consumer protection.* Consumer Advisory Council, London, Jan 1962, p. 19 (includes a bibliography).
37. Standardization in the service of consumer. *ISO Bulletin,* June 1970, p. 3.

38. Consumer questions (Report of a meeting of ISO/TC 73 Committee on Consumer Questions). *BSI News*, Jan 1971, p. 15.
39. Wilhelms, Fred T. and Heimerl, Ramon P.: *Consumer economics.* McGraw-Hill, New York, 1960, pp. x + 534.
40. Britt, Steuart Henderson: *The Spenders – where and why your money goes.* McGraw-Hill, New York, pp. xiii + 293 (includes a bibliography).
41. Millar, Robert: *The affluent sheep,* Longmans, London, 1963, p. 203.
42. Standards for the American home. *Magazine of Standards,* New York 1970, vol 41, no 2, pp. 54-57.
43. Lundberg, John: *In our own hands.* Kooperativa forbundet (The Cooperative Union and Wholesale Society), Stockholm, 1969, p. 56.
44. Radetzki, Marian: *Economics of consumer cooperatives.* International Co-operative Alliance, New Delhi, 1965, p. 27.
45. Blank, G., Bykov, A. and Gukasyan, B.: *The consumer cooperative in the USSR,* Centrosoyus, Moscow, 1966, p. 72.
46. Hesselbach, Walter: *Cooperative enterprises in West Germany – The contribution of the trade unions to a consumer-oriented economic policy.* Europäische Verlagsanstalt, Frankfurt am Main, 1966, p. 130.
47. *British cooperatives – a consumers' movement?* Research Institute for Consumer Affairs, London, 1964, p. 38.
48. Smith, Augustus H.: *Economics for our times.* Fourth edition. Webster Division, McGraw-Hill, New York, 1966, p. 628.
49. See pp. 107-111 of reference 25 above.
50. Warne, Colston E.: Protecting the consumer. *Britannica – book of the year 1971,* pp. 502-504.

Chapter 19

Problems of Developing Countries

19.0 The term "developing country" would appear at first sight to be a misnomer. All countries, in fact, are developing – some at a faster rate than others. Indeed, it is the so-called developed countries that are developing at quite a rapid rate. Some of them, like Germany, Japan, USA and USSR, are developing at a tremendous pace, much faster than many others in the developed class.[1] That means that the developing countries in the modern sense of the word are those that stand most in need of development or perhaps those whose economies are in early stages of development. At one time this group was called "underdeveloped," which was perhaps nearer the truth. But certain objections were raised to this nomenclature, which led the United Nations to adopt the new term and 77 of its member nations voluntarily chose to be classed among what are now known as the "developing countries." Since this group was formed, other developing countries have joined the UN and then there are several countries still outside the UN family of nations, which could also be classed as developing. On the whole the group may well exceed one hundred.

19.1 Both the existing stage of development as well as the pace of development among this group of countries vary quite widely and so do the nature and character of their problems. Even their sizes show terrific disparities – from a million or less of population to several hundred millions. In respect of the stage of development, some of them, like India, South Korea, Malaysia and Singapore, may be taken to be quite well advanced in the industrial sphere, while others, like Bhutan, Afghanistan and Nepal, are still largely dependent on agriculture, forestry and handicrafts. Some among them, like Saudi Arabia and other near Eastern countries, may have predominant interest in one resource only, namely petroleum, while still others, like Iran and Venezeula, are attempting to develop all-round balanced economies. Obviously, no generalization of their standardization problems could possibly be attempted, except that the range of variation may be noted and a general method of approach for carrying out a detailed individual study of each case may be indicated. To a large extent the requirements and peculiarities of developing countries have already been discussed in earlier chapters in connection with the specific subject matter dealt with in each chapter. Here the attempt will chiefly be to present a brief résumé, giving appropriate references.

COMMON PROBLEMS

19.2 The common problems which beset most of the developing countries to a varying extent are those relating to the lack of industrial and management know-how, shortage of trained and experienced manpower, inadequate finances and sometimes due to the smallness of their size, the limited range of essential raw material resources. Another serious hurdle they have to overcome in early stages of development is the lack of ready availability of industrial raw and semi-finished materials, parts and components; for example rolled, flat, cast and forged metal products – both ferrous and non-ferrous – screwed and other fasteners, wires, conductors and cables, relays, motors and control gear, fertilizers, pesticides and other agricultural inputs, and many products and services which in an advanced country are taken for granted to be available around the corner or within the range of a telephone call. The important thing in this context is not so much the lack of ready availability of such goods, which could be remedied to an extent by free imports, as the basic need of any self-sustaining economy for these goods to be in conformity with a uniform set of standards. Well-established and regular sources of supply of standardized goods are most essential to ensure interchangeability and fitness for purpose, and to avoid disparities, delays and wastages. Pace of development in many developing countries is retarded due mainly to these factors.

19.3 *Know-how* Lack of know-how is often made up by establishing collaborative enterprises or simply by purchasing plans and blueprints from overseas for establishing new enterprises. Alternatively the establishment of new plants in a new country may directly be undertaken by overseas interests. All these devices bring in their wake the problem of multiplicity of standards and practices imported from a number of different countries.[2] This situation creates a form of chaos which not only complicates supply and procurement operations but also raises considerable hurdles in the way of creating uniform national and industry standards. The problem could assume serious proportions, because, on the one hand, all the developing countries would wish to force their pace of development by buying or borrowing the know-how from outside, and on the other hand, for very good economic and political reasons, such countries cannot always be choosers of the countries from which to buy or borrow. In each particular case, they are inclined to accept assistance from any source which may be considered most economical and otherwise acceptable. The question of alignment of standards is hardly raised, if ever, in such negotiations. It must, however, be pointed out that for some countries, where there exists a modicum of standards movement, it should be quite possible for the government to stipulate that overseas interests in establishing their enterprises within the country would be required to follow the national standards wherever available and, wherever they were not available, they would assist and cooperate in the task of establishing such standards. Even in fairly advanced developing countries, this stipulation, though quite simple and straightforward, is often overlooked in the general rush for finalizing the collaboration agreements and establishing immediate production.

19.4 Know-how is also exchanged through free flow of literature between the National Standards Bodies from one to the other, and between them and the international organizations like ISO and IEC. This includes not only the free exchange of published standards but also draft standards during the course of formulation. Sometimes it so happens that a developing country may circulate to several NSBs abroad an enquiry on a moot technical point which may not have been clarified either in any published national standard or other literature. Experience has shown that sister bodies have often gone out of their way to furnish the necessary data and clarification. Standards literature has also been made freely available to countries which may not have in operation a National Standards Body but where it might be in the planning stage. Thus ANSI, BSI, ISI and many other bodies have presented complete sets of their standards entirely free of charge to several such countries. Usually, such presentation is either made at government-to-government level or by the NSB concerned through diplomatic channels. This helps to build up a standards library in new countries which is the basic requirement for the success of any newly emerging standards programme.

19.5 *Manpower and Technical Assistance* Shortage of trained and experienced manpower is a problem which can only be resolved with time – through education and training. The subject of training has been treated separately in a subsequent chapter (Chapter 25). But for immediate relief, certain other means are available by which this shortage can be alleviated to some extent. Through bilateral and international assistance, services of technically qualified and experienced personnel can often be procured for undertaking the actual tasks of planning, organizing and operating the standards institutions in newly emerging countries. This overseas assistance personnel can also undertake the training of the seconds in command who in due course would have to take charge of the actual operations of running the standards institutions. A very large number of developing countries has taken advantage of such international arrangements, among which the most prominent one is the Special Fund of the United Nations.[3] In the field of standardization, the United Nations agency that carries the executive responsibility for the UN Special Fund projects and for other UN assistance schemes is the United Nations Industrial Development Organization (UNIDO) with headquarters in Vienna. Many regional and bilateral agencies also exist, such as the USAID and the Colombo Plan, but it is hardly pertinent to deal with them all in any detail in this chapter. Governments of all developing countries are quite familiar with the various plans under which they may receive technical assistance and the procedure required to be followed for the purpose.[4]

19.6 A few words may not, however, be out of place in regard to taking the best advantage of the services of technical assistance personnel. First of all, it must be recognized that the advisers, experts and consultants come from different backgrounds of experience derived from differing social and economic environments. In a new country, it is necessary, therefore, to allow them an adequate period for acclimatization, to get adjusted to the new environments and to pick up adequate knowledge of the new background in which they have been called upon to function. Secondly, they should be given every assistance to perform their duties effectively and efficiently, without avoidable loss of time and undue expenditure of effort, such as requisite subordinate manpower and suitable equipment. Thirdly, to take the best advantage of their presence in the country, which has necessarily to be for a limited period of time, it is most advisable to attach to them, at a very early stage, one or more understudies or counterparts, to learn as much as possible of the methodology and techniques essential for carrying forward the work initiated with their assistance. While the practice of attaching counterparts to overseas specialists is not at all uncommon, it is rarely that adequate thought is given to the choice of suitable personnel for this purpose. Apart from their suitability for the job and knowledgeability of the relevant subject, it must also be borne in mind that, after the specialist withdraws, the counterpart would be entrusted with the full responsibility and given the necessary authority to carry the job forward from the point the specialist leaves it.

19.7 It may also not be out of place here to add a word of advice to the overseas personnel. As indicated earlier, such advisers, consultants and experts usually come from relatively well advanced countries, where the traditions that prevail and environment in which one works are usually quite different from those in the relatively newly developing countries. In the latter, the educational accomplishments of the peoples and their standards of living would be rather low. Their pace of life may be more leisurely and they may not be quite used to the idea of hustle and bustle of an industrially advanced society. In spite of the pressures generated by a forced pace of planned industrial development, the people may prefer to take life in their stride. Under such circumstances the overseas personnel will be well advised to exercise a great deal of patience and try to take the people along with him. This may have the effect of reducing his own pace of work somewhat and of setting back his original ideas of quick accomplishment. But, if he is to attain any results at all, he must realize that he cannot do it all by himself – he has necessarily to take his associates and the people of the country with him.

19.8 There may be occasions when he may be up against an indigenous variety of red tape, with which he is quite unfamiliar. Whatever its shade or length, the red tape of a developing country, it must be admitted, cannot be cut offhand any more readily than that of a developed country. One has patiently to discover its ramifications and learn how to get around it. It might even happen that one may feel that one has succeeded in one's endeavours by getting one's recommendations generally accepted, only to discover that no action follows. Of course, an outsider need hardly concern himself with the whole of the responsibility for accomplishing all the goals of the institution to which he is attached. He need not, therefore, feel unduly frustrated when things do not go all his own way. On the other hand, he should use such occasions for reflection and self-analysis to find if he himself has not gone wrong. In short, there may be many hurdles and pitfalls to overcome, which may be considered as a challenge to his acumen and professional ability.

19.9 In addition to the assignment of personnel through international and bilateral arrangements for long or medium terms, there exists a constant flow of visits of officials of one NSB to another. The developing country personnel can and do take full advantage of such opportunities for learning from relatively more advanced countries. Even for countries equally advanced, such visits hold the potential advantage of knowing what is going on elsewhere that may be new or useful to them. The occasions of international meetings where standards personnel gather together from many countries can be particularly productive and instructive in this regard. The developing country delegates have wonderful opportunities here of not only learning by direct exchange of views with delegates from other countries but also of arranging visits to several NSBs and associated institutions abroad, without much extra effort or expense.

19.10 *Finance* As far as the common problem of availability of funds is concerned, there is little that can be put forward in the form of general guidelines, except that each country must resolve these problems in its own particular context. Given forceful leadership, forward looking policy and pragmatic approach to the standards movement, financial problems can always be resolved by patient application and soundly based programmes which would satisfy the clientele that the standards being prepared and implemented are in their own direct interest.[5, 6] Moreover, in developing countries it is the governments which must share the greatest portion of the financial burden. All this requires careful handling of every situation in which tactful persuasion and quiet education of top level management in governments, in industry and in trade have to be carried out without any one being conscious of being pressurized. As some one said, the motto should be: "Men must be taught as though you taught them not."

19.11 *Raw Materials* In respect of raw materials, again, patience coupled with assiduous research and development are highly essential. The objectives of such research and development would include, among others, the following tasks:

(1) To determine the adequacy of the indigenously available materials and explore ways and means of removing their drawbacks, if any, or finding other methods to meet the situation.

(2) To explore new sources of materials through planned geological surveys and to develop them for extensive exploitation.

(3) To investigate the possibility of substituting a material that may be in short supply with one that may be more readily available.

(4) To explore the possibilities of substituting imported materials with indigenous materials, without sacrificing the quality and performance of the product.

(5) If in any of these cases, quality and/or performance has to be somewhat moderated, to determine the extent to which this may be considered tolerable, without in any way affecting economy or the pace of development.

19.12 Having found the solutions to such problems, the important and more difficult part of the task begins. The newly developed ideas have now to be sold to those who are immediately concerned with and responsible for putting them into effect. They have to be convinced of the efficacy of the solutions and the substitutes so found and be imparted the necessary know-how for making the change in their production programmes.

19.13 A few examples of tackling the materials problem may be cited from Indian experience, which may be of interest to other countries. One of the earliest hurdles that had to be overcome was in respect of the phosphorus content of Indian structural steels, which according to the British specification then prevailing had to be limited to 0.060 percent. However, the nature of available raw materials in the country, such as iron ore and coking coal, made it difficult to maintain this limit and at the same time expand production. No economic method could be found to reduce the phosphorus content of the raw materials before they went into steel production. On further investigations, it was discovered that the 0.060 percent phosphorus limit need only be adhered to in making steel which was to be used for dynamically loaded structures, such as bridges and railway rolling stock parts. For welded structures, the same phosphorus limit together with lower carbon content was found necessary. So far as statically loaded structures were concerned, a relaxed phosphorus limit of 0.065 percent was found to be adequate and later on even 0.070 percent was considered acceptable. It was, therefore, decided, as early as 1950, to divide structural steel in two grades[7] which later were increased to three[8] – one for the dynamically loaded structures, another for welding purposes, and the third for the statically loaded structures and other ordinary applications. It was thus made possible to make the best use of the indigenous resources of raw materials to multiply the production volume of steel, without in any way compromising the technical requirements of fitness for the intended purpose.

19.14 Similarly, in the case of portland cement, it was found difficult to adhere to the 5 percent limit for magnesium oxide content mainly because of the high magnesium content of the Indian limestone deposits. Here again it was fortunately found that the soundness of cement was related more to the size and dispersion of periclase crystals in cement rather than to the total magnesium oxide content as determined by chemical analysis. After prolonged investigations it was found possible to raise the minimum magnesium oxide limit from 5 to 6 percent, with the proviso that a rigid control should be kept on the soundness requirements of the specification. These changes were found helpful in promoting better utilization of the available raw material deposits for the manufacture of cement, without in any way affecting its quality.[9]

19.15 A good example of conserving materials in short supply by other more readily available ones is perhaps that of the promotion of the use of tamarind kernel powder in place of maize starch for the purpose of sizing fabrics produced by the cotton and jute textile industries.[10, 11] This led to conservation of maize for food in a country suffering from food shortage. In an agricultural country, it should generally be possible to find substitute sources of supply of many raw agricultural materials which may either be in short supply or needed for more urgent use in other applications.

19.16 As regards import substitution, a considerable amount of work has been carried out in India in a great many different fields of industry, over the past 10 years or so.[12] Perhaps the best example in this category of work is that of replacing copper as an electrical conductor by aluminium.[13] It may be noted that Indian resources in copper and certain other non-ferrous metals are extremely limited. On the other hand, bauxite deposits are plentiful. Electrical energy being abundantly available from multi-purpose river valley projects, it was possible to establish a number of aluminium producing plants. An intensive programme of utilization of aluminium as raw material for electrical products and codification of practices made it possible to do away with the need for importing copper in large quantities. This has meant an estimated potential saving of foreign exchange of something like 300 million rupees yearly.

SPECIFIC PROBLEMS

19.17 So far we have been discussing the common problems which afflict most developing countries. Another quite serious set of problems would be those that are specific to each developing country, emanating from its own peculiar socio-cultural background and politico-economic set-up. Thus, in certain countries, for example, the political climate is such that voluntary standards would have little relevance, as in Cuba. Then, there are other countries in which mandatory standards would generally be quite inappropriate, as, for example, in Singapore (*see also* Chapter 14, paras 14.1–14.7). Furthermore, in each country, the existing institutions, their objectives, scopes and patterns would have to be taken into account in determining the shape of the standards institution and its modes of working. For instance, in Indonesia, a large number of technological institutions exist which are equipped to undertake industrial testing and research work; these could be utilized to great advantage by their being integrated into the structure of the National Standards Body. In a relatively newer country like Algiers, on the other hand, it may be necessary to think of creating a laboratory wing within the structure of the National Standards Body from the very beginning. Yet another set of circumstances may make both these lines of approach necessary. For example, in India, in view of the volume of work involved, ISI, besides operating its own testing laboratories and establishments for certification marking, farms out considerable amount of testing and research work to national and other laboratories.[14] Similar considerations would prevail in regard to deciding the assignment of the responsibility for controlling weights and measures in a developing country to a National Standards Body or to some other organization, existing or specially created for the purpose (*see* Chapter 4, paras 4.13–4.14 and Chapter 13).

19.18 In view of their differing economic structures, each country has to emphasize certain fields of work in preference to others. In a country like Singapore, for example, which has no natural resources except its very advantageous geographical location, industrial standardization with a good deal of commercial content would have to be given predominance with special emphasis on maritime commerce and small- and medium-scale industries. On the other hand, in a country like Thailand it would be relatively more important to give high priority to agricultural needs of the economy, though industry standards could not be overlooked, for industrialization also is taking place to an extent. Then there are countries like Burma and Malaysia, whose most important economic assets are the minerals and forest resources, for which grading, sampling, analytical and materials-handling standards should receive high priority. It would thus be clear that in each developing country, the pattern of detailed priorities for standardization in different fields would have to be individually determined in the light of economic needs and plans for national development. Reference is also invited to the discussion of priorities as presented briefly in Chapter 4 (*see* paras 4.11-4.16).

19.19 Another set of specific problems of developing nations would be concerned with the evolutions of technical and scientific terminologies in the local language or languages. This question has also been dealt with earlier in Chapter 2, and reference to paragraphs 2.23 to 2.27 is invited. Then there might be other problems of social, cultural or economic nature peculiar to a given country, which must be dealt with individually. It is hardly possible or appropriate to generalize in this respect, except to suggest that each problem after careful delineation should be subjected to close study before a solution is attempted.

NATIONAL STANDARDS BODIES

19.20 The varied character of the existing National Standards Bodies in many of the newly developing countries has already been indicated in Chapter 9 in which reference may be made particularly to the general pattern described in paragraph 9.6. The method of approach to the setting up of new National Standards Bodies in developing countries has also been discussed in that chapter (*see* paras 9.21-9.28). In this regard reference is also invited to the discussion of a desirable location of the headquarters for a standards body given in paragraphs 9.29 and 9.30. The concluding paragraphs of that chapter, namely 9.31 and 9.32, may also be of interest in this context, which deal with standardization in certain countries as a concern of several other bodies besides the National Standards Body.

19.21 In certain particularly small newly developing countries, it has been found uneconomic to organize an independent National Standards Body (*see* Chapter 12, para 12.1) and where multi-national standards bodies have been formed to serve a group of countries having geographical contiguity and economic similarity. In theory, the solution is worthy of commendation to countries where such a need may be felt. But it is strongly suggested that great care be taken in working out the *modus operandi* of such a multinational institution before actually deciding to launch one. Experience in respect of two such bodies, namely ICAITI and SACA, (*see* Chapter 12, para 12.1), that have so far been created, has not been entirely free of problems. It would, therefore, be rewarding to study the details of this experience and try to provide safeguards to avoid the possible pitfalls. The main difficulty seems to arise from a factor which affects most intergovernmental bodies, namely the need for preservation of the sovereignty of each participant. Being sovereign states, each member would look at the working of such an organization from its own individual viewpoint and attempt to have subjects of its own interest given the highest priority. The usual question arises as to what extent each participant would be prepared to surrender its sovereignty for the common cause. This should be carefully delineated in the charter of any multinational standards body.

19.22 Then there may be very small nations, smaller even than those which can afford to create a multinational standards body. Such nations may not even aspire to become industrial powers and may thus not be concerned with the establishment of national standards. In such cases it may be quite adequate for standards of other countries or international standards, wherever available, to be utilized, both for internal as well as external purposes. Even here some sort of authority would be useful to be created, even if it is a one man cell in an appropriate government ministry. This cell would look after the standards needs of various sectors of the national economy and be authoritative enough to be able to declare the adoption of such external standards as may be considered desirable for application to the country. The cell could also perhaps make a positive contribution to promoting external trade by organizing quality control and pre-shipment inspection of the country's exports, which are always of important economic significance to any developing country no matter how small.

GOVERNMENTS' RESPONSIBILITY

19.23 Whether a developing country is small, medium sized or large, whether its development programmes are carried out in accordance with a preconceived plan or otherwise, and whether it would emphasize industrial or agriculture activity or both, the responsibility of its government for promoting standardization goes far beyond that of the government of a developed country. In a country of the latter class the people are mostly educated and well enough informed to look after their own interests. They are also in a position to be able to move their governments to undertake the responsibility to initiate the kind of action which a government alone could successfully undertake. In a developing country, on the other hand, it is usually the government which carries the major responsibility for all the social, educational and economic programmes and for bringing up its people to share in due course a part of this responsibility. But standardization, being an urgent prerequisite for most development programmes, has necessarily to receive the earliest attention of the governments of all developing countries.

19.24 It is in this context that the observations made in Chapter 9, para 9.6(3), may be explained concerning the fact that a majority of the existing National Standards Bodies of the developing countries are either governmental or jointly-run organizations. Governments of all developing countries where NSBs do not exist would, therefore, be well advised to consider the question and decide to take the initiative in legislating for the establishment of National Standards Bodies and to participate actively in running them. Wherever an NSB exists and where some spadework has been accomplished, it would be desirable to consider early introduction of company standards departments within government ventures and thus to set an example for others in the private sector to follow. As described in Chapter 14, governments of developing countries have also to carry a special responsibility for adopting manifold policies in the interest of implementing standards on a wide scale. Among other matters, legislative measures would need to be taken for the purpose of enforcing conformity with standards in such fields as safety of persons and property, purity of foods and drugs, export promotion and other matters of public interest.

STANDARDIZATION POLICY

19.25 In Chapter 10, the procedures and practices for preparing national standards and their underlying guiding principles have been discussed. These apply universally to all countries of the world – whether developed or developing. But while sticking to these principles and procedures, it is hardly necessary for developing countries to start the work on every topic from grass roots, namely starting from research, investigations and surveys to compilation and analysis of data for drafting standards, and so on. Good deal of this preliminary work up to the drafting stage could often be abbreviated or simplified by the adoption of a guiding policy based on certain principles of specific importance to the developing nations in general, which may be enunciated as follows:

(1) A prestigious approach of trying to be original, unique or different from other countries should be avoided at all events. It is not only costly and time-consuming but also likely to work against the best interests of the nation.

(2) Taking the pragmatic view, it must be recognized that the ultimate objective of all national standardization effort should be to work towards international harmonization of standards, which process is considerably facilitated by reducing to a minimum the divergences between national standards, that may either be already in existence or that are likely to be brought into existence by fresh independent action.

(3) In view of the limitations of available resources, particularly of manpower and technical know-how, full advantage should be taken of the

readily available experience gained in other countries, particularly in those in which the stage of development and economic and other conditions happen to be similar to the country in question. Such experience is usually available in the form of national standards of other countries as also to a more effective degree in the international standards.

19.26 If these principles are generally accepted, then the step-by-step approach to the policy to be adopted by NSBs of the developing countries for the preparation of their national standards on each approved topic may be stated briefly as follows:

(1) Examine the ISO and IEC Standards and Recommendations for standardization on the topic, if such exist, and adopt them in their original form, or with such minor modifications or amplifications as may be dictated by local considerations. Such changes sometimes become necessary because it does happen that in certain cases the international documents do not take care of the detailed requirements of the developing countries and often enough they are too brief for their adoption *in toto*. But in making these changes great care should be exercised to safeguard the country's interest in export trade, which would largely be guided by the international documents.

(2) If this solution fails to apply, examine the standards of other countries, particularly of those with which there exist close ties of trade and industry, and whose circumstances and development happen to be of a parallel character; and adopt the standard of one such country, either as such or after adapting it to the peculiar needs of the country in question.

(3) In case neither of these two courses lead to fruitful results, then proceed to prepare national standards by pooling the experience of several countries considered appropriate, as may be reflected in their respective national standards and in other relevant publications of importance both indigenous and foreign. In pursuing this course great care must be taken not to introduce latent contradictions or other factors which might create difficulties in implementations. It would be best, therefore, that in such cases the draft standard be subjected to more than the usual intensive practical tests and trials, before being issued.

(4) Only as a last resort and in exceptional cases, it should become necessary to undertake the preparation of an entirely original standard. This process should be preceded by collecting and analysing all the necessary and relevant data from commercial, industrial and other institutional sources, within and without the country. In many such cases developmental research and investigational work, including surveys, may

have to be conducted to collect the requisite data and to devise appropriate methods of test.

Whichever of the various courses of action is chosen – adoption, adaptation, compilation or original spadework – it should always be requisite that the needs of the country's own industry, trade and consumer be kept in view and the experience of the indigenous producer, user and engineer be fully reflected in the national standards.

19.27 As regards the formal procedure itself for the processing of standards, the NSBs of the developing countries stand to benefit considerably by a close and detailed study of the work carried out in several other countries on the time taken for preparing standards (Chapter 10, paras 10.36 to 10.47). Based on the experience elsewhere of reducing the overall time involved, these NSBs could simplify their own procedures to a great extent. But any such simplification must be done rather carefully, taking due account of the prevailing economic and other conditions in the country, and keeping in view the necessity for securing ready implementation of standards.

19.28 In conclusion, it may be pointed out that the problems of developing countries are not basically different from those of advanced countries. The chief difference lies in their relative urgency and comparative ability to resolve them. The developing nations have a great deal to learn from the experience of industrially advanced nations, yet, if the gap between them happens to be too large, the flow of useful expertise from one to the other is not always quite smooth. It is much better, therefore, that when a developing country is seeking help from abroad, it should rely more on those developing countries, which are relatively more advanced than itself, and among these latter, preferably on those in which somewhat similar conditions prevail in respect of economic resources, climatic factors and social, cultural and linguistic backgrounds.

REFERENCES TO CHAPTER 19

1. Verman, Lal C.: Why does India participate in international work? *Markets, standards and profits* (Proceedings of the Fourteenth National Conference on Standards, Washington, DC). American Standards Association (now ANSI), New York, 1964, pp. 64-66.
2. *Overseas collaboration and standardization in India – 13 papers for Session 7 of the Ninth Indian Standards Convention* (Bangalore). Indian Standards Institution, New Delhi, 1965:

S-7/1 Yogeshwar, R. and Bir R. S.: Conversion – the link between collaboration and standardization.
S-7/2 Rao, K. Sitharama, Majumdar, A. and Mathews, P. E.: Overseas collaboration and the role of Indian Standards Institution (ISI).
S-7/3 Rao, C. S.: Overseas collaboration and national standards in instrument industry.
S-7/4 Bulgrin, H.: Collaboration needs better company standards in India.
S-7/5 Balakrishnan, T. V. and Bhagowalia, B. S.: Influence of overseas collaboration on standardization in India.
S-7/6 Ramajayam, N.: An approach to overseas collaboration and standardization.
S-7/7 Tolpadi, S. G. and Sathyanarayana, K.: Overseas technical collaboration and standardization – a realistic outlook.
S-7/8 Mehta, V. N.: Overseas collaboration and standardization.
S-7/9 Mitra, H. K.: Multiplicity of refractories specifications – a bane on the Indian scene.
S-7/10 Parikh, R. D.: Effects of overseas collaboration and need for standardization of plastic pipes.
S-7/11 Toshniwal, G. R.: Standards and collaboration in industry.
S-7/12 Mukherjee, B. K.: Overseas collaboration *vis-à-vis* standardization.
S-7/13 Krishnamurthy, S.: Impact of overseas collaboration on Indian standards.

3. *Manual on the use of consultants in developing countries* (UN Publication No E.68.II.B.10) United Nations Industrial Development Organization, Vienna; UN, New York, 1968, p. 158.
4. Martin, Edwin M.: *Development assistance – 1969 review.* Organization for Economic Cooperation and Development, Paris, 1969, p. 325.
5. Appeal for ISI building fund. *ISI Bulletin,* 1954, vol 6, pp. 88-89.
6. Annexe to Manak Bhavan – appeal for building fund. *ISI Bulletin,* 1963, vol 15, p. 285.
7. *IS:226–1950 Specification for structural steel.* Indian Standards Institution, New Delhi, p. 8.
8. *IS:226–1958 Specification for structural steel (Second revision).* Indian Standards Institution, New Delhi, p. 8.
9. *IS:269–1967 Specification for ordinary, rapid-hardening and low heat portland cement (second revision).* Indian Standards Institution, New Delhi, p. 11.
10. *IS:189–1956 Specification for tamarind kernel powder for use in the cotton textile industry (revised).* Indian Standards Institution, New Delhi, p. 9.
11. *IS:511–1962 Specification for tamarind kernel powder for use in the jute textile industry.* Indian Standards Institution, New Delhi, p. 8.
12. *Standardization for import substitution – 34 Papers for Session S-1 of the Tenth Indian Standards Convention (Ernakulam).* Indian Standards Institution, New Delhi, 1966–67:

S-1/1 Indian Standards and import substitution.
S-1/2 Riebensahm, Hans E.: Contribution of standardization to the substitution of scarce materials.
S-1/3 Banerji, A. P.: Integrated standards programme for accelerating import substitution.
S-1/4 Krishnamurthy, S.: Role of standardization for import substitution.
S-1/5 Kidao, T. V. N.: Quality assurance for import substitution with particular reference to heavy duty commercial vehicle industry.
S-1/6 Chatterjee, A. K. and Moorthy, S. R.: Role of standardization in indigenization of imported spares for defence vehicles.

S-1/7 Sethy, V. A. S., Handa, M. K. and Rao, K. Nagesha: Import substitution in Hindustan Machine Tools Limited.
S-1/8 Ramarao, C. R.: Standardization for import substitution in the field of water supply fittings and builders' hardware.
S-1/9 Arora, K. L. and Gokhale Y. C.: Suggestions for substitution of imported asbestos in asbestos-cement products.
S-1/10 Singh, Jatindra: Standardization for import substitution with particular reference to river valley projects and agricultural development.
S-1/11 Chari, R. K.: Standardization for import substitution with aluminium.
S-1/12 Subramanian, V. R.: Metal arc welding industry and import substitution.
S-1/13 Chakraborty, R. N., Saxena, K. L. and Chattopadhaya, S. N.: Studies on recovery of zinc from viscose rayon waste.
S-1/14 Rao, A. S. N.: Some aspects of rationalization of boiler steels.
S-1/15 Mukherjea, D. C.: Review of progress made by the mints on import substitution.
S-1/16 Shah, P. S.: ISI's role in import substitution.
S-1/17 Sen Gupta, P. K.: Import substitution of raw materials used in wire and cable industry.
S-1/18 Import substitution in the field of overhead transmission lines.
S-1/19 Bhatt, N. M.: Amendment of existing specifications leading to greater use of aluminium and steel in manufacture of ACSR and AAC.
S-1/20 Ajwani, M. B. and Subramanian, V.: Role of standardization for import substitution.
S-1/21 Rao, C. E. B.: A new series of economic electric motors with aluminium windings.
S-1/22 Venkatasubbu, L. and Krishnaswami, A.: Hot dip aluminizing of transmission line hardware.
S-1/23 Rao, C. Koteswara and Nayudamma, Y.: Indigenous vegetable tanstuffs *vis-à-vis* import substitution.
S-1/24 Lodh, Dalpat R.: Role of talc in import substitution.
S-1/25 Sinha, R. K. and Gupta, N. R.: Progress in the substitution of minerals in the world and their possibilities in India.
S-1/26 Menon, A. G.: Research leading to import substitution in petroleum industry.
S-1/27 Rao, Thirumala S. D. *et al.*: Indigenous substitution for imported lecithin, lanolin and palm oil (tinning oil).
S-1/28 Rao, V. K. and Bhattacharyya, B. N.: Mineral based industries and import substitution.
S-1/29 Nair, S. R. and Thampy, R. T.: Role of research in import substitution with particular reference to plastics and other synthetic high polymers.
S-1/30 Gopalachari, A. S.: Investigations leading to import substitution of some metals and minerals.
S-1/31 Das, B. N.: Role of metallurgical research for import substitution.
S-1/32 Pillai, N. R.: Possible avenues of research leading to import substitution.
S-1/33 Thomas, K. S.: Research leading to import substitution.
S-1/34 Dash, Bhagwan and Bedi, Ramesh: Indigenous drugs for import substitution.

13. See Papers S-1/11, S-1/17, S-1/19 and S-1/21 listed in reference 12 above.
14. *Twenty-five years of ISI.* Indian Standards Institution. 1972, p. 123 (*see* p. 95).
15. *Development plans and programmes* (Studies in development no 1). Development Centre of the Organization for Economic Cooperation and Development, Paris, 1964, p. 219.
16. Sen, S. K.: Experience in standardization efforts in a newly industrialized country (India). *UN Inter-Regional Seminar on Promotion of Industrial Standardization in Developing Countries. Vol 1 Papers presented at Seminar.* Danish Standards Association, Copenhagen, 1966 pp. 195-217.
17. Shourie, H. D.: *UNCTAD-II – a step forward.* Indian Institute of Foreign Trade, New Delhi, 1968, p. 407.
18. Standardization in developing countries (Jamaica and Nigeria). *BSI News,* May 1971, pp. 16-17.
19. *Industrial standardization in developing countries* (UN Publication sales No 65.II.B.2). United Nations, New York, 1964, p. 136.
20. Industrial development and standardization:
Part I Importance of industrial standardization for economic development of developing countries.
Part II National standardization – a survey of present conditions.
Part III International standardization.
Part IV Company standardization.
Part V Implementation of standards.
Standards Engineer, 1969, vol 3, pp. 7-14 and 24-30.

Chapter 20
Planning

20.0 The USSR was perhaps the first country to adopt planning as an instrument of state policy for achieving systematic economic and social development of its people. That was soon after the First World War. Since the Second World War, national planning has become more universally accepted. Practically every socialist country and most developing countries today have come to depend on it for a coordinated approach to advancement. It has been generally recognized that planning helps to avoid waste of time and effort and achieves maximum possible results with the available resources at the disposal of a country. A great deal of literature has accumulated on basic principles and techniques of planning and a good deal of practical experience has been gathered in many countries, which comprise a large number of success stories as also an occasional failure.

20.1 Any national plan must be built up of several component plans each dovetailing into the other. For example, the industrial component of the plan must be designed, on the one hand, to meet the developing needs within the industrial sector itself as well as those of all other sectors of economy and, on the other, it must be assured of adequate inputs from other sectors in the form of investments and operating capital, infrastructure services, skilled and unskilled manpower, and raw materials and other requirements to enable it to meet its own obligations. Similar situation would exist in relation to every other sector of the overall plan – each one must be so designed as to meet the needs of the society as projected and to be provided with all the wherewithals for that purpose. Every sector being concerned with practically every other sector, or at least with several of them, it would be clear how complicated becomes the problem of planning accurately and rationally on nationwide basis. Many mathematical models have been developed to handle the problem and many computers are being pushed into service to find the answers.

NATIONAL PLANNING AND STANDARDIZATION

20.2 As far as standardization is concerned, it should be considered as a part of the infrastructure required for national development.[1] But standardization differs considerably in character from other forms of infrastructure, such as a transport network, or a telecommunication system. Standardization in its broad sense may be said to provide the essential infrastructure required for supporting and regulating every sector of national activity – be it economic, educational, social or scientific.[2] Furthermore, furnishing as it does the very basis for the most economic utilization of resources it can be a most powerful constituent of any well-conceived national plan of development.

20.3 In most socialist countries it has come to be generally recognized that standardization can give maximum return if it is made a constituent part of the socialist economic system;[3] it should, therefore, be taken into account in all economic planning in such detail as may ensure the supply of suitable quality and quantity of raw materials as also those of the finished products.[4] In the five-year plan of the USSR for 1966–70, "questions of standardization occupy an important place in every branch of industry and agriculture. Included in this plan are qualitative characteristics which must be adhered to in the five-year period in concrete terms. . . . In this way, standardization is implemented in strict accordance with a plan."[5] In East Germany, interesting discussions have taken place extending the interrelationship of national plan and standards to include commercial contracts[6, 7] which have a direct bearing on the execution of plans.

20.4 It would thus be obvious that standardization should form an integral part of the national plans wherever they are framed. Furthermore, in order that they be effective, standardization should proceed in parallel with planned development in all sectors. Wherever possible, it would even be better if standardization could remain somewhat ahead of developments. Every plan project requires several years to be processed – from preliminary survey and feasibility study to full-scale production. There is usually ample opportunity for standards to be prepared and adopted before the production stage is reached. But oftentimes it is advantageous and sometimes necessary to have agreed standards available even at the stage of initial planning of a project, so that appropriate processes of production could be adopted, correct machinery and equipment ordered and suitable auxiliaries and stores provided for in the project plan. A good example would be the need for a standard series of voltages for the planning of an electric power project. Early decisions on product standards are also very pertinent, for these would form the basis of all considerations for design details of a project. It is obvious, therefore, that any national plan to bring forth the best possible results should be accompanied by a parallel standardization effort and that the latter should be so organized that standards solutions are made available to planners and those responsible for executing the plan, in good enough time for planning all new ventures – be they industrial, agricultural or otherwise.

PLANNING FOR STANDARDIZATION

20.5 In order to meet this criterion it would appear essential that any country adopting planning as an instrument of development should give high priority to strengthening its standardization movement at all levels, particularly at the national and company levels. Furthermore, to be able to derive maximum benefit from standardization effort, it would be most useful to adopt a regular plan for the evolution of standardization and related activities, which may run parallel to but be somewhat in advance of the national development plan. It is most likely that in certain socialist countries, where national planning has been adopted as a state policy for many decades, standardization movement has also been subjected to formal long-term planning. But little information is available in published literature for review or comparative assessment.

20.6 Normally, every organized body, whether it is a standards institute or otherwise, undertakes planning to some extent. A yearly budget is always prepared well in advance of the year to which it pertains and its financial estimates have necessarily to be based on the activities, new and old, which the institution plans to initiate and to continue during the particular year. The need, the justification and the extent of development of such activities, whether explicitly stated in a separate document or not, constitute the plan for the year and furnish the basis for the budget provisions being made. Sometimes, as in Japan,[8] a national standards institution may prepare a five-year programme for the preparation of standards, which would go somewhat beyond the annual budgetary planning, but which may or may not be related to a national plan. While planning on a yearly basis may be, in fact must be, a feature of every standards institution, it is the long-term planning on lines parallel to national planning, which is under consideration here. Such long-term plans may spread over varying periods of time, depending on the number of years covered by the national plan. A span of 5 years appears to have become generally adopted in a majority of countries. This period has come to be considered as satisfactory, for it is long enough to permit certain projects to assume concrete enough shape for being assessed in the light of the original targets and not too long to entail an undue delay in making a review of the performance of the plan as a whole before starting to prepare for the next one.

PLANNING AT DIFFERENT LEVELS

20.7 These considerations would apply equally to the planning of national standardization activity but as far as company and industrial level standardization is concerned, it would appear that a yearly plan at the budgeting stage would perhaps be preferable. Changes in technology and consumer tastes and market demands take place quite rapidly, and these are bound to affect the company and industry standardization programmes significantly. Any long-term planning would only stand in the way of flexibility that is desirable to be maintained at the industry and company levels. As far as international standardization goes, planning can apply only in a limited sense, because international work is really dependent on and caters to the needs of national standardization. It would be most interesting to formulate a plan for international standards development, if one could get a large enough number of constituent NSBs to frame their own individual long-term plans. Furthermore, such national plans, in order to be useful for international planning, should synchronize with one another and stretch over the same general period of years. This is, of course, hardly feasible in the background of the present stage of development of the concept of "One World." The International Organization for Standardization (ISO) has, nevertheless, tried to plan its activities in advance in certain restricted spheres of its work. For example, it is presently in the process of reorganizing its technical committees in a new divisional structure,[9-11] with a view to securing better coordination and expeditious disposal of work. Very recently, ISO has set up a long-range Planning Committee with a view to planning for meeting worldwide requirements of international standards during the 1980s.[16]

PLANNING FOR NATIONAL STANDARDIZATION

20.8 In advanced countries, where the practice of standardization has come to be generally adopted and national standardization has been well enough organized, there would perhaps be little need or incentive to evolve or adopt a long-term plan for the development of national standards and allied activities. However, to this group also belong those advanced countries where national planning has become a state policy. Among such countries the National Standards Bodies have necessarily to take cognizance of the plan requirements and shape their own programmes and policies to help meet the targets laid down in the national plan.[3-5] For this purpose it may not always be necessary for such NSBs to have an elaborate and formal long-term plan of their own, because their organization would generally be well enough developed to take care of the demands of the national plan as a part of their normal activity, though some advance programming may be called for. The problem, however, assumes a serious proportion among the developing countries which pursue a policy of planned development but whose national standards movement may not have had a chance to strengthen itself adequately for being able to meet the national-plan needs expeditiously. It was to meet just such a situation that the practice developed in India of formulating regularly the successive Five-year Plans for the development of the Indian Standards Institution and its activities. A similar course of action, as far as is known, has not been adopted anywhere else. The Indian experience has, however, yielded valuable results[1,2] and it may be of interest to review it here in brief. It could prove useful to other countries also. To some developing countries particularly, it may even indicate useful guidelines to explore and pursue.

ISI PLANS

20.9 Planning for standardization in India began in 1951, starting with the First Five-Year National Plan of India, when the Indian Standards Institution was still in its infancy. At that time, ISI was in its early stage of development and engaged primarily in working out its procedures and getting its staff together. Today, when it is pursuing its Fourth Five-Year Plan, it can well be said to have come of age. This may be said to be largely the result of successful completion of its three Five-Year Plans, during which it managed almost always to exceed its targets and meet the needs of the national plan well in advance of schedule. For the same reason, ISI had no difficultly in securing the necessary finances for its development and growth. Both the government and the industry were fully convinced that the contributions made to ISI were being utilized to the best of advantage, as had been demonstrated by its performance and accomplishments, in both qualitative as well as quantitative terms.

20.10 The method of approach to the formulation of the First Plan was somewhat different from the succeeding plans. Since at that time there was very little prior experience available, the First Plan was based mainly on broad lines of the National-Plan needs which placed great emphasis on agricultural development at the cost of industrial growth. Industry was expected to find a good proportion of its resources from savings brought about by "modernization of plant and machinery, reorganization of uneconomic units, standardization of production and scientific management."[12] This meant reliance on the best utilization of every available resource – in other words, on an all-out standardization effort. The first ISI Plan, therefore, provided for the creation of all the necessary facilities to make the Institution a viable body and provide the services required for as many branches of industry as possible, giving priority to machine-building industries, building construction, chemicals and textiles.

20.11 While framing the succeeding Plans, it became possible to study and analyse the performance of the previous Plans in quantitative terms. This study revealed, among other things, the trends of development of the demand for standards, the capacity of the Institution to meet this demand, the possible projected demand of the future years, and whether the Institution at its then existing stage of development would be able to meet the expected level of demand or whether its facilities would have to be expanded, and, if so, to what extent. The framework of the new National Plan that was being prepared at about the same time would give its own targets and indicate trends of the developments which would have to be taken due account of in framing the ISI Plan. Integrating all these data together, it was possible to form a rough idea of the extent to which the output of the Institution would need to be expanded so as to be able to meet the probable demand of the country for standards and other associated services.

20.12 But these estimates could not always be adopted as targets for achievement during the forthcoming plan periods, for it was also necessary to take account of the wherewithals that could possibly be marshalled for achieving the targets; for example, the requisite staff, office accommodation and laboratory facilities. The matters were also studied in the light of the past plan performance, and the quantum of financial support received from industry and government was objectively assessed. It was largely, but not wholly, in the light of the quantum of support that could reasonably be expected to be received in the future that the size of the forthcoming plan was determined. The other important consideration was to ensure an organic growth of the Institution without letting itself become an ineffective bloated organism. The size of the plan thus determined was usually much below what could be justified by the volume of rising demand for standards and the associated services. But the plan had to be realistic and, therefore, cut to reasonable size. For example, while framing the Second Plan, it was estimated that a two-fold increase in ISI services would be justified in the light of the increasing demand, but in view of the financial and other limitations mentioned above, the Plan had to be restricted only to a 50 percent expansion. For similar reasons, the Third Plan aimed at only a two-fold expansion, in spite of the indication that a three-fold expansion was needed to meet fully the demands of the country, which at that stage were going up steeply with the increased pace of industrialization. Subsequently, when framing the Fourth Plan, it was observed that, as a result of having always exceeded the previous plan targets, the capacity of the institution had grown to sizable proportions. The expansion volume indicated as necessary at that stage was roughly 80 percent, which could be accepted as such for the plan target. Later, however, it had to be revised downward for other extraneous reasons, when the Fourth Plan was reviewed.[12]

20.13 Such a rapid and sustained rate of expansion of ISI activities, as demanded by the high rate of development of the industry and other sectors of economic life according to the national plans, could not have been achieved but for the adoption of a training scheme in late 1955. With the help of that scheme, ISI could recruit bright young graduates of technical institutions and offer them specialized training in standardization, which would enable them to chalk out a professional career for themselves in this newly emerging discipline. In most developing countries, and India was no exception, the technically trained and well experienced personnel in different disciplines is always in short supply, as it is in constant demand by every section of the newly developing economy, where new enterprises are daily coming into being (*see also* Chapter 19, paras 19.5-19.9.) Under these circumstances, any standards institution would have to face quite a stiff competition from private as well as public sector employers. It was, therefore, thought that newly graduated engineers and technologists who could more readily be attracted would perhaps meet the ISI needs after two years of intensive on-the-job training. This training programme added yet another branch of activity to the already complex set-up of ISI. But for this planned approach to the development of national standardization, which demanded a forced pace of personnel recruitment, the need for the training project of ISI might never have been felt so acutely. (*see also* Chapter 25).

20.14 Having the means for facilitating staff recruitment, it was essential to provide also for adequate accommodation both for laboratories and offices. This led to the proposal for constructing a multistoried structure, well equipped with laboratory facilities. No direct solution could be found for financing this project except to plan for going out with a begging bowl to the government and industry for substantial financial subvention. Both industry as well as the government came handsomely to the aid of ISI[13,14] and each contributed almost equally towards the total cost of about 2.1 million rupees of which about 15 percent was met from savings from the ISI budget itself. The building project envisaged in the First Plan (1951–56) was completed early during the Second Plan period. The accommodation of 6,200 square metres included, besides the committee and conference rooms and office space for committee secretariats, a library, a laboratory, executive and other essential auxiliary services, such as certification marking, implementation, publication, public relations and accounts. A novel feature which contributed its own quota to the success of ISI Plans was a prominent wing housing a standards museum, which has since been serving a very useful function of enabling one to explain graphically to the layman the objectives of standardization and how they are attained.[15]

20.15 Towards the end of the Second Plan period, it became evident that if the Third Plan (1961–66) objectives were to be realized, the building accommodation would have to be more than doubled. This was duly planned for and a second multistoried building of 10,000 square metres area was constructed and occupied towards the close of the Third Plan period. It took care of the overflow from the first building and in addition accommodated the ever-expanding library and a number of new testing laboratories which had become essential for looking after the considerably increased demand for certification marks. The funds for this building (nearly 3.5 million rupees) were contributed mostly by the government, a small fraction coming from rental income of a part of the floor area. The reason why industry was not called upon at this time for funds was because it had already contributed during this period substantial amounts to ISI for running two large international conferences of IEC and ISO. Rates of subscriptions of ISI members had also been raised considerably to meet the financial needs of the Third and Fourth Plans.

MAIN ELEMENTS OF ISI PLANS

20.16 The above bird's-eye view of the ISI Plans gives the general background and an idea of the ways and means adopted to accomplish the objectives. The main elements of the Plans, however, covered all the major activities of ISI, namely:

(1) preparation of standards;
(2) certification marking;
(3) implementation (including company standardization);
(4) institutional organization (committees and councils);
(5) directorate organization (divisions and sections);
(6) publicity and conferences;
(7) international standardization (including regional level activity);
(8) building (offices and laboratories);
(9) staff (including training);
(10) staff welfare and housing; and
(11) finances.

Activity concerning each element was reviewed in the light of the experience gained during the previous Plan period. An analysis of the quantitative data pertaining to this period revealed many factors which could be made good use of in preparing the next Plan. It would hardly be appropriate in this discussion to deal with all of these elements individually and indicate how they were actually analysed and planned for from one Plan period to the next.* It might, however, be interesting to give here a couple of examples concerning the preparation of standards and development of certification marks activity. Both of these relate to the stage at which the Fourth Plan (1966–71) was being prepared, namely in the fiscal year 1965–66.

20.17 *Preparation of Standards* First of all the data concerning the demand for standards over the previous three Plan periods were assembled. This was done in terms of the total number of subjects that had been approved for standardization at different stages. It may be noted that the approval of subjects is the result of investigations and consideration by the Institution of the various requests and proposals for preparing new standards received from different quarters. In addition, the data concerning the total number of standards prepared and issued during each preceding year were collected. The number of revised standards issued every year was also indicated separately. All these data are shown plotted

* If details are required, they may be obtained on request from ISI.

in Fig. 20.1, each curve being extrapolated over the Fourth Plan years in dot-dash lines, which indicate the trends of development that might be expected. In addition, certain dotted line curves are drawn asymptotic to

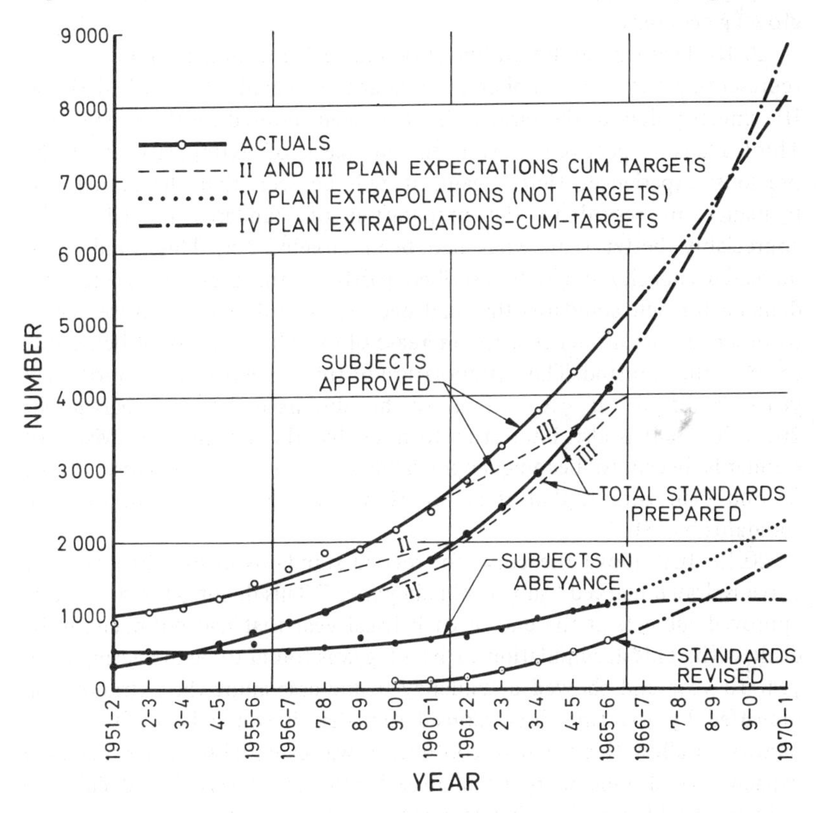

Fig. 20.1 Expectations, targets and actuals of the four five-year plan periods of Indian Standards Institution in respect of subjects approved and standards prepared.

the top two curves. Those marked II and III, radiating from the curve for subjects approved, represent the extrapolations of earlier data pertaining to previous Plan periods, showing the trends which could have been expected for the demand of new subjects during the Second and Third Plans respectively. From the difference between these dotted curves and the full line curve representing the actual performance, it would be apparent that the demand for new standards had increased appreciably more than what might have been expected on the basis of extrapolations of the previous performance. This had occurred during both the Second and Third Plan periods, showing that the value of national standards was being generally appreciated more and more in different sectors of the growing economy.

20.18 Two similar dotted line curves radiating from the solid line curve representing the total number of standards prepared, also marked II and III, similarly denote the targets that had been adopted for the Second and Third Plans respectively, again on the basis of extrapolations of the previous experience. Here, again, it will be seen that the actual performance in regard to the preparation of standards proved to be appreciably better than what had been targeted for. This build-up of increased capacity might be ascribed partly to the pressure of excessive demand for new standards that had been registered and, of course, partly to other factors influencing the increase of overall efficiency of being able to meet this demand. The extrapolation of this curve over the Fourth Plan years is based on calculations to be discussed later in paragraphs 20.21-20.23. It is also interesting to note that the demand for revision of standards began to become appreciable only towards the close of the Second Plan period, that is about 10 years after the issue of the first standards by ISI.

20.19 Just above the curve for revised standards in Fig. 20.1 another curve is drawn marked "subjects in abeyance." This denotes the number of approved subjects at the end of each fiscal year that had not begun to be dealt with. This accumulation of backlog was found to be increasing from year to year and showed a tendency to increase along the extrapolation indicated by the fine dots. Considering that it would be most unsatisfactory to allow this trend to continue, it was essential to create increased capacity for dealing with the increased demand. It was thus decided to hold this backlog of the subjects in abeyance at a constant maximum value of something like 1,200, as indicated by the dot-dash extension of this curve in Fig. 20.1.

20.20 Another important statistic which has been found useful in planning for the preparation of standards is that expressing the average number of standards resulting from one approved subject, averaged over the course of a Five-Year Plan period. This was determined for all the previous plan periods and found to be 2.0 during the First Plan, 1.6 during the Second and 1.3 during the Third. For the purpose of the Fourth Plan this continuously decreasing tendency of this statistic was taken into account and it was assumed that during this period it might be reduced, on the average, to 1.1 standard per subject. The explanation of this reducing trend appears to lie in the observation that as standards consciousness in a new country increases, the interested parties making proposals for new standards become more and more precise in specifying the subjects for which they require standards to be prepared. For example, at earlier stages of the standards movement, it would be quite likely for an NSB to receive a request for standardization of "electrical motors," a very wide field which would take several standards to cover; whereas at a later stage of development, similar requests would become much more specific, for example, it might be for a "squirrel-cage motor."

20.21 Having thus studied the performance during the previous plan periods and ascertained the relevant statistics, the targets for the Fourth Plan could now confidently be calculated on the basis of these studies. It will be seen from Fig. 20.1 that the total number of subjects which could be expected to be approved by the end of the Fourth Plan on the basis of the then existing trends could exceed 8,200. Although there was no guarantee for the ever-increasing trend to cease as observed successively during the past three Plans, yet for the purpose of planning one had no option but to assume the existing trends as the basis. Furthermore, assuming that it would be possible to restrain the backlog of subjects in abeyance from exceeding the 1,200 mark, and noting that the number of subjects approved at the end of the Third Plan period was 4,900, it follows that there would be 8,200 – 4,900 = 3,300 new subjects to be dealt with during the Fourth Plan period. This amounts to roughly 660 subjects on the average to be disposed of every year, which under the assumption that each subject would lead to 1.1 standard on the average would mean 726 new standards to be prepared per year on the average.

20.22 It was also noted that during the last year of the Third Plan the estimated number of new standards to be prepared would be 465 out of a total of 625, the difference representing the revised standards. This meant an increase of 726 – 465 = 261 standards on the average per year to be prepared during the Fourth Plan period. Obviously, this increase could not be brought about immediately during the very first year of the Fourth Plan but had to be spread over the whole period and built up gradually. Thus assuming a linear increase over the five years, the increment of 261 would have to be achieved during the middle or the third year of the Plan. This would lead to an increment of 261 divided by 3 or 87 per year. Adding this to the existing production capacity of 465 during successive years, the number of new standards to be produced per year during the Fourth Plan period comes out to be as follows:

Year	*No. of standards*
1966–67	552
1967–68	639
1968–69	726
1969–70	813
1970–71	900
Total:	3,630

20.23 This aggregate of 3,630 new standards for the whole period of the Fourth Plan added to the number of 3,470 standards expected to be issued by the end of the Third Plan would make an overall total of about 7,100 standards by the end of the Fourth Plan period. For the determination of overall production capacity of the institution, to this workload for preparing new standards must be added the number of standards expected to be revised, which would, however, not add to the aggregate of the total number of published standards. According to the existing trend as indicated by the curves of Fig. 20.1, this number would amount roughly to 250 revised standards to be prepared during the last year of the Fourth Plan. Consequently, an aggregate capacity required to be aimed at by the end of the Fourth Plan for the preparation of new standards plus revised standards would amount to 900 plus 250 or, say, 1,150. This implied a minimum increase of capacity for the preparation of standards in the ratio of 1,150 divided by 625, that is, by a factor of 1.84, or in other words, by 84 percent.

20.24 In addition to the quantitative targets as dealt with in the preceding paragraphs, qualitative features of the standards to be prepared were also included in the plan. These took the form of priorities to be given to the various subjects on lines parallel to the overall national plan for development of the country (*see also* Chapter 4, paras 4.11-4.16). Appropriate precautions were also taken during the execution of the plan to ensure that, in attempting to achieve the quantitative targets, the quality and adequacy of the standards did not suffer in any way. In fact, in a planned approach of this type, quality had always to be safeguarded most jealously and at times even at the risk of missing the quantitative targets.

20.25 *Certification Marks* The certification marks activity of ISI started towards the close of the First Five-Year Plan and began to become effective during the Second Plan period. In Fig. 20.2 are plotted the data concerning the number of applications received and the number of licenses

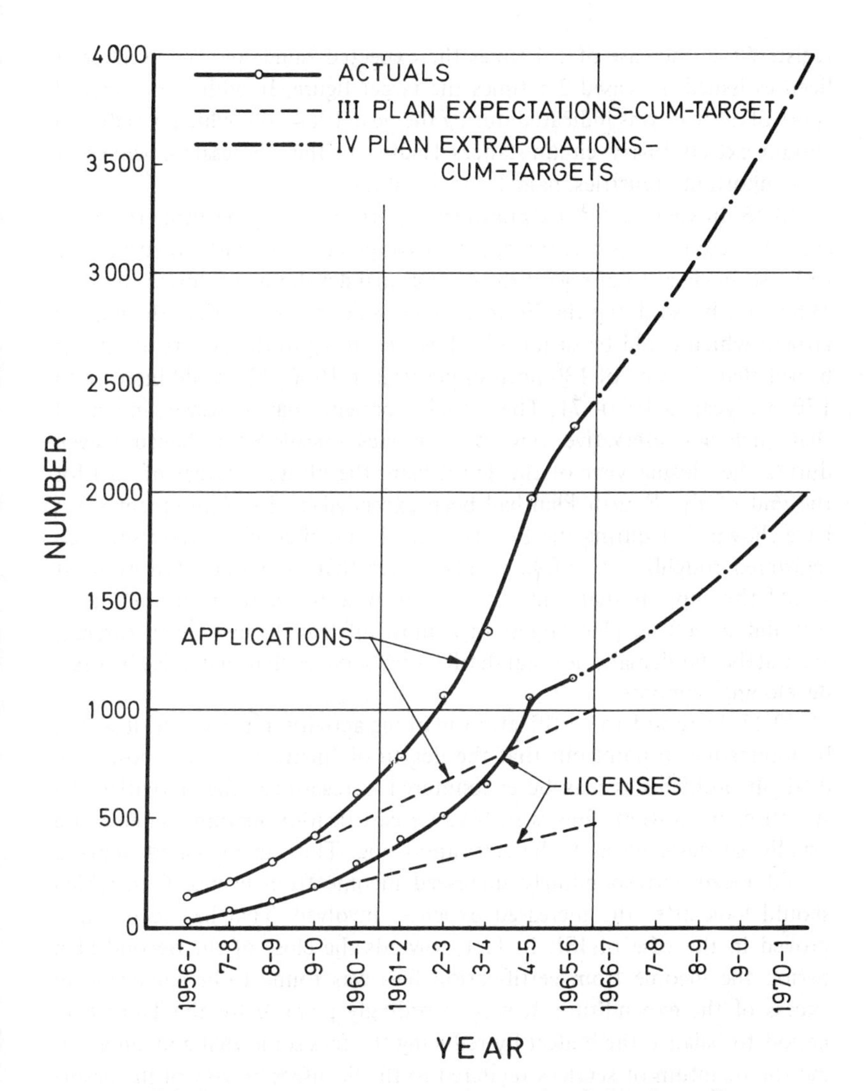

Fig. 20.2 Expectations, targets and actuals of the three five-year plan periods of Indian Standards Institution in respect of applications received and certification marks licenses issued.

issued during the Second and the Third Plan periods, on the basis of which the Fourth Plan was framed. It will be seen that during the Third Plan period, there was a considerable spurt in certification marking activity which went way beyond expectation. The number of applications registered an increase of 2.4 times the expected value, and the number of licenses issued increased 2.5 times the target figure. In addition, a special spurt of activity was indicated during the year 1964–65, which was due to certain extraordinary circumstances leading to the wholesale coverage of two important industries, namely steel and jute.

20.26 In view of this extraordinary spurt of activity resulting from the special circumstances mentioned, the extrapolation of the two curves had to be so made as to disregard the exceptional development during the year 1964–65. By so doing the Fourth Plan was designed to reflect the normal growth which could be sustained. Thus, according to the targets set, it was hoped that the rate of 130 licenses per year in 1966–67 would increase to 170 per year in 1970–71. The actual experience has, however, indicated that such a conservative view was perhaps uncalled for, because even during the closing year of the Third Plan, the ultimate target of 170 for the end of the Fourth Plan had been exceeded by 15. Subsequent years have shown that during the Fourth Plan the number of licenses issued has registered roughly a two-fold increase over that of the targeted value. It would thus appear that while it may be wise to exercise an amount of restraint in setting plan targets, it is most difficult to be able to forecast accurately the demand for standardization services that may arise in a fast developing economy.

20.27 In regard to certification marking activity in particular, it would be interesting to point out that the dearth of finances does not pose such difficult problems as may be encountered in regard to other activities of a standards institution. This is so because certification marking schemes are usually so designed as to be self-supporting. Thus an increased demand would mean correspondingly increased income from license fees, which should look after the increased expenses involved. That has been amply proved in the case of ISI. In fact, towards the close of the Second Plan period the income from certification fees was found to be somewhat in excess of the expenditure. It was accordingly planned for the Third Plan period to balance the budget by reducing the fees somewhat and augmenting the quantum of services rendered to the licensees by way of increasing the frequency of inspection and instituting a more active consultancy service. Both these steps led to a reversal of the financial situation towards the end of the Third Plan period. It thus became necessary to reverse the process somewhat during the Fourth Plan so as to restore the balance. It is hardly necessary to describe how the actual adjustment between income and expenditure was arrived at. But it may be pointed out that it was the planned approach that made it possible to control closely the finances of such an operation as certification marking so that it might be maintained at self-sustaining level.

CONCLUSION

20.28 Without going into further details of all the other elements entering into the Plans, it may safely be concluded that the technique of planning can prove to be a powerful instrument in the hands of an NSB for advancing the cause of national standardization movement in all its ramifications. Particularly in countries where national planning has come to be adopted as a state policy, planning in the field of standardization in one form or another has to be considered as most essential. In the developing countries, where national standardization work is in early stages of development, the only way to meet the standards requirements of the national plan effectively would appear to be to adopt the practice of parallel planning in the realm of standardization as well. By so doing the NSB of the country would not only facilitate the achievement of national goals but also be able to attain its own objectives most economically.

20.29 Long-term planning enables an NSB to set its targets in both quantitative and qualitative terms so as to fulfil the requirements of the national plan. This is done on the basis of the quantum of the expected demand for its services on the one hand and its capabilities on the other, as revealed by a close study of its past performance, existing resources and the new resources that could possibly be marshalled in the near future. Resources, of course, include not only the financial but also the manpower, the know-how, the equipment, the accommodation, the organizational and so on. In defining its targets it is advisable for the NSB to be realistic and base its calculations on the actual data as collected, but a degree of boldness in setting these targets is also most essential. While too conservative an approach could cripple growth, an over-optimistic outlook might lead to disappointment and discouragement. It must also be recognized that there is an optimum size to which an organism can grow safely, before it becomes too unwieldly and no NSB can allow itself to grow in that manner. This tendency can readily be checked, for planning provides ample opportunities for the NSB to review its operations and accomplishments at convenient periodical intervals, when the plan provisions could be revised and readjusted upward or downward, depending on the indications of the review. Solid organic growth can thus always be ensured.[16]

20.30 It is a pity that details of experience of planning national standardization activity in other countries are not readily available for comparative study and analysis. It is hoped that the Indian experience briefly described here might prove of some value.

REFERENCES TO CHAPTER 20

1. Verman, Lal C.: Standardization as infrastructure for development of ECAFE region. Paper for the Asian Conference on Industrialization – Second Session, Tokyo. ECAFE, Bangkok, 1970, p. 25.
2. Verman, Lal C.: *Standardization – a triple point discipline* (Presidential Address, 57th Indian Science Congress, Kharagpur). Indian Science Congress Association, Calcutta, 1970, p. 9. Also printed in adapted form in *ISI Bulletin*, 1970, vol 22, pp. 47-50.
3. Merbach, Horst: Standardization, planning and management in the work of setting up cooperative system in socialism. *Standardisierung,* 1969, vol 15, pp. 102-106 (*in German*).
4. Beirau, E.: Perspective plan and standardization. *Standardisierung,* 1965, vol 11, p. 2. (*in German*).
5. Standardization in the USSR (contributed by GOST). *International Standardization* (issued by ISO), 1967, vol 1, pp. 7-11.
6. Schade, Hans-Joachim: Plans, standards and commercial contracts – relations between commerical contracts and standards (I). *Standardisierung,* 1966, vol 12, pp. 17-19 (*in German*).
7. Schade, Hans-Joachim: The specific tasks of plans, standards and commercial contracts and their relations to one another (II). *Standardisierung,* 1966, vol 12, pp. 66-70 (*in German*).
8. *JIS Yearbook 1971.* Japanese Industrial Standards Committee, Tokyo, p. 181 (*see* p. 11, para 3.1).
9. International standards need of the day – ISO Council gears up machinery – four technical divisions constituted. *ISI Bulletin,* 1969, vol 21, pp. 473-474.
10. Technical divisions start work. *ISO Bulletin,* vol 2, no 3, p. 3.
11. ISO technical division 2. *ISO Bulletin,* vol 2, no 4, p. 3.
12. *Twenty-five years of ISI.* Indian Standards Institution, New Delhi, 1972, p. 123 (*see* p. 22).
13. Appeal for ISI building fund. *ISI Bulletin,* 1954, vol 6, pp. 88-89.
14. Annexe to Manak Bhavan – appeal for building fund. *ISI Bulletin,* 1963, vol 15, p. 285.
15. Visvesvaraya, H. C.: Standards museum at Manak Bhavan. *ISI Bulletin,* 1962, vol 14, pp. 13-15.
16. New ISO committee set up for long-range planning. *ISO Bulletin;* Dec. 1971, vol. 2, no. 12, p. 1.

Chapter 21

Economic Effects of Standardization

21.0 By its very definition one of the basic aims of standardization is to achieve optimum overall economy (*see* Chapter 2, paras 2.10 and 2.12). By overall economy is meant the overall economy of the particular group of interests belonging to that level to which a given standard may relate. Thus a company standard is designed for the overall economy of various divisions and departments of the particular company concerned and a national standard for that of the nation as a whole involving every possible national enterprise and endeavour. At the international level of standards the situation becomes a little more complex, but the basic principle remains the same. This means that any international standards should be aimed at the optimum overall economic benefit of all the nations of the world considered as one entity. Whether such an assessment can or cannot be made quantitatively and objectively is besides the point. Nevertheless, the aim of standardization should remain valid in any attempt at preparing standards for any and all levels.

21.1 Economy of any kind is a complex enough affair to assess under any circumstances, but the overall economy of a given standardization effort is even more so. This is mainly due to the many-faceted nature of standards and to the fact that some of the economic effects resulting from the introduction of standards are quite intangible and difficult of quantization. How, for example, is one going to reduce to economic terms the gains accruing from a better all-round understanding of well-defined terms, leading to smoother flow of work; or for that matter, the economic effect of standards on the efficiency of personnel with the psychological atmosphere of general security brought about by the orderliness resulting from standardization? Even the assessment of savings accruing from conservation of natural resources is not quite a simple matter.[1] However, many quantitative studies have been made, particularly in the company standardization field, which show that considerable economic benefits are brought about by standardization. Some of these have been indicated in Chapter 7 and a considerable number of references have been cited[2] (*see* paras 7.10-7.11).

21.2 The chief difficulty in making any quantitative assessment of standardization effort is that an accurate assessment has, first of all, to be made of the economic situation before the introduction of any standards, or let us say, a new or a revised standard, and then an assessment of the new situation in terms of the same variables after the introduction of these standards. These steps are to be followed by an inter-comparison of the two sets of values, so as to determine the economic gains achieved. But while the change from the old to the new practice is being introduced, many other changes may also take place. Furthermore, practices are oftentimes subjected to influences extraneous to standardization, the effects of which are most difficult to eliminate. But such elimination is necessary to enable the effects of standardization alone to be taken into account for the economic assessment.

ECONOMIC GAINS AND LEVELS

21.3 In the restricted field of individual and company level standardization, where one decision-making agency alone is involved, the quantitative assessment of economic gains does not present so many difficulties. In individual level standardization effort the economic gains are relatively simple to calculate, as the variable factors involved are practically absent. Even at the company level, the matter is not so complicated either, as is evidenced by the numerous useful studies referred to earlier.[2] But as one enters the field of higher levels from company to industry and through national to international, one is faced with increasing difficulties in being able to make quantitative estimates which could be considered reliable enough as regards their accuracy, universality or applicability over long periods of time. This is not to say that positive gains do not exist or cannot be demonstrated to an adequate degree of satisfaction of sceptics. It only means that whatever attempts have so far been made are in the nature of early explorations in a most complex realm, and that a great deal of work is still required to be done by way of further research and investigations, both in relation to the development of methodology as well as the assessment of actual economic effects in specific cases.

21.4 In adjudging or assessing economic gains from standardization at a given level, it should be borne in mind that these gains emanate not only from the specific standardization effort made at that particular level, but also from efforts made at all other levels, particularly at those higher than the given level. It is apparent that this must be so, for standardization at every level, while feeding experience and know-how to the next higher levels, is also closely related to and dependent on the decisions made at these higher levels (*see also* Chapter 3). AFNOR (the French National Standards Body) has reported, in an ISO document,[3] the interesting results of a survey of gainful effects of standardization in 84 percent of the enterprises in France among the 669 replying to one of their questionnaires. Among those replying, 14 percent indicated no noticeable effect and 2 percent pointed out difficulties only, while none appeared to have suffered any form of loss. In this document, the most pertinent remark made deserves reproduction:

> Some investigators would have liked the replies to distinguish amongst results ascribable to:
> national standardization;
> international standardization (ISO or IEC);
> plant standardization effected internally in each company.
> In point of fact, however, such a distinction would have been completely arbitrary and of no practical significance. It would scarcely be making a caricature to compare this kind of distinction to separating what, for a given level of employee diligence in a company, can be ascribed to social laws, the work of the ILO, the personnel director, etc.

Thus benefits assessed as a result of standardization effort at a given level in a particular case might only appear to be flowing from that effort. In actual fact they would represent, as pointed out earlier, the collective result of the related work done at all the other associated levels.

POSSIBILITY OF ADVERSE EFFECT

21.5 It may be pertinent to sound a word of caution at this stage. Beneficial effects from standardization can only be expected if the standardization process has been carried out in a rational manner and implemented correctly. Uncalled-for over-standardization, gross under-standardization and wrong application of standards may lead to negative effects on the economy of production and/or utilization. In this connection, an American authority has sounded a precautionary note[4] which may be regarded as over-strong, but would be of interest to reproduce here:

> Standards can be costly when you have blind adherence to a standard; where you are lulled to sleep by an aura of security because something has been standardized; where you adopt standards just for standardization's sake; where you accept standards as a solid permanent solution and do not realize that standards are constantly on trial; where standards breed such product sameness as to lose sales and customer appeal; where wide horizons are not considered in the full cost picture; where standards curtail inventiveness and superior design and overlook customer needs; and where standards are not related to your specific production, engineering, and purchasing problems.

ASSESSMENT AT INDUSTRY LEVEL

21.6 Perhaps the very first economic assessment of the effects of standardization on record was that made by the Federated American Engineering Societies in 1920 in a survey conducted at the initiative of Herbert Hoover, its President[5, 6] (*see also* Chapter 1, para 1.13). The survey, in spite of the negative character of its conclusions, proved extremely valuable at the time for bringing to light the extent of wasteful use of materials and wasteful operations in industry. "Twenty-five percent of the cost of production could be eliminated, the report disclosed, without affecting wages or labour. In six typical industries, wasteful practices accounted for almost 50 percent of materials and labour. If waste was most prevelent in industry, industry had no monopoly on it. Owing as much to long-established custom as to the wake of war, it was to be found throughout the economy." This was the state of affairs not only in the USA but no doubt everywhere, at the stage when importance of standards in eliminating wastage was beginning to be realized and the standardization movement was just becoming organized among the industrially advanced countries of the world.

21.7 During the succeeding decades, though standardization had made quite some headway, no comprehensive industry-wise study seems to have been made either of the wastage resulting from lack of standardization or of the positive economic gains achieved from their application. After the close of the Second World War, however, several industry-wise studies were sponsored by the Foreign Operations Administration of the US Government.[7] Their alleged objective was to provide guides for the rehabilitation of industry in European countries and they concentrated mainly on case studies of individual company operations indicating particular advantages accruing from different steps taken for introducing standardization, simplification and specialization. In a sense, therefore, these studies dealt with economic benefits gained by individual companies through standardization at the company level, within of course the orbit of certain specific industries. Though constituting a very valuable contribution to furthering the cause of standardization they cannot strictly be regarded as studies for assessing the advantages brought about by standardization to a given industry as a whole.

ASSESSMENT AT NATIONAL LEVEL

21.8 Though no exhaustive industry-wise study appears to have been carried out, yet certain attempts at quantitative assessment of benefits of standardization made at national level are of particular interest. In a sense, the Herbert Hoover survey of 1920, referred to in paragraph 21.6 above, could be considered as a national level study. In another sense, the French survey,[3] referred to in paragraph 21.4, was also intended to be a national level survey. But neither of these attempted to bring out explicitly the quantitative estimates of gains accruing to national economy as a result of a general or a specific standardization effort. An interesting quantitative estimate of this type was made for the hides and skins industry of Ethiopia, which at that time had adopted no standards for the grading of these products for export. This particular estimate was intended to serve as an argument in favour of introducing suitable standards. For the purpose of the study, the basis of standardization was assumed to be the East African standards for grading hides and skins in respect of weights, defects, etc. Knowing the total quantity of the commodity exported from Ethiopia, and its approximate distribution pattern among various East African grades, it was possible to assess the value which it could have commanded in the export market. Comparing this value with what the ungraded Ethiopian hides and skins were presently earning, the price differential could readily be calculated, which Ethiopia could have additionally earned in the export market.[8] This amount aggregated to some 3.5 million Ethiopian dollars additional income for the producers and 5.1 million for the exporters, which in terms of US dollars comes to a total of some 3.5 million yearly – quite a sizable sum for a developing country's income from one export item only.

21.9 Perhaps one of the most interesting subject-wise economic studies carried out in the field of national gains, which could accrue from a comprehensive standardization effort was the one made in India. This study had its origin in the exhaustive programme of work undertaken in 1954 by the Indian Standards Institution for preparing a comprehensive set of standards to deal with different phases of production and utilization of structural steel in the country.[9] This so-called steel economy project of ISI covered standardization and rationalization of steels, hot- and cold-rolled sections and tubes for structural use; formulation of codes of practice for the design, construction and maintenance of steel structures in accordance with the latest advances in theory and practice; and the preparation of codes for welding of structures together with specifications for welding equipment and welding accessories. When, after 10 years of concerted effort, an adequate number of standard specifications and codes had been prepared and some of them actually applied in practice, it was felt that an economic study of the overall national gains to be secured through a wholesale adoption of these standards would be most useful in securing their prompt implementation. The task was formally entrusted to the National Council of Applied Economic Research of India which brought out its report in 1965.[10, 11, 26-33]

21.10 The various steps that were followed in carrying out this study included the following:

(1) Assessment of savings on the basis of direct substitution of old sections by those newly standardized, which came to approximately 19.1 percent on the average.

(2) Assessment of savings on the basis of redesigning typical steel structures of different categories such as bridges, factories, buildings and warehouses, in accordance with the new design codes using the new standard sections, which came to about 4.2 percent.

(3) Assessment of savings that could be brought about by the use of welding instead of riveting and bolting, which came to some 14.6 percent.

(4) Assessment of savings by the use of tubular sections and cold-formed light gauge sections instead of hot-rolled sections, which came to about 30.4 and 42.9 percent respectively.

Having made these various assessments, the next major step was to collect data on a nationwide basis concerning the indigenous production and import of various sections, types, numbers and tonnages of different kinds of structures being fabricated in the country; and to estimate the extent to which it would be possible to introduce tubular and cold-formed sections in the structural field generally.

21.11 Two sets of data were thus made available. One of them gave the actual percentage savings that could be made in specific applications of steel by the use of new specifications and codes, and the other the relative distribution of the total quantity of steel used and usable in various sectors of the national economy. Putting these two sets of data together led to the conclusion that in the overall picture of the economy a clear saving of 23.6 percent of steel could be made if all the newly issued Indian standards were fully implemented. Surprisingly enough this came pretty close to the rough estimate of 25 percent made earlier by the specialists who had been engaged in the steel economy project for the previous ten years, but who had not made any detailed economic assessment on the basis of national statistics.

21.12 This study was concluded at about the time when India's Fourth Five-Year National Plan for economic development was being prepared and projections for the Fifth Plan were being made. Applying the above calculations of steel savings to the expected production of structural steel during the forthcoming Plans, it was revealed that quite a substantial national economy would result from this single effort at comprehensive standardization in this field of endeavour. The figures were:

Plan	*Years*	*Production*	*Savings*	
		10^6 tonnes	10^6	10^6 rupees
Third	1961–66	3.9	0.92	644
Fourth	1966–71	7.6	1.79	1,253
Fifth	1971–76	12.6	2.97	2,079

The aggregate of the estimated savings of some 3.33 billion rupees over the ten-year period (1966–76) would amount, in terms of the present-day rising prices of steel, to the equivalent of some 820 million US dollars to the nation. This is quite a substantial amount for a developing country's economy, especially when it is compared with the relatively insignificant amount of direct expenditure of the equivalent of less than 100,000 dollars which was involved in completing the ISI steel economy standardization project. Of course, since these estimates were made, production figures have undergone some changes, but these changes in no way influence the general character of the conclusions. This means that, while the aggregate savings of steel and its monetary value may change with changes in production and utilization patterns and in price levels, yet the percentage saving of steel as a material of construction would be of the order of 23.6 percent.

21.13 However, neither the gains nor the expenditure incurred to secure the gains as indicated above include all items of the gains and expenditure that are actually involved. Among the gains are to be included those normally associated with standardization when applied to production processes and to distribution and utilization of a commodity, including effects of variety reduction. Among the expenses would have to be included those to be incurred by the producers, designers, erectors and others to switch over their operations to the new standards. Change of rolls for rolling hot-rolled sections, provision of new equipment for cold-forming and welding are but a few of the important items of expenditure involved in the case of steel. To arrive at a comprehensive overall balance sheet for the whole national endeavour would be quite a complex task, in the tackling of which the greatest hurdle would be the collection of reliable basic data. Some detailed thinking has, however, been done in Germany to work out a methodology for assessing such a national balance sheet.

21.14 Kunow[12] and Guenther,[13] who have taken the initiative, are of the view that national economic benefits fall under two major categories, namely in-factory benefits and out-factory benefits. Each item of benefit is to be calculated by subtracting the cost after standardization from that before standardization, but in doing so it is necessary that every effect of standardization is reduced to quantitative monetary terms. This has not always been found possible. But those effects which can be so reduced would, according to the authors, include the following cost items:

In-factory cost items:

(1) materials;
(2) wages and salaries;
(3) power;
(4) machine and tool wear;
(5) risks in production, stocking, handling and transport;
(6) general costs (administration, supervision); and
(7) other costs.

Out-factory cost items (part of which may be incurred within the factory):

(8) freight, transport, excise;
(9) storekeeping;
(10) maintenance;
(11) depreciation;

(12) improvement of performance;
(13) improvement of health; and
(14) other improvements.

21.15 Having discussed in detail the manner in which each one of these factors could be translated into quantitative monetary terms so that benefits from each item could be estimated, Guenther comes, at the end, to the disappointing conclusion that "no one can indicate even approximately the benefits to national economy brought about by standardization. This benefit can be considered as an intricate mosaic consisting of innumerable pieces of stones. The in-factory benefit can at least be grasped to some extent, but the benefits outside the factory, though real, are very difficult to gauge. The value obtained from statistics (of this type) may safely be regarded as minimal values." If the last statement about the minimal values is to be granted, there is no occasion to be despondent over the difficulties to be faced in assessing benefits to the national economy. Having detailed, as Guenther and Kunow have done, the various items of costs, and derived the monetary figures for the in-plant and out-plant benefits, the question arises: can we not find a method to make a weighted summation of all these benefits to obtain a figure of some sort for national assessment? It should not be difficult to answer this question in the affirmative, except that the fact would still remain that such an attempt has not yet been made, in the overall manner in which Guenther and Kunow envisage it should be made. In this sense, the Indian assessment of steel economy described above (paras 21.9-21.12) would appear to be quite restricted in its scope, covering as it does only one of the 14 items listed above, namely, the materials saving. If the national benefits, accruing from the remaining 13 items, were to be assessed, added together, and found to be insignificant, even then the results of the materials item alone would remain quite remarkable and well worthwhile.

21.16 So far we have been discussing the possibility of assessing economic benefits of national standardization in an individual sector of industry or a restricted field of endeavour. If such an assessment could be extended to cover a number of more important economic fields, then only could the overall national benefits from standardization be correctly gauged. In spite of all the difficulties indicated in developing a proper methodology to compute the national economic benefits from national standardization, it would appear that such a comprehensive attempt has been successfully made in the USSR. The State Committee on Standardization, Measures and Measuring Instruments of the Council of

Ministers – the National Standards Body of the USSR – has recently published some official estimates of the annual savings from standardization accruing to the country's economy.[14] The aggregate total savings of some 1,500 million dollars annually are said to include the following more important items:

ITEMS	US $ *(in millions)*
Machinery and small tools	412
Power generating equipment	250
Chemistry	200
Measuring instruments, computers, etc.	83
Radio, telephone, telegraph and electronic components	66
Foodstuffs	55
Mining	47
Civil engineering	31
Transportation	19
Domestic appliances, sports equipment, etc.	19
Petroleum products	15

Similar overall estimates of national gains have been published in France,[15] which run up to 2,000 million francs annually as against an overall national expenditure of some 100 million francs.

21.17 Though no indication has been given as to the exact method followed in making these estimates, yet it must be admitted that the size of the task involved must have been quite as enormous as the aggregate total itself. It is very likely that the methodology adopted was perhaps not very complicated. But whether complicated or not, it would be extremely interesing to know more about its details. Furthermore, it must be recognized that any great exactitude or extreme accuracy in such estimates is quite unnecessary. Orders of magnitude of the benefits derived should suffice for most purposes for which these estimates are generally required, namely, for emphasizing the justification of standardization work in terms of cost-benefit analysis and for determining national plan programmes and priorities.

ASSESSMENT AT INTERNATIONAL LEVEL

21.18 It will be seen from the above discussion that as one ascends the ladder along the level axis of standardization space, complexities of the task of assessing the economic effects and benefits of standardization become more and more intricate. It was also hinted in the opening paragraph of this chapter that the very concept of optimum overall economic benefits for international level of standardization was difficult of definition. Every nation is an independent economic entity and it is not infrequent that the economic gain of one would constitute an economic loss for another. The only manner in which nations are related to one another economically is through the ties of international trade. In recent years, of course, this relationship has become further extended to include the ties of mutual economic aid which includes technical and other forms of aid. Thus the effects of international standardization, beneficial or otherwise, would have to be assessed mainly in terms of international trade and mutual economic aid.

21.19 In addition, international standardization has certain indirect effects on the economies of individual nations, insofar as it influences and guides national standardization. The very existence of an international standard constitutes a benefit as it eliminates the need for the preparation of national standards on the subject and saves a great deal of the resources of National Standards Bodies. Furthermore, once an international standard becomes nationally adopted, it assumes the same role as all other national standards in conferring benefits on the national economy. It must also be recognized that, in the past, international effort has chiefly been directed towards harmonization of the existing national standards. In certain countries, such harmonization has sometimes necessitated serious changes being made in national practices, which have meant additional expenditure and effort to bring these practices in line with the newly adopted international recommendations. This would constitute a cost item in the red to be set against the gains which might be expected to accrue from the adoption of the newly harmonized international standard. More recently, however, the emphasis in international standardization is tending to shift towards the preparation of original international standards in fields not fully covered by national standards. Such international standards could directly be adopted as such by the different National Standards Bodies for application to their own national economies. In such cases, it is obvious that the National Standards Bodies would stand to gain a great deal financially and in other ways by not having to prepare separately national standards of their own and also by not having to follow them up for getting them coordinated internationally. However, such an approach can only be considered feasible in the fields of newly developing technologies for which national standards have not yet emerged in any large number.

21.20 The whole broad question of economic effects of international standardization was taken up for study by ISO as recently as 1965. The ISO Standing Committee for the Study of Scientific Principles for Standardization (STACO) has been seized of the problem ever since. Several proposals and suggestions have been made and discussed. A significant contribution to STACO thinking has been made by Aubrey Silberston,[16] a British economist, who, after discussing the various economic effects of international standardization on different sectors of the world economy, concludes that "any economic evaluation . . . might be difficult to make, both because forecasts over a considerable period of time might be needed and because the effects in one field would have consequences in other fields." According to Silberston, it would not be possible to put money value on many of the costs and benefits – for example, those associated with emergence of greater competition. Nor would it be easy to evaluate welfare benefits to consumers. He further goes on to express the hope that it would be important to attempt some estimation of welfare benefits, because these might be the source of one of the greatest benefits to emerge from standardization.

21.21 As regards a possible methodology that might be followed in making such evaluation, Silberston has no definite suggestion to make; nor indeed has any suggestion come from any other source. It must, however, be said to the credit of STACO that it continues to persist in its search for appropriate methods of quantitative and qualitative evaluation of benefits to be derived from international standardization.[17] In addition, it is studying other related matters such as the economic problems relating to standards evolved on the basis of international standards, the role of international standards in international economic relations and evaluation methods to determine priorities for preparing international standards. It must be admitted that the economic benefits of international standardization as such are generally recognized and appreciated by all concerned. That is why so many nations are cooperatively working towards its advancement. The difficulty arises only in their quantitative evaluation. In view of the fact that every nation is an economic entity in itself and that economic interests of some often conflict with those of others, the question that comes to the forefront is whether we can define a common international economic interest, in terms of which the estimations envisaged could be expressed. Furthermore, one is tempted to enquire whether it would not be more pertinent to concentrate first of all on the evaluation of national benefits from international standardization as well as national benefits from national and other levels of standardization. Once certain dependable methods have been evolved, accepted and tried out in actual practice, the time would be ripe to further extend the search for developing the evaluation techniques for international benefits.

MATHEMATICAL APPROACH

21.22 While the question of assessment of economic benefits has of recent years begun to attract considerable attention, it cannot be claimed that a great deal of success has been achieved either in regard to the development of an acceptable methodology or in the form of actual results. The whole subject bristles with difficulties and would appear to be in very early stages of development. It must be recognized, however, that any rational system that might be evolved should be founded on sound logic and mathematics, if possible. Mathematical approach to such problems, if successful, should enable one to predict the magnitude of gains and the nature of benefits that might be expected to accrue from a given standardization measure. Progress in this direction has been extremely handicapped mainly because of the difficulties of quantization of a large number of variables involved and the comprehensive nature of standardization istelf. However, a definite beginning towards the introduction of mathematics would appear to have been made in recent years, though in quite a limited sphere of application.

21.23 One of the important aspects of standardization is variety reduction which makes possible relatively long runs of production of a given type and size of an item. It is in this field that attempts have been made to introduce mathematical analysis. Frontard[18 19] refers to Albert Caquot's suggested law applicable to the cost of manufacture of a unit as a function of the length of run or the number of units manufactured at one time. It is admitted that the law has not been derived mathematically, yet in actual experience it appears to have been found approximately valid. Simply expressed, the law states that the cost of production per unit will vary in inverse proportion to the fourth root of the number of units produced in a given series. Expressed as an equation:

$$K = \frac{1}{\sqrt[4]{N}} \text{ or } N^{-0.25}$$

where K = relative cost per unit, and
N = number of units in the series.

21.24 In Fig. 21.1, the Caquot equation is shown plotted together with the actual cost in terms of man-hours required for the production of tankers, collected by Rone Fould, chairman of a French shipbuilding firm.

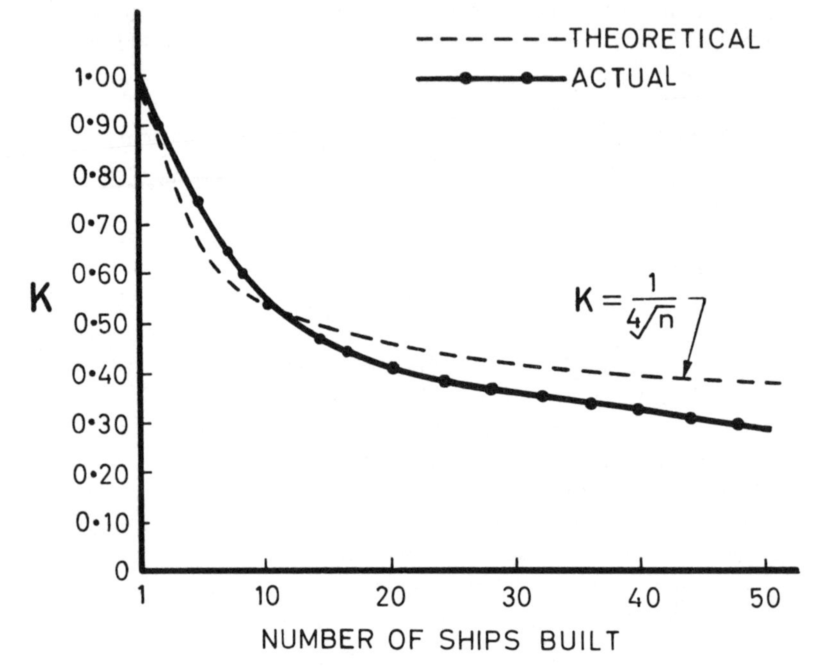

Fig. 21.1 Correspondence between Caquot's Law and actual cost in terms of man-hours required for the production of tankers.

The correspondence between the theoretical and the experimental curves appears to be remarkable. G. Maxcy and A. Silberston[20] have also obtained similar correspondence between this law and the cost data on the model T Ford. It would be most interesting to have further examples of similar experimental data obtained under controlled conditions from other industries and in other countries for further confirmation of Caquot's law. Simply stated, the law implies that, if the volume of production is doubled or, which is approximately the same thing, the variety of products is halved, the cost may be expected to be reduced by approximately 16 percent. If the volume is tripled or variety is cut down to one-third, the cost reduction would be of the order of 24 percent and so on, as follows:

Relative volume of production	0.10	0.25	0.5	1	2	3	4	5	10
Relative variety reduction	10	4	2	1	0.5	0.33	0.25	0.20	0.10
Unit cost, percent	177.8	141.4	118.9	100	84.0	76.0	70.7	67.0	56.2

It will be seen that decreasing the volume of production increases the cost much more steeply than the cost reduction obtained by increasing the volume. The same thing applies to variety reduction in the reverse direction. For example, one-tenth the production volume or ten times variety proliferation would increase the cost by 78 percent, while ten times the original volume of production or a variety reduction to one-tenth the original value would reduce the cost by only 44 percent. This feature alone should be enough to emphasize the importance of standardization, specialization and variety reduction as against a *laissez-faire* approach to production policy.

21.25 Matuura[21,22] in Japan has also been interested in reducing to mathematical formulae the calculations for benefits to be derived from variety reduction and has arrived at somewhat similar results to that which Caquot intuitively reached. Matuura concentrated on deriving his mathematical formulations empirically on the basis of actual cost data which he assiduously collected from many different types of industries. The data related to the figures of cost before and after the introduction of variety reduction of products. Some of these industries were the manufacturers of the products, while others were users and consumers. When the cost data was plotted in terms of percentage cost against the ratio of variety reduction (*see* Fig 21.2), it was found that most of the results lay bounded between two curves, namely:

$$Y_1 = X^{-0.25} \text{ and } Y_2 = X^{-0.50}$$

where Y_1 and Y_2 correspond to C_1/C_0, the percentage of cost ratio *after* to *before* variety reduction, and
X corresponds to P_0/P_1, the variety reduction ratio *before* to *after* standardization.

It will be seen from Fig 21.2 that most of the producer industry points lie quite close to the Y_1 or the inverse fourth root curve while most of the user or consumer industry points lie close to the Y_2, or the inverse square root curve. Some interesting details of Matuura's work, including the questionnaire he used for collecting the industrial data and the data thus collected are given in an ISO document[23] which might be of value to those interested to pursue the subject further.

21.26 In the USSR literature also a similar equation has been used for expressing the economic gains as a function of variety reduction ratio. Thus, Ratner[24] and Tkachenko[25] have given the formula $Y = X^{-0.3}$, which differs only in the value of the exponent from those of Caquot and Matuura. To show how little the various formulae differ from one another

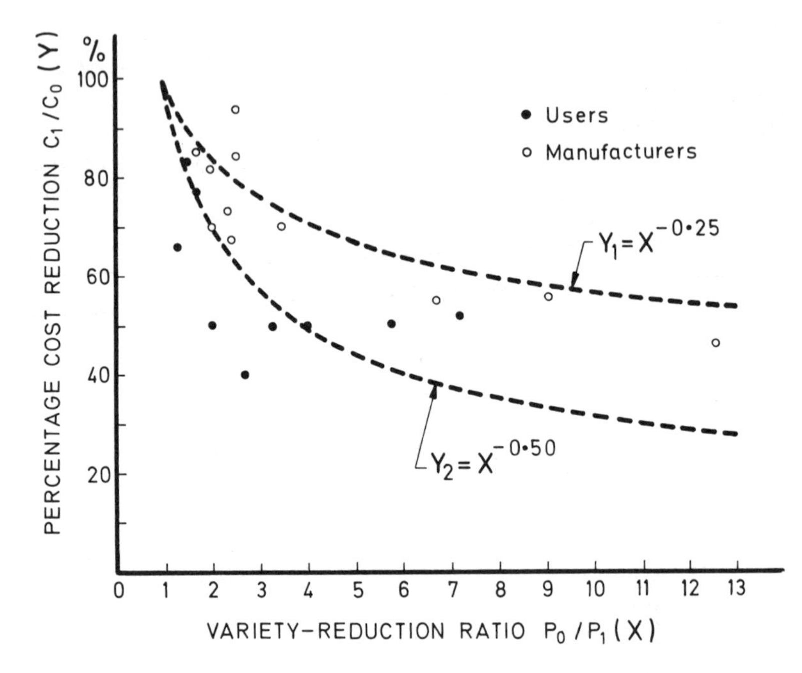

Fig. 21.2 Dependence of cost reduction C_1/C_0 on variety-reduction ratio P_0/P_1.

they have been plotted together in Fig. 21.3. Since the experimental data show such a scatter from industry to industry as may be seen from the plot of Fig. 21.2, it would appear that there is perhaps little hope of obtaining a single mathematical expression of a general character applicable to all the diverse industries. Matuura has been attempting to discover just such a single expression for covering both the producer and the user data of Fig. 21.1, but he still feels diffident about its general applicability. It would appear that he is perhaps much nearer the truth in seeking a region bounded by two curves which will envelope most of the industrial data. At the same time, as seen from Figs. 21.1 and 21.3, it would also appear possible to obtain a good fit between a single theoretical expression having an appropriate exponent and the industrial data pertaining to one particular industry. Thus, for each industry it may become possible to work out separately an expression of the type $Y = X^{-b}$, with a specific value of exponent b applicable to the cost data of the industry in question.

To investigate this point would require a considerable amount of work of collecting the relevant data and their detailed analysis.

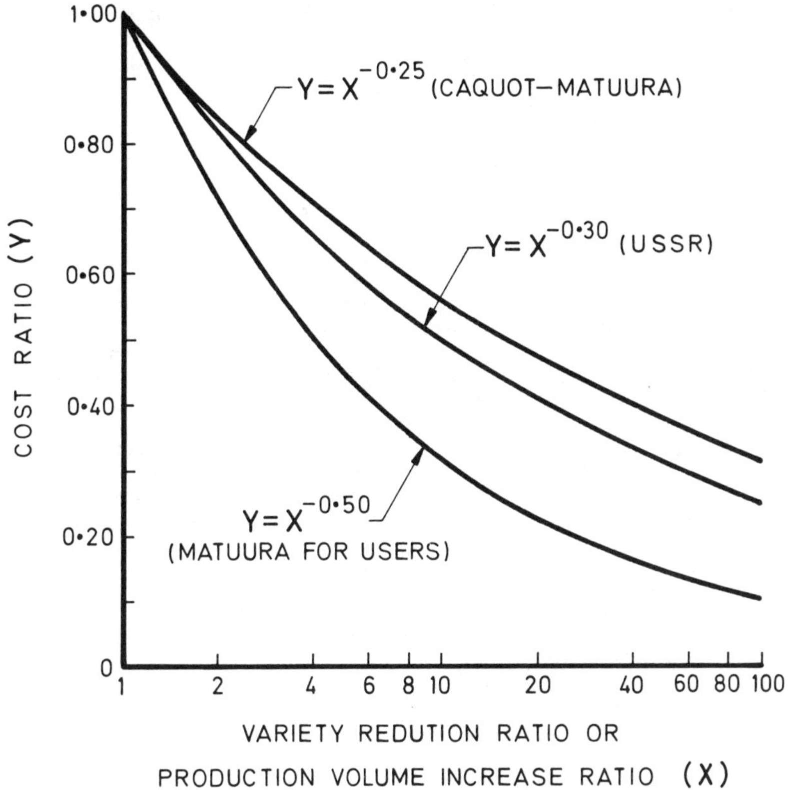

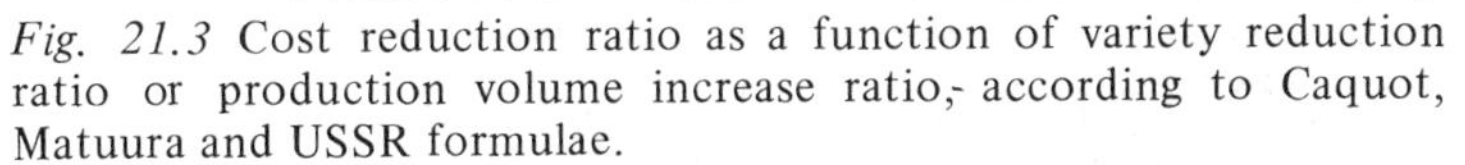

Fig. 21.3 Cost reduction ratio as a function of variety reduction ratio or production volume increase ratio,- according to Caquot, Matuura and USSR formulae.

REFERENCES TO CHAPTER 21

1. Brady, Robert A.: "Standards and resource conservation." Pp. 335-342 *in* Reck, Dickson (Ed.): *National standards in a modern economy.* Harper, New York, 1956, p. 372.
2. See references 11 to 14 and 39 to 54 of Chapter 7.
3. *Document ISO/STACO/WG 10 (France–16) 378E. Economic effects of national standardization.* International Organization for standardization, Geneva, 1971, p. 7.
4. Berry, Richard B.: When does standardization cost money? Proceedings of tenth annual meeting of Standards Engineers Society, 1961, p. 86 (*see* p. 71).
5. See reference 10 of Chapter 1.

6. Cochrane, Rexmond C.: *Measures for progress – a history of the National Bureau of Standards.* US Department of Commerce, Washington, 1966, p. 703 (*see* p. 253).
7. *Cost savings through standardization – simplification – specialization* in:
   electrically operated household appliances, p. 172;
   the building industry, p. 182;
   materials handling equipment, p. 125;
   the clothing industry, p. 57; and
   containers, p. 42.
   Five separate publications: United States Department of Labour, Bureau of Labour Statistics, Washington, 1952.
8. Johnson, T. J.: Standardizing hides and skins brings profit. *Herald,* Addis Ababa, 18 March, 1969.
9. Krishnamachar, B. S.: Steel economy through standards. *ISI Bulletin,* 1964, vol 16, pp. 163-167 and 189.
10. *Saving in structural steel through standardization.* National Council of Applied Economic Research, New Delhi, 1965, p. 100.
11. India can save over 23 percent structural steel – NCAER assessment of steel economy through standardization. *ISI Bulletin,* 1965, vol 17, pp. 307-313.
12. Kunow: Determining the economic advantages of standardization. *Standardisierung,* 1959, vol 5, pp. 1/559-1/562 + 1/575 (*in German*).
13. Guenther, Otto: Benefits of standardization to national economy and their assessment. *Elektro-standard,* no 4, 1961, pp. 1-5 (*in German*).
14. Standards save the Russians millions of roubles a year. *Standards Engineers* (India), 1971, vol 5, pp. 45-46.
15. See p. 23 of reference 3 of Chapter 7.
16. Silberston, Aubrey: Economic consequences of international standardization. *Document ISO/STACO (Secretariat–54) 389E.* International Organization for Standardization, Geneva, 1967, p. 7.
17. *Document ISO/STACO (WG 10–13) 490E – Report.* International Organization for Standardization, Geneva, 1971, p. 4.
18. Frontard, R.: Economic aspects of standardization. *ISI Bulletin,* 1966, vol 18, pp. 441-446.
19. Frontard, R.: Standard and profit – reasearch on the economic effects of standardization. *Courier de la Normalisation,* 1967, vol 34, no 194, pp. 211-221 (*in French*).
20. Maxcy, George and Silberston, Aubrey: *The Motor Industry.* Allen and Unwin, London, 1959, p. 245.
21. Matuura, S.: The economic effect of simplification. Researches and studies applied to standardization and quality control. *JSA Technical Report No 5.* Japanese Standards Association, Tokyo, 1970, pp. 30-34.
22. Matuura, S.: Measurement of the effect of standardization. *Document ISO/STACO (Japan–9) 396E.* International Organization for Standardization, Geneva, 1967, p. 13.
23. *Document ISO/STACO (Japan–10) 424 – Report.* International Organization for Standardization, Geneva, 1969, p. 11.
24. Ratner, M. L.: Economic evaluation of the effect of manufacturing standardized products. *Standardization* (USSR), 1963, vol 27, no 5, pp.3-4 (*in Russian*).
25. Tkachenko, V. V.: *Method and practice of standardization.* Komitet Standartov, Gosudarstvenniy Soveta Ministrov SSSR, Moscow, 1967, p. 612 (*see* Chapter 10, pp. 168-182) (*in Russian*).
26. Paulssen van Lunaeboerg, O. G. F.: The problem of steel economy. *ISI Bulletin,* 1951, vol 3, pp. 6-9 and 35-39.

27. Steel saving and standardization. *ISI Bulletin,* 1951, vol 3, pp. 31-32.
28. Llewellyn, Fred T.: The rational design of a series of steel sections. *ISI Bulletin,* 1952 and 1953, vol 4, pp. 97-102 and vol 5, pp. 6-13.
29. ISI launches national project for steel saving. *ISI Bulletin,* 1954, vol 6, pp. 47-49.
30. Saving steel with Indian Standards – ISI specifies improved structural sections. *ISI Bulletin,* 1957, vol 9, pp. 149-158.
31. Verman, Lal C.: Steel economy through Indian standards. *ISI Bulletin,* 1958, vol 10, pp. 193-198.
32. Mehta, Ashoka: Economy in use of steel would achieve higher physical targets. *ISI Bulletin,* 1965, vol 17, pp. 443-444.
33. National standards and economy of steel fabrication. *ISI Bulletin,* 1970, vol 22, pp. 114-116.

Chapter 22

Presentation of Standards

22.0 Standardizing the detailed manner in which a standard may be drafted and published should naturally be the first of the few tasks to attract the attention of any newly organized standards issuing authority. It would be contradictory to the spirit of the discipline of standardization, if different standards emanating from the same source were allowed to differ in style and presentation. Such a situation would not only constitute a source of inconvenience to the users of standards and hinder their ready implementation, but may even lead to unnecessary misunderstanding. Most standards authorities from company level up to the international level, therefore, adopt a formal format or style for the presentation of their standards.[1-15] Reference has already been made (*see* Chapter 7 para 7.17) to examples of a few company standards formats. In the present chapter it is intended to discuss some of the important points of general interest to be borne in mind when adopting a standard format or compiling a standard style manual. These points should be of particular interest for work at the national level, but would also be generally useful at other levels – company, industry and international. Individual level standardization, being not so much of a repetitive process, does not require standardization of formats in any formal manner, though a systematic presentation can be very useful even in this case.

22.1 Normally, one format or style manual applicable generally to all types of standards is found adequate; it is suitable for most subjects and most aspects. But in certain cases, a large National Standards Body, which issues hundreds of standards a year and covers a great many sectors of industry, may find it necessary to issue additionally, a special set of instructions for guidance in drafting standards in specific fields of endeavour.[16-18] In all such cases it is important to ensure that the coverage in these guiding instructions in no way conflicts with the general format or style of presentation. Such a guide need repeat nothing of what has already been covered in the general document, but the two sets of publications should be so compiled as to be supplementary to each other.

SIZE

22.2 In standardizing the size of published standards and other publications issued by a standards body there is today no great difficulty – indeed there is very little choice. One of the first few recommendations issued by ISA and subsequently adopted by ISO was the one which dealt with paper sizes.[19] Most National Standards Bodies have adopted this recommendation as a national standard. Though in many countries the ISO paper sizes have not become generally adopted for publications and other uses, yet most NSBs have standardized the sizes of their standards and other publications on the basis of ISO recommendations. Thus the most popular sizes that have come to be used are the so-called A4 (210 mm. × 297 mm.) and A5 (148 mm. × 210 mm.). In most countries one of these two sizes is adhered to by the NSBs for their standards, but in some countries both sizes are used by the same organization. The main reason for letting both sizes prevail is that while the smaller A5 size is handy for a pocket-size booklet, the larger A4 size is more convenient when a standard happens to contain large tables, drawings and charts. It is up to the standards issuing authority to make a decision whether it would select one size or the other or adopt both.

NUMBERING OF STANDARDS

22.3 Standards documents are so frequently quoted in other standards and many other types of documents, contracts, business transactions and so on, that it would be most inconvenient if they had always to be referred to by full title, source and date of publication. By numbering the standards with a code number, the standards authorities usually provide the facility to refer to them in a convenient manner. For example, a client may write to a supplier that "the installation drawings for the equipment ordered should conform to BS 308:1965." This common method of numbering standards consists of three important elements, meant to indicate the name of the authority issuing the standard, the title of the standard and the year in which it was issued, respectively. In the above example:

BS stands for British Standard issued by the British Standards Institution;

308 is the serial number in the series of standards issued by BSI, which according to the BSI Catalogue or Handbook of Standards would indicate the exact title of the standard, in this case: "engineering drawing practice;" and

1965 denotes the year of publication of this particular version of the standard.

22.4 In some cases the date of publication is not made an integral part of the number of the standards, but is separately given and printed close to the number. In DIN standards of Germany and several other continental countries, this happens to be the practice. But it is extremely important that the year of publication be given prominently, though the month is not very important because standards are seldom if ever revised or amended within a few months of their original publication. The year of publication has always to be cited in referring to a standard, because the latest version would be the one currently in force. The BS 308 mentioned above for instance was first issued in 1927 then successively revised in 1943 and 1953. Every edition contains the latest decisions and may differ materially in many respects from the former. Thus if the year is made a part of the number of the standard, it simplifies the task of reference.

22.5 Some institutions for their own convenience introduce a code letter in the number indicating the classification in which the subject matter of the standard falls (*see also* Chapter 4, para 4.4). Thus, in the Japanese Standard numbered:

JIS Z 8301 – 1957

the letter "Z" indicates the class to which the standard entitled "Style manual for Japanese industrial standards" belongs, the class "Z" being assigned to "Fundamental and General" subjects.[20] Similarly, certain standards bodies differentiate between the ordinary run of standards and codes of practice; for example:

BS CP98:1964 Code of practice for preservative treatment for constructional timber.

The British Standards Institution, in fact, has a number of distinct series of publications[21] other than the normal British Standards, including AU (automobile), MA (marine), PD, Aerospace series, etc. But certain other institutions, like the Indian Standards Institution, simplify the practice and designate almost all their publications including codes of practice as standards, bringing them under one overall series of numbers and leaving only a few emergency standards and the so-called special publications (SP), which, though related to standards, cannot possibly be called standards as such; for example:

SP: 5–1969 Guide for the use of international system (SI) units.
or SP:6(1)–1966 Handbook for structural engineers – structural steel sections (revised).

22.6 It must be recognized that the main object of adopting a numbering system is to provide a simple and concise method of referring to standards for the convenience of all concerned, particularly the users of standards. It should, therefore, be adequate to adopt the system, typified by the example given in paragraph 22.3, which contains three main elements, namely, the source, the serial number which is linked with the title, followed by the year of issue. Any other additional element introduced in the number may facilitate internal operations of the standards authority concerned, but does not materially add to the facility of outside users. Punctuation marks most appropriate to the language of the standard should be used to separate the three elements.

FORM OF PRESENTATION

22.7 Two distinct practices prevail in the world in respect of the form in which standards are usually presented, which for convenience may be referred to as:

(1) the sheet form, and
(2) the discursive form.

In the sheet form of standards, the material is presented in a compact manner mostly through drawings, tables, charts and graphical statements, resorting to descriptive passages only when it becomes most essential or unavoidable. In the discursive form, the matter is expressed mostly in descriptive style, using tables, charts and graphical representation as adjuncts to the main discursive text. When the sheet form is used, it is often, but not always, possible to condense the presentation of a whole standard on one or two printed sheets. In the other case, the ancillary material itself, such as the title page, the foreword and the appendices may take a few pages, which in the sheet form is usually omitted. The sheet form of presentation is generally preferred for company standards (*see* Fig. 7.1 in Chapter 7). It is also quite popular among most National Standards Bodies of the continental countries of Europe. In English-speaking and English-using as well as some other countries the discursive form is preferred for the presentation of national standards. But company standards even in these countries are cast mostly in sheet form.

22.8 Obviously, there are merits and demerits in both systems and each standards authority must decide for itself the form it may adopt. As far as company standards are concerned there appears to be little to recommend the discursive form. It is indeed the merits of the sheet form which have led to its widespread adoption at this level. At other levels, however, different considerations arise, such as the frequent use of industry-level and national-level standards for legal and contractual purposes, where textual clarity is so essential. Even the differences in the meanings of words like "shall," "will" and "may" can influence legal interpretation and decision. In a sheet form standard, where textual matter is reduced to the barest minimum, it is hardly practical to bring out such differences in shades of meanings of words and expressions. But for internal company use, refinements of this character are hardly necessary. All concerned within the organizational framework of a company understand everything given on the sheet has to be adhered to as far as any company operation is concerned. Furthermore, the sheet form facilitates filing and also addition and replacement, which are bound to take place more frequently at company level than at other levels.

22.9 For developing countries, it would appear that the discursive form is more suited for national standards, even though it may involve an amount of extra expenditure in printing. Standards in such countries, besides being used for legal and other purposes, serve as guides for the development of new industries which are almost always in need of a great deal of fresh technical information about new products and processes. For example, certain manufacturing details, which in complicated cases are sometimes included in standards, are often quite useful. Besides, the introductory matter contained in forewords to standards can be most informative for users. Such material may cover a wide range of information, such as the genesis of the standard; the reasons for including certain unusual and difficult items of requirements; correspondence of the contents of the standard with ISO or other international standards, if any; instructions for the use of the standard and methods of interpretation of some of its contents. Another element which is sometimes introduced in national standards of a developing country is a complete list of members of the committees responsible for preparing the respective standards and/or the interests the members represent. All such information, besides being interesting, can be of great service to users, particularly to those who are comparatively recent entrants in the field of production and distribution of goods and services. For users of this class, who predominate in developing countries, the compact presentation of requirements in sheet form could present serious problems for the understanding of the exact implications of the provisions of standards.

22.10 In the international field also, the discursive form has been generally accepted; both ISO and IEC and most other international agencies issuing standards adhere to this form. In the sixth Commonwealth Standards Conference this question of the form of presentation received some collective attention of representatives of several countries.[22] The following comment recorded by the Conference would be of interest.

> Consideration was given to whether compact (sheet form) standards would be suitable for contract or legal purposes, about which there was some doubt. It was thought that the compact form might be useful in particular cases, or might be partially adopted by the use of more tables in traditional (discursive form) standards. It was agreed that the possibilities deserved further consideration and that countries should exchange views if their practices developed in this direction.

The recommendation implicitly contained in this statement is that a synthesis of the two practices may be attempted to advantage. No significant development along these lines appears to have so far been reported from any country. Perhaps ISO/STACO could usefully take up the problem and find a widely acceptable *via media* which may help to unify the practices in all countries. In the meantime newer NSBs would do well to adopt the discursive form of presentation of their standards, using tables, charts, figures, symbols and formulae to the maximum possible extent consistent with ready comprehension by their clientele.

NUMBERING OF ELEMENTS OF STANDARDS

22.11 A systematic style of numbering of paragraphs, tables, illustrative figures and appendices within the body of standards is just as important as the numbering of standards themselves, and for the same purpose, namely that of providing facility for referencing and cross-referencing. In either style of presentation, whether sheet or discursive, a standardized method of numbering these elements is considered equally important.

22.12 The most pervasive of these elements is the paragraph and hence the system for paragraph numbering should receive special attention. The generally adopted method used by most standards authorities is the point system, in which each major item or important constituent idea is given a whole number: 0, 1, 2, 3, etc. Primary subdivisions of a major item are assigned numbers starting with the major item number followed by a point or full stop and another number corresponding to the ordinal position of the subdivision. For example, 3.1 and 3.2 would be the first and second subdivisions under the major item 3. Similarly, the next level of subdivisions under a primary subdivision would be designated by 3 numbers separated by two stops; for example, 4.3.1 and 4.3.2. Any further subdivision requiring four numbers is usually discouraged, for it may appear unnecessarily complex and may even create avoidable confusion. In extreme cases, four numbers may, however, be justified but certainly no more than four should be tolerated.[23]

22.13 It must be pointed out that this system of paragraph numbering is not to be confused with the normal decimal system of notation exemplified by the Dewey decimal classification. The decimal system is limited to a maximum of 10 subdivisions, or, in other words, only to as many sub-items as may involve the use of a single digit number from 0 to 9. In the present system of paragraph numbering described here, no upper limit can be placed on the number of subdivisions of an item or sub-item. This is necessary because it is quite in the nature of standards that a given idea or item may require more than ten subdivisions to be fully covered. Thus, it should not be uncommon to find a standard paragraph numbered 7.15 or 5.0.11 or 0.3.10, and so on. Such numbering in the decimal system would have an entirely different meaning if used at all. For example, 3.15 in decimal system would mean 3.1.5 in the present system. For further examples and useful hints, reference is invited to IS:12–1964.[9]

22.14 Similarly, for tables and figures, it is advisable that a uniform convention be adopted by each standards authority. Normally, a simple system of serially numbering them is found adequate, such as Table 1, Table 2, etc. and Fig. 1, Fig. 2, etc. There should be no need to distinguish between the different types of illustrative material; for example, diagrams, drawings, graphs, charts, maps, photographs, etc., could all be lumped together under one serial order of figures, with suitable captions for each. Another style of numbering may have to be adopted for appendices and annexes to standards. A common practice is to number them alphabetically but there can be no serious objection to numbering them differently. It is, however, suggested that only one class, either appendix or annexe, may generally be recognized. There may be occasions when both may have to be used,[24] but such occasions will be rare.

SPECIFYING NUMERICAL VALUES

22.15 A great deal of thought has been given during the last couple of decades to this question which had previously been taken for granted. Properties of materials and characteristics of products have necessarily to be expressed in terms of numerical quantities and it is essential that it be done in a consistent manner and as accurately as demanded by circumstances. For example, when the acid value of an oil is specified not to exceed 6, is it meant that it should be so exactly 6 that it should not exceed the least amount beyond this value, say, not even by as much as 0.0001? But this would be absurd. There is always an ascertainable degree of uncertainty in any determination and it is only within that limit that the test result can be guaranteed to be accurate. Neither could such a high degree of accuracy as 0.0001 be justified from the point of view of utility of the oil. In this example, for instance, if the uncertainty of test result were as much as ± 0.02, then any test leading to the conclusion that the acid value was 6.01 would mean that it is actually somewhere between 5.99 and 6.03. The question would naturally arise whether under these circumstances the consignment of oil would be taken to pass the test or not. Many a dispute has arisen on just such borderline cases – the inspection agency maintaining that the acid value of 6.01 exceeds the specification value and the vendor claiming that the excess is well within the error of determination. The argument may appear to be trivial at first sight but what it really amounts to is that the existence of a standard has not served one of its main objectives, namely the elimination or minimization of the chance of dispute.

22.16 Oftentimes values are specified along with their tolerances; this is particularly the case in dimensioning a drawing. But even in engineering drawings, it is not rare for rough dimensions to be given without a tolerance. In such cases, the interpretation of the specified untoleranced dimensions is left to the good sense of the user of the drawing. In certain cases, however, companies have adopted for their own use, certain conventions by which the implied tolerance of an untoleranced dimension is to be interpreted. No such national or international standards exist. The British Standards Institution[25] and the American Society for Testing and Materials[26] have issued two very useful standards discussing this question and giving useful instructions for standards writing. But a more definite convention has been proposed and is being used with advantage by the Indian Standards Institution[27] in writing Indian Standards which should be particularly useful to developing countries where fresh traditions in the field of standards are being extablished. The ISI system is based on the assumption that whenever a value is given without any tolerance, it is implied that it is to be regarded as correct to half the value of the unit in the last place of the specified value. For instance when the distance between two towns is stated to be 48 km., it may be taken that it is nearer to 48 than it is to either 47 to 49. For if it was otherwise, it could not have been stated to be 48 km. Now both commonsense and the rules for rounding off of numbers formally adopted by a number of NSBs[25 26 28 29] dictate that the actual number should be anywhere between 47.5 and 48.5, if it is to be rounded off to 48. Thus one may safely conclude that the accuracy of the distance specified as 48 km. may be safely taken to be ± 0.5 km. In the case of the acid value example, the interpretation would be that 6 would mean 6 ± 0.5 or a variation between 5.5 and 6.5 would be acceptable. But if such a variation is not to be considered acceptable, the standard should have specified a value of 6.0 or 6.00, in which case the unstated tolerance limits would have been ± 0.05 and ± 0.005 respectively.

22.17 Thus the first rule for specifying values emerges, namely, that untoleranced values should be so stated that they may be taken to be correct to within half the unit in the last significant place. The second rule emerges from the common practice of writing large values in such a manner that several non-significant zeros are unnecessarily retained and significant zeros are sometimes omitted from decimal fractions. For example, a bacterial count of 1,000,000 per millilitre obviously cannot be accurate to ± 0.5 or even one bacterium. It may be accurate, say, to 1,000, in which case it should be written as 1,000 ± 10^3. A dimension intended to be accurate to, say ± 0.05 mm. may not be written as 12 mm.; strictly speaking, it should be written as 12.0 mm. Thus emerges the second rule for stating values: Retain all significant zeros and drop all non-significant zeros, so as to remain consistent with the first rule. A third rule, which emerges as a corollary to the above two, is that in any toleranced value, retain as many significant places as are consistent with the tolerance itself. Thus the interpretation of the result of a test would be facilitated, for it would have to be rounded off to the same number of places as the specified value before being compared with the latter.

22.18 In addition, there are several other rules of minor importance for specifying values which should be provided for in any style manual or a guide for writing standards: whether to use a stop or a comma for decimal point, whether a zero should always precede a decimal fraction smaller than one, whether to group digits in threes before and after a decimal point, and so on. All decisions of this type would depend largely on the language and usage of a country. But in making and adopting such rules for a new NSB, a great deal of assistance can be drawn from the existing practices elsewhere,[3–15, 25–29] but particularly from ISO and IEC documents on the subject.[1, 2]

SPECIFYING METRIC VALUES

22.19 For countries which are presently engaged in introducing the metric system and those which may follow, many considerations of conversion of older values to metric values arise for deciding upon the actual magnitudes of specification values to be adopted as standard. Here the questions that arise are numerous and complex. For example:

(1) what the present prevailing practice is in the industry and to what extent it could conveniently be altered, without the industry suffering serious economic consequences;

(2) to what extent the direct conversion of values from the existing system of measurement to the metric system[27] would meet the situation or whether certain readjustments would be called for (*see* Chapter 23, paras 23.10–23.14);

(3) whether advantage could or should be taken of the opportunity offered by the overall metric change to introduce a greater degree of rationalization than had existed before.

These and similar other considerations have been discussed in an Indian Standard[30] and a line of approach suggested.[36] It is not proposed to go into details here, except to say that the 10-year metric conversion programme of India was greatly facilitated by this approach[31] – particularly in respect of writing new standards and rewriting old ones in metric terms. Developing countries would find it useful to study this matter carefully before formulating their own programmes for metricization of their standards.

ABBREVIATIONS

22.20 Standardization of abbreviations to be used in standards is another concern of style manuals, which must be attended to separately for each language. In the more popular world languages like English and French, certain usages have come to prevail from which selections can be made. But even here several choices may exist for each case and the decisions made in the most commonly occurring cases should be so listed as to be handy for reference wherever required (*see*, for example, Appendix C of IEC Guide,[1] Appendix B of SAA MP 15–1971,[4] or Appendix A of IS:12–1964[9]).

SUBSTANTIVE MATERIAL

22.21 So far the discussion has been confined to matters of form and general interest concerned with the presentation of standards. The substantive material which is the most important part of any standard, in fact its very *raison d'etre*, has also to be systematically organized and presented in such a manner as to provide the maximum of facility for the user. Many styles of presentation have come to be adopted[3–15] by several authorities for their own internal use and a great deal of common ground has been covered between them. No attempt appears to have been made for arriving at an international agreement on the layout of national standards, except on a "regional basis," by the Commonwealth Standards Conferences.[32] The IEC[1] and ISO[2] guides on the subject are also relevant efforts in the same direction. Though these latter guides are intended for assisting the drafting of IEC and ISO standards publications respectively and not any other standards, they do represent international accords, though in limited spheres of application. Inasmuch as ISO and IEC publications are often adopted directly in their original form as national standards, it should be in the fitness of things to use the ISO and IEC guides as models for compiling national standards style manuals. Of these two, the ISO guide would perhaps be preferred because it is designed to cater to all varieties of subjects and, incidentally, deals in fair detail with the question of how generally to organize the substantive material in the case of most standards.

22.22 Without going into fine details, it would be useful to reproduce here the main elements that may be necessary to be included for covering the substantive matter dealing with a given subject and the possible sequence in which these elements could be presented to best advantage. Following is the list reproduced from the ISO Guide:

Preliminary elements

(a) Title page
(b) Brief history

(c) Foreword
(d) Contents

Body of standards

(e) Title
(f) Introduction
(g) Scope
(h) Field of application
(j) Definitions
(k) Symbols and abbreviations
(l) Classification (grading)
(m) Terminology
(n) Manufacture
(o) Characteristics and tolerances
(p) Sampling
(q) Methods of test and inspection
(r) Designation
(s) Marking, labeling and packing
(t) Annexes

Accessory Elements

(u) Footnotes
(v) Appendices

22.23 It must be recognized that not all the elements listed above might be required in each standard, nor may their order of listing be considered sacrosanct in any way. Indeed many style manuals and guides differ from one another in several such details, without in any way losing much of their effectiveness. Furthermore, it must be recognized that whatever aspect of a given subject may be under consideration, this ISO list of elements would be found adequate to deal with it. Obviously, when a standard is required to cover only one or two aspects, its substantive material would hardly need to include all the elements mentioned in the list. For example, a code of practice would have to be cast in quite a distinctive manner, particularly suited to the subject matter in hand rather than in the generalized manner represented by the list. Similarly, a standard on terminology would need to deal with only a few of the listed elements. A specification of a product or material would perhaps cover more elements than almost any other aspect.

22.24 It is also to be remembered that in special cases new elements not mentioned in the list may have to be introduced. Particularly, in matters of detail, some style manuals emphasize certain elements or subelements which could more often be elaborated in a standard to advantage than they usually are in general practice. For example, the item on "Characteristics and tolerances" in the ISO list, which is termed "Specification clauses" in IS:12–1964,[9] is divided into three subgroups, namely, "Obligatory clauses," covering general and specific requirements, "Optional clauses" and "Informative clauses." Similarly, under sampling, an element, which is often overlooked, could usefully be included, namely, "Criterion for acceptance – method of interpretation of test results for deciding the conformity of the lot as a whole to the requirements of the specification." In countries where national certification marks facilities exist, the marking clause of standards may be required to mention that such and such certification mark may also be applied to the product covered by the standard. In such cases it is useful to append a note explaining what the certification mark means and how the license for its use could be obtained.[33]

22.25 Thus in every country the National Standards Body must compile its own style manual for drafting national standards in such a manner that the numerous experts engaged in writing standards within the committee structure and the directorate of the NSB are properly and uniformly guided in their work. In this way the task of editing the drafts before publication would be greatly simplified and require attention to be given only to linguistic and grammatical details. It will be appreciated that if standards writing among a large group of authors is not so guided, it may often lead to the necessity of some of them having to be rewritten in their entirety by the editorial staff. In any such case the possibility of inadvertently changing the sense of the technological requirements of a standard cannot be ruled out. But technological requirements can only be accurately interpreted by the technical experts and not by the editorial personnel. Such a procedure is, therefore, bound to entail unnecessary delay. Furthermore, no internationally accepted style manual could be worked out in a comprehensive enough manner to be universally applicable. It could only cover broad requirements in general terms, because each country has its own language or languages to cater to. Linguistic and grammatical details and other questions of local usage would have to be looked after by the individual countries in the light of the genius of their own languages and backgrounds.

COPYRIGHT

22.26 The question of copyright of standards differs in no way from that of ordinary books and other publications. There are two international conventions under which copyright obtained in one country becomes automatically effective in other member countries which are signatories to the convention. The Berne Convention[34] for the Protection of Literary and Artistic Works and the Universal Copyright Convention (UCC)[35] are well known, the former having 55 signatories and the latter 53, with 26 countries being common signatories to both. This still leaves 46 member countries of UN which are not signatories to either of these conventions. This latter group includes the following 14 countries which have National Standards Bodies of their own:

Albania	Iran	Sudan
Burma	Iraq	USSR
China	Jamaica	UAR
Colombia	Malaysia	Venezuela
Ethiopia	Singapore	

22.27 The need would hardly arise for copyrighting individual standards and/or company standards which are private and internal documents respectively, but standards at other levels are usually copyrighted because they are publicly sold to users and have an economic value for the publishers. So far as the ISO and IEC standards and publications are concerned, the copyright, besides vesting in the parent organization, is freely and openly assigned to member bodies and national committees. It is the latter organizations which are responsible for looking after this copyright within their own respective jurisdictions. Certain National Standards Bodies have also adopted the practice of bilaterally allowing other sister NSBs to use freely the textual material from their own publications. Such a practice prevails quite generally among member countries of the Commonwealth Standards Conference. Other NSBs also make such a facility generally available to sister bodies on request. It is universally appreciated that this practice promotes and contributes to the coordination of national standards and facilitates the evolution of international standards.

REFERENCES TO CHAPTER 22

1. *Guide to the drafting of IEC documents.* International Electrotechnical Commission, Geneva, 1961, p. 25.
2. *Guide for the presentation of ISO recommendations.* International Organization for Standardization, Geneva, 1971, p. 15.

3. *Style manual for world food standards.* Joint Committee FAO/WHO Codex Alimentarius, Poland, 1965, p. 29.
4. *SAA MP 15–1971 SAA Style manual.* Standards Association of Australia, North Sydney, 1971, p. 46.
5. *PD 6112–1967 Guide to the preparation of specifications.* British Standards Institution, London, 1967, p. 17.
6. *Style manual for CSA standards.* Canadian Standards Association, Ontario, 1961.
7. *Procedure manual for the development of CSA standards.* Canadian Standards Association, Ontario, 1971, p. 15.
8. *DIN 820 Standardization procedure:*
   Blatt 1–1960. Fundamentals, principles, system, p. 11 (*in German*).
   Blatt 2–1969. Arrangement of standard sheets, p. 32 (*in German*).
   Blatt 2, Beiblatt 1–1966. Arrangement of standard sheets. Notes on the composition of DIN designations, p. 6 (*in German*). English version p. 7.
   Blatt 2, Beiblatt 2–1969. Arrangement of standard sheets. Examples for application, p. 20 (*in German*).
9. *IS:12–1964 Guide for drafting Indian standards (second revision).* Indian Standards Institution, New Delhi, p. 57.
10. *JIS Z 8301–1957 Style manual for Japanese Industrial Standards.* Japanese Standards Association, Tokyo, p. 29.
11. *ICAITI 4 001–1968 Preparation of standards,* p. 6 and *ICAITI 4 002-1963 Formats of standards,* p. 2. Central American Research Institute for industry, Guatemala (*in Spanish*).
12. UNCO 2–1–0.001–1959 Presentation of Colombian standards. Instituto de Normas. Universidad Industrial de Santander, Bucaramanga (Colombia), p. 2 (*in Spanish*).
13. *NC 00–01–1967 Guide for the presentation of standards,* p. 5 and *NC 0.0–1960 Cuban standard for drafting standards,* p. 3. Department of Standards, Ministry of Commerce, Havana (Cuba) (*in Spanish*).
14. *KS A 0001–1962 Style manual for Korean Industrial Standards.* Korean Bureau of Standards, Seoul (South Korea), p. 12 (original in Korean, English translation available).
15. *ASA PM 117a–1960 Style manual for American standards.* American Standards Association (now American National Standards Institute), New York, p. 28.
16. *PD 1992–1954 Guidance for committees drafting building standards.* British Standards Institution, London, p. 19.
17. *PD 2436–1956 Memorandum on the preparation of British Standards for non-ferrous metals.* British Standards Institution, London, p. 32.
18. *PD 2879–1957 Guidance for committees drafting British standard specifications for electrical materials and apparatus.* British Standards Institution, London, p. 22.
19. *ISO/R 216–1961 Trimmed sizes of writing paper and certain classes of printed matter.* International Organization for Standardization, Geneva, p. 8.
20. *JIS yearbook 1971.* Japanese Industrial Standards Committee, Tokyo, pp. iii + 181 (*see* p. ii).
21. *BSI yearbook 1971.* British Standards Institution, London, p. 823.
22. *Commonwealth Standards Conference 1965.* British Standards Institution, London, p. 35 (*see* p. 18).
23. See p. 19 of reference 22 above.
24. See p. 7 of reference 2 above.
25. BS 1957:1953 The presentation of numerical values. British Standards Institution, London, p. 15.

26. ASTM Designation: E29–67 Recommended practice for indicating which places of figures are to be considered significant in specified values. *1969 Book of ASTM Standards with related material, Part 27.* American Society for Testing and Materials, Philadelphia, p. 907 (*see* pp. 805-809).
27. *IS:787–1956 Guide for interconversion of values from one system of units to another.* Indian Standards Institution, New Delhi, p. 19 (*see* paragraphs 2.1, notes and 4.3.1.1).
28. *IS:2–1960 Rules for rounding off numerical values (revised).* Indian Standards Institution, New Delhi, p. 12.
29. *ASA Z 25.1–1947 Rules for rounding off numerical values.* American Standards Association (now American National Standards Institute), New York, p. 7.
30. *IS:1722–1970 Guide for specifying metric values in standards.* Indian Standards Institution, New Delhi, p. 15.
31. Verman, Lal C. and Kaul, Jainath: *Metric change in India.* Indian Standards Institution, New Delhi, 1970, pp. xxviii + 529.
32. See pp. 19-21 of reference 22 above.
33. See p. 14 of reference 9 above.
34. Berne Convention. *Encyclopaedia Britannica,* 1961, vol 6, p. 1005.
35. *Universal Copyright Convention.* UNESCO, Paris, 1952, p. 24.
36. Interconversion and coordination of values in metric and inch systems. *ISI Bulletin,* 1961, vol 13, pp. 227-229.

Chapter 23

Mathematical Aids for Standardization

23.0 Mathematics is the basis of all sciences, and inasmuch as standardization discipline is closely related to physical as well as biological sciences, its dependence on mathematics is, to an extent, inevitable. It must, however, be admitted that though this dependence was recognized long ago, intensive exploitation of mathematics for standardization purposes dates back only to the last couple of decades. It was, no doubt, late in the nineteenth century that Renard introduced his geometric series of preferred numbers, and even statistical quality control techniques were discovered during the twenties of the present century. However, it was only with the extensive utilization of sampling inspection during the Second World War that the real application of mathematics to standardization began to develop. Since then considerable progress has been registered in many directions by extending the use of mathematics in resolving different kinds of standardization problems and actually introducing mathematical concepts, or more specifically statistical concepts, in the actual texts of standard specifications and methods of test.

23.1 Standardization, as a process, being most readily applicable and conducive to mass production, in which large numbers of similar products are involved, it would be obvious that in establishing and assuring their conformity with the requirements of standards, statistical methods would play an important part. This indeed happens to be the case. While most branches of mathematics may be called upon to help in standardization,[1-3 63] statistics has come to play the greatest role. At the instance of the Polish National Standards Body, ISO has taken up the study of the question of application of mathematics to standardization. The ISO Committee on Scientific Principles of Standardization (ISO/STACO) had organized, at Warsaw, in 1969, the first seminar of the kind on these problems. The Report[4] of this Seminar should be helpful to those interested in pursuing the subject further.

23.2 It is not intended here to enter into a detailed discussion of the various aspects involved, of which there are many – some of them interrelated to one another. It is proposed only to touch briefly upon a few of the aspects, quoting references to relevant literature in order to facilitate further study. Though the references are fairly copious, they are by no means to be considered exhaustive. The literature is admittedly scattered and worthy of being collected and collated in book form by some mathematician, especially interested and experienced in standardization.

23.3 A few of the mathematical techniques which have proved helpful to standardization have already been referred to in the earlier pages of this book. See, for instance,

Chapter 5, for:

(1) inventory control (para 5.28)
(2) programme evaluation review technique (PERT) (para 5.29) and
(3) linear programming (para 5.30)

Chapter 21, for:

(4) evaluation of economic effects of variety reduction (paras 21.22–21.26)

Chapter 22, for:

(5) specifying numerical values (paras 22.15–22.18).

The topics that are proposed to be touched upon in this chapter would include preferred numbers, interconversion of values, statistical quality control, several other statistical tools and some related miscellaneous items.

PREFERRED NUMBERS

23.4 One of the commonly occurring situations is that in which a standard engineer is called upon to define a series of sizes of a product in relation to one or more of its properties. These may be dimensional properties, which by and large, is quite frequently the case, or it may be certain performance characteristics, such as the output power of a line of electric motors or diesel engines. Such a situation may arise in the case of a new product, for which no traditional or customary sizes may have existed before, or it may be an occasion where an existing product line may have become uneconomical because of an undue proliferation of sizes which could reasonably be rationalized. In earlier days, a tendency prevailed to fix the sizes, rising step by step, at first by a small fixed amount and then as the size grew larger, by increasing amounts; for example, 1, 2, 3, 4, 6, 8, 12, 16, 20 and so on. This tendency to mix up arithmetic with geometric progression exists even today to an extent. But with the advent of preferred numbers, it is gradually coming to be realized that this approach to standardization of sizes is not only *ad hoc* and irrational but also uneconomical in the long run. It stands to reason that as the size of a product increases the step from one to the next larger size must be significantly larger than that adopted at the lower range of sizes, and that this should happen not only after certain intervals but continuously throughout the range from each step to the next. This is the most obvious way to meet the demand for various sizes at an economic level. But the question is how the incremental steps should be devised and what should be their magnitude.

23.5 This was perhaps the first standardization problem to be tackled with the help of mathematics. It was in the seventies of the last century that Charles Renard, a well-known French balloonist, hit upon the solution and proposed the geometrical progression of numbers which have come to be know as "preferred numbers" or the "Renard Series."[5-7] The basic series is derived by the use of integral powers of the fifth root of 10, such that the expression

$$(\sqrt[5]{10})^n$$

yields the series, where n is any integer. Thus we have for $n = 0, 1, 2, 3, 4, 5$, the series

1, 1.58, 2.51, 3.98, 6.31, 10

This so-called R5 series provides a geometrical increment of approximately 60 percent between steps. To obtain finer incremental steps other sister series of the same type are adopted, namely:

R10 series – $(\sqrt[10]{10})^n$ – increment 26 percent
R20 series – $(\sqrt[20]{10})^n$ – increment 12 percent
R40 series – $(\sqrt[40]{10})^n$ – increment 6 percent

Any of these series can indefinitely be extended in either direction – beyond 10 or below 1 – by simply multiplying or dividing the numbers by integral powers of ten.

23.6 The preferred number series have been extensively used in France from as early as the latter part of the nineteenth century, and their use has now extended to many countries all over the world. They were first adopted as standard by ISA and later taken over by ISO[8] in 1953. Subsequently, almost all leading National Standards Bodies have incorporated them in their national standards.[9–13] Besides giving complete tables of the numbers, their derivation and properties, some of these standards include useful hints indicating how actually to select a suitable series of numbers for a given purpose. Several standards bodies have issued separate standards to guide the standards engineer as to how to use the preferred numbers.[14–16] National and other standards in several countries have used these numbers in specifying and rationalizing the sizes of products: electrical current ratings in amperes by IEC, speeds of machine tools in revolutions per minute in Belgium, water tank capacities in litres in Germany, lifting capacities of cranes in tonnes in Norway, and so on.[14]

23.7 The practical use of preferred numbers has attracted a great deal of attention of standards engineers in the world and a considerable amount of literature has grown around this question,[17–22] but it still remains to be demonstrated that the solutions offered by their use in resolving standardization problems are the most economical. It has often been stated and it stands to reason also, that, in practice, these solutions have proved to be quite economical, but this cannot be taken to mean that they are the most economical and that no better or more economical solution is possible. What is really necessary here is a strictly mathematical proof. With the advent of the computers, advance of mathematical techniques, and active interest in the subject among leading mathematicians, such a proof may one day become available. In the meantime, other series of numbers have occasionally been proposed[23] and sometimes even utilized[7] in standardization, but the preferred number series continue to be the centre of attraction for standards engineers.

MODULAR COORDINATION AND PLANNING

23.8 Somewhat opposed to the concept of geometric progression of sizes is the philosophy of modular planning, which insists on uniform arithmetical increments. The conditions that give rise to situations where only arithmetical increments can lead to economy are those in which increments have of necessity to be arithmetical such as in the case of coinage and weights series of 1, 2, 2, 5, 10, and where a large number of similar components or parts have to be assembled or put together for creating systems or products of similar character but intended to serve somewhat differing aims. Such situations arise in the layout of factories and machine shops,[24] in the assemblage of electronics equipment for telecommunications systems, industrial controls and other purposes. But the greatest use of modular planning techniques is perhaps made of in the field of building construction industries.[25-29] Here, besides designing the buildings and preparing their layouts on a modular basis with repeatable units, the building components themselves, such as bricks, blocks, windows, doors and sanitary fittings, lend themselves to being advantageously dimensioned on the basis of a fixed module. The idea and the technique of modular coordination of building components and the modular planning of buildings arose purely out of economic necessity. Modular components help to save time and labour by eliminating the need for cutting and adjusting their sizes and shapes while putting them together at the site. On the other hand, the modular planning of buildings itself facilitates the fitting together of modular components and simplifies the design of large building complexes having several repeat elements.

23.9 The modular theory and practice may hardly be called a mathematical development. Nevertheless, being in direct contrast to the geometrical progression approach of the preferred numbers, it is important to be taken note of as an arithmetical device which also contributes richly to achieving economy through standardization. Numerous studies have been made on the subject and standards for basic modules have been adopted in many countries. The present tendency is to adopt 100 mm. as the basic module for component standardization and 3 and 6 times this size as multimodules for larger dimensions used in building planning.[30-35] ISO has also issued two Recommendations on the subject[36,37] and thus put a seal of approval on the modular movement which is picking up momentum all over the world. Considerable progress has been made, but there are many problems of detailed character yet to be resolved before it becomes universally adopted as a general practice in the building construction and other industries.

INTERCONVERSION OF VALUES

23.10 The existence of two major systems of units of measurement in the world (inch-pound and metric) and the exchange of technologies and trade between countries using the two systems had made it necessary to evolve a suitable method of interconversion of values from one system of units to the other. Normally, ordinary tables of conversion factors for the various units issued by standards bodies and other organizations would have sufficed to serve most of the needs, for all that was required was a simple operation of multiplication of a given value in one system with the appropriate conversion factor to obtain the requisite value in the other system. It did not much matter how many places of decimal were retained or how many were dropped – a rough and ready rounding-off operation sufficed. This was so because most of the quantities requiring such conversion pertained to measurements of bulk materials, in terms of large units, such as mineral ores in tonnes, timber in cubic metres and so on. But when it came to more sophisticated manufactured products requiring dimensional interchangeability, more precise conversions were called for and certain agreed methods had to be evolved and adopted.

23.11 One of the first attempts at evolving a procedure for such interconversion was made in 1933 by the American Standards Association, which issued a standard on the American practice for inch-millimetre conversion for industrial use.[38] Though this standard was later revised in 1947, its approach remained *ad hoc*. The reliance placed in the standard is mainly on a prescribed number of decimal places to be retained in the converted values, without any regard to the precision of measurement or the exact value of tolerances involved, stated or implied. One of the well thought-out contributions of importance bearing on this subject was that by Nickols,[39] who took full account of precision of the original value and the tolerances associated with it. Nickols also considered the statistical variability of measurements and proposed a method which was later incorporated in British and other standards.[40–42] The method tackled the problem of interconversion of linear inch-millimetre dimensions quite satisfactorily. It ensured interchangeability between parts made to the original given dimensions and those made to converted dimensions. It maintained the precision of the original value within 5 percent, which was adequate for interchangeability. But having been designed to deal with the specialized problem of linear sizes only, it did not lend itself for use in a generalized manner to deal with the interconversion of quantities other than linear or for that matter even with linear quantities other than those expressed in inch or millimetre units.

23.12 Nickols' method was based on tables giving ranges of tolerances of given values in inches and millimetres. Corresponding to each range was indicated the unit to which the converted value was to be rounded off. The method was thus too specific and could in no way be extended to other units except by the device of compiling similar sets of tables for each pair of units for each quantity separately. Furthermore, if more than just two systems of units were to be dealt with, it would require fresh sets of tables for each possible pair of units of interest in the various systems. This would involve a tremendous amount of work which could perhaps be avoided, if a more generally applicable method could be evolved for use with any pair of units from any pair of system of units. The author was faced with this problem of precise interconversion when in 1955 plans were being made for the wholesale adoption of metric system in India which included the task of rewriting the existing standards in metric terms and also of preparing new metric standards to help expedite the conversion of existing industrial practices which were mostly in non-metric units of measurements, including inch-pound as well as indigenous. A solution was finally found resulting in a generalized method which was duly incorporated in an Indian Standard.[43] This method is presently under consideration of ISO for worldwide adoption.

23.13 The method depends upon the definition of a new mathematical concept, namely, the significant part of a number, which consists of the significant digits occurring in a given value written down as an integral number without a decimal point and without the non-significant zeros. For example:

Given value	*Significant part*
0.0591	591
2.110	2110
4.1×10^3	41

The procedure for conversion then simply reduces itself to the determination of the appropriate fineness of rounding to be used for rounding off the conventionally converted value. This is determined as follows:

(1) Write down the significant part of the original value for using as standard of comparison. Call it S_o.

(2) Write down the significant part of the converted value and drop from it one digit at a time, starting from the right, until the significant part assumes, for the first time, the same order of magnitude as S_o. Call this S_1.

(3) Drop another significant figure from S_1 and observe that the resultant value S_2 will also have the same order of magnitude as S_o.

(4) Of S_1 and S_2, choose one as S which bears the least ratio to S_o. That is:

$$\text{if } \frac{S_1}{S_o} \leqslant \frac{S_o}{S_2}, \text{ choose } S_1 \text{ as } S;$$

$$\text{if } \frac{S_1}{S_o} > \frac{S_o}{S_2}, \text{ choose } S_2 \text{ as } S.$$

(5) The fineness of rounding should then be taken as unity in the last place retained in S.

(6) Using the fineness of rounding thus determined, round off the calculated converted value and retain in the final converted value all the significant zeros and drop from it all the non-significant zeros.

23.14 An example is perhaps called for to illustrate the use of the above procedure. Suppose a specification calls for the weight of a woven fabric to be 12.5 ± 0.3 ounces per square yard, which is required to be converted to metric units of grams per square metre. First of all, the limits of the given value may be written as 12.2 and 12.8, which have to be converted separately. Looking up in tables or working out from separate tables of conversion for ounces to grams and from square yards to square metres, one finds the appropriate conversion factor as being 33.906 g. per sq. m./oz. per sq. yd. This factor should be taken to at least two significant figures more than those in the given value, so that an adequate margin of accuracy is maintained for proper rounding off. Now multiplying the given limiting values with this factor one gets the converted values as 413.6532 and 433.9968 g. per sq. m. The significant parts of the given and converted values may be written as –

Given values	*Converted values*
122	413 6532
128	433 9968

By striking off one digit at a time from the first converted value according to step (3) of the procedure detailed above, one finds that

$$\frac{S_1}{S_o} = \frac{413}{122} = 3.39$$

$$\text{and } \frac{S_o}{S_2} = \frac{122}{41} = 2.98$$

which means that

$$\frac{S_1}{S_o} > \frac{S_o}{S_2}$$

According to step (4) of the procedure, S_2 or 41 should be chosen as S. The last place retained in 41 corresponds to a fineness of rounding of 10

g. per sq. m. Thus the converted value would be 4.1×10^2 g. per sq. m. Notice that it should not be written as 410, because that would imply a degree of accuracy of ± 0.5 g. per sq. m., which is much greater than that given in the original value (*see also* Chapter 22, paras 22.15–22.18). Similarly, the upper limiting value would become 4.3×10^2 g. per sq. m. The two could now be written together as $(4.2 \pm 0.1) \times 10^2$ g. per sq. m.

23.15 This method of precise interconversion to ensure interchangeability has been further simplified and reduced to a tabular form by Sen.[99] In using this table, reproduced as Table 23.1, one must determine the error of measurement or the precision of the original given value and with the knowledge of the conversion factor read off the fineness of rounding from the appropriate column and row of the table.

The converted values can then be rounded off directly by the use of this fineness. For toleranced values the two limits are converted and rounded separately. If it is desired that the converted limits must not exceed the original value, the upper and lower limits of the converted value should be rounded downward and upward respectively.

RULES FOR ROUNDING OFF

23.16 In all such calculations and in many other situations, values have to be rounded off to the desired fineness of rounding retaining only the requisite number of places and dropping the rest. The procedure for rounding off has been standardized in most countries, which is more or less uniform.[44–46] It is generally agreed that when there is an equally good choice for rounding to be done upward or downward in the last place retained, the rounding should be done so that the retained figure remains even. Thus 4.675 and 4.685 would both be rounded to 4.68. In adopting such a rule, it should not really matter whether the even or odd figure were chosen, to which to round off, because statistically either would lead to a sound solution. But it is most essential that everyone should use the same procedure, and now that the rule of even number has been generally adopted, it should be universally maintained. It is understood that in ISO also the same practice is being adhered to.

STATISTICAL QUALITY CONTROL

23.17 When, way back in 1924, Walter A. Shewhart first proposed the application of statistical methods for the control of quality of mass produced telephone components, it could hardly be suspected that

TABLE 23.1

FINENESS OF ROUNDING (FR) FOR INTERCHANGEABLE CONVERSION

Error of Measurement											
Above →		0.000000158	0.00000158	0.0000158	0.000158	0.00158	0.0158	0.158	1.58	15.8	158
Up to →		0.00000158	0.0000158	0.000158	0.00158	0.0158	0.158	1.58	15.8	158	1581
Corresponding Precision / Conversion Factor ↓		± 0.0000005	± 0.000005	± 0.00005	± 0.0005	± 0.005	± 0.05	± 0.5	± 5	± 50	± 500
Above	*Up to*	FR	FR	FR	FR	FR	FR	FR	FR	FR	FR
31.6 x 10^6	31.6 x 10^7	10^2	10^3	10^4	10^5	10^6	10^7	10^8	10^9	10^{10}	10^{11}
31.6 x 10^5	31.6 x 10^6	10	10^2	10^3	10^4	10^5	10^6	10^7	10^8	10^9	10^{10}
31.6 x 10^4	31.6 x 10^5	1	10	10^2	10^3	10^4	10^5	10^6	10^7	10^8	10^9
31.6 x 10^3	31.6 x 10^4	0.1	1	10	10^2	10^3	10^4	10^5	10^6	10^7	10^8
31.6 x 10^2	31.6 x 10^3	0.01	0.1	1	10	10^2	10^3	10^4	10^5	10^6	10^7
316	3160	0.001	0.01	0.1	1	10	10^2	10^3	10^4	10^5	10^6
31.6	316	0.0001	0.001	0.01	0.1	1	10	10^2	10^3	10^4	10^5
3.16	31.6	0.00001	0.0001	0.001	0.01	0.1	1	10	10^2	10^3	10^4
0.316	3.16	0.000001	0.00001	0.0001	0.001	0.01	0.1	1	10	10^2	10^3
0.0316	0.316	0.0000001	0.000001	0.00001	0.0001	0.001	0.01	0.1	1	10	10^2
0.00316	0.0316	0.00000001	0.0000001	0.000001	0.00001	0.0001	0.001	0.01	0.1	1	10
0.00316	0.00316	0.000000001	0.00000001	0.0000001	0.000001	0.00001	0.0001	0.001	0.01	0.1	1

statistical quality control (SQC) methods would assume such enormous importance as they have today in the field of industrial production in general and standardization in particular. Statistical methods of quality control furnish a most effective and economical tool for ascertaining the quality level which a given product may be achieving during production. From its results, it can readily be judged whether a given standard is being adhered to. In case it is not, the extent of the deviation can be gauged and corrective action instituted. Furthermore, the results obtained from quality control operations can advantageously be used as feedback for further improvement of standards.

23.18 It is obvious that the employment of SQC methods is most essential for an effective implementation of standards, but quality control process can well proceed without the existence of a standard. Though the quality control experience is often reflected in the improvement of existing standards, it has not so far been considered as an essential prerequisite for the initial establishment of new standards. The experience of the past several years has, however, indicated that greater concern with quality control processes during the initial establishment of standards would lead to much more practical and sound standards.[47] This has been especially emphasized in the context of the newly developing economies of the world.[48] Furthermore, a new concept of total quality control[49, 50] has been developed during the past few years which, besides standardization, includes within its purview the fields of market research, product design and consumer satisfaction. From a very modest beginning in the early twenties, SQC has thus come to assume an extremely important position in world economy. It is not proposed here to deal with all aspects of quality control except to indicate briefly the rudiments of the basic methods involved.

23.19 To determine the quality characteristics of a batch of products or a continuously running production line, it is hardly feasible to examine each unit. For one thing, the tests involved may be destructive, for another, they would be rather time-consuming and expensive. Though in certain cases 100 percent inspection becomes necessary and is found feasible, yet a statistical approach is quite often adequate, whereby an appropriate size of sample is taken and examined, and the results obtained are used to deduce the quality of the original batch or the output of the production line. The manner in which samples are taken and the results of tests run on them interpreted and reflected back into the production line constitute the SQC methods. These methods, which have been highly developed and refined during the past four decades or so, help in improving the economy and quality of production in several ways;[51-55] for example by:

(1) leading to the reduction of defects in the product and consequently of the percentage of rejections;

(2) predicting the impending trouble before much wider and unacceptable deviation of quality occurs;

(3) saving in raw materials, machine time and labour;

(4) reduction in cost of inspection;

(5) furnishing better basis for establishing specifications and improving the existing quality levels without any basic changes in manufacturing techniques and processes employed; and

(6) helping to narrow down tolerances and variations from the normally specified values of the characteristics and thus ensuring a more uniform product quality.

23.20 The statistical quality control methods are based chiefly on the assumption that under normal conditions variations in product quality would be distributed according to the Gaussian law of chance and any undue deviation outside the expected limits should be ascribed to a non-chance cause which should be ascertained and eliminated. Though distribution patterns other than Guassian may sometimes occur, they are quite rare. In any case, if the shape of the distribution curve is known, the problem can readily be handled by widening the control limits appropriately in one direction or the other. But in most cases Gaussian assumption can be and is adhered to. The basic procedure for setting up quality control involves the establishment of control charts on which periodic inspection data is plotted continuously as production proceeds and the trend of the plot watched carefully in order to detect any undue deviations which may not be ascribable to chance causes. If one such deviation is detected, it is a signal for alarm and investigations are instituted to determine the exact cause and to eliminate it. Sometimes the trouble may be serious enough to have led to the production of an appreciable size of batch of defective product in between two sampling and inspection intervals. In such cases, further inspection of the particular batch involved would be called for to determine whether it could be freed of defectives by 100 percent inspection, reprocessed, scrapped or otherwise disposed of.

23.21 Pioneering work for the standardization of methods and procedures for SQC operations was carried out in the UK and USA and several standards were published[56-58] during the thirties and early forties.

In later years, these methods were adopted, adapted and further developed within these countries and elsewhere, and several other national standards were published on the subject.[59-62] All these standards are quite self-contained and include all the necessary statistical constants and other data

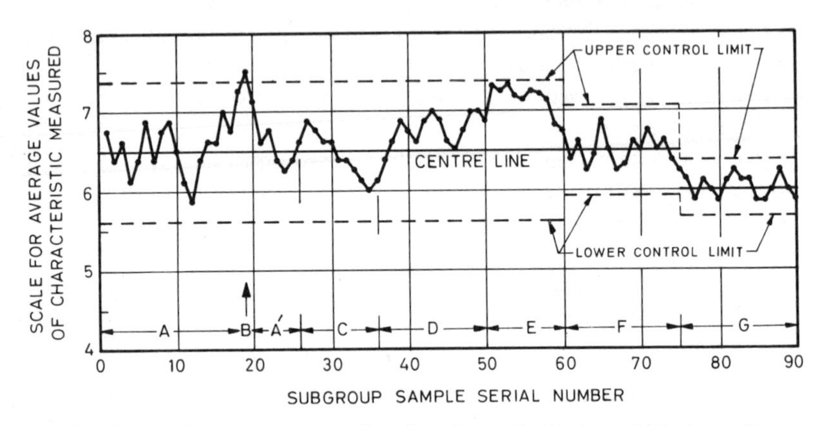

Fig. 23.1 Illustrative example of an hypothetical statistical quality control chart.

required for guiding a non-statistician in setting up quality control procedures for controlling quality during production. The main steps necessary for this purpose include:

(1) the selection of the characteristics of the product whose production is to be controlled;

(2) the determination of the size and frequency of the subgroup or sample to be drawn for inspection and testing, and designating the point or stage of production from which the samples may be drawn;

(3) designation of the tests to be run on the sample and the mode of collection and recording of the relevant data;

(4) determination of the types of control charts to be maintained, which may cover any one or sometimes two but rarely more of the following statistical measures:

Quantitatively measured quality characteristics

(a) $\bar{x}$, the average or arithmetic mean of the observed values for the subgroup,

(b) σ, standard deviation or the root mean square deviations of the observed values,

(c) R, the range or the difference between the largest and the smallest observed values;

Qualitatively ascertained characteristics

(d) p, fraction defective or the ratio of defective units to the number of units inspected in the sample, sometimes expressed as percent defectives or 100p,

(e) c, the number of defects in a sample of constant size;

(5) preparing the layout of data sheet and control chart;

(6) collection of preliminary data on 10 to 25 samples;

(7) calculation of the central and control lines of the chart and plotting the actual data; and

(8) as data accumulate and occasion may demand, redetermining fresh central line and control limits.

23.22 During the actual operation of quality control, the observational data is continuously calculated and plotted on the control chart, which is carefully watched to detect any undue variation from random behaviour and to detect any trends that may be indicative of non-chance or assignable causes coming into play. In Fig. 23.1 a hypothetical control chart is given in which it has been attempted to illustrate the various dispositions which the plot of control chart may assume and which would indicate the various courses of action to be taken. In region A of the chart, the observed averages lie well within the upper and lower control limits and are randomly scattered above and below the central line. The process in this disposition may be taken to be well under control. At B is shown the point corresponding to the average of subgroup sample serial number 19 as having gone beyond the upper control limit. This is the warning signal that a certain assignable non-chance cause has come into play in the process which should immediately be investigated, detected and removed. The next few points in region A′ indicate that the disturbing cause has perhaps been removed and the process has come again under control. In region C, however, a decreasing tendency of the average values is clearly to be observed. This too may be due to some ascertainable cause such as a progressive deterioration of the machine which would be worth looking into. The next region D illustrates a situation in which most of the points lie above the central line and if this tendency continues for an appreciable length of time there would be justification for recalculating and shifting the position of the central line and perhaps that of the control limits. Similarly, in region E the points lie too close to the upper limit. Here again if the tendency continues to persist there would be justification for action. Region F shows that the deviations in the observed values have been considerably reduced which may justify recalculation of control limits and tightening them accordingly. Region G illustrates a situation quite similar to D and E in which both the central line as well as the control limits have been recalculated and altered. Another situation that may well arise and which has not been illustrated in the figure, may be that without any change in the spread of the control limits, the position of the central line may justify a shift upward or downward. Generally, an $\bar{x}$ chart is accompanied by either a σ chart or an R chart, which further helps in detecting the changes in variability.

23.23 It is clear that the control chart method of quality control furnishes a sensitive means for keeping one's fingers on the pulse of the production process and helps one readily to diagnose significant variations in the process, which might require corrective action. Apart from keeping a tab on production and product quality, the method furnishes, as stated earlier, a positive means for ensuring that a particular specification is being adhered to.[60] Furthermore, the data so accumulated provides a sound basis for improving or relaxing the previously agreed specification in the light of the process capabilities discovered. Regions D, E, F and G, in Fig. 23.1, illustrate situations where such an action may be considered justified. It would thus be clear how standardization and quality control techniques are closely interrelated and how they depend on each other. In this brief description of SQC methods, no attempt has been made to go into mathematical details and operational directives, which will be found abundantly described in the literature and the standards cited.

STATISTICAL SAMPLING

23.24 SQC techniques described earlier in this Chapter have not yet been adopted universally enough to serve in all cases the object of proving conformity to standard specifications and to eliminate the need for sampling and inspection of consignments for determining their acceptability. Even if a product does come out of a plant in which satisfactory operation of SQC exists, there may be a need for its subsequent inspection at the point of its transfer to the purchaser or user. A purchaser may, for instance, be over-anxious to ensure his own safety or have inadequate faith in the validity of the producer's quality control operation. Also, situations may arise where a producer himself may be desirous of double-checking his own operation or re-inspecting his stock before offering it for sale, for the simple reason that it may have been kept in store for an undue length of time. Obviously, it is uneconomical, unless justified by good and valid reason, to resort to inspection of every unit of the product with the object of determining its conformity to specifications. Statistical methods of sampling and statistical criteria for acceptance or rejection on the basis of inspection results are, therefore, extremely useful in this context.

23.25 In the recent couple of decades a great deal of attention has been paid to the development of statistical methods of sampling and inspection, particularly helpful in the process of ensuring compliance with standards; and many standards, national and others, have been issued on the subject.[63–75] These standards, besides describing the general principles, present in simple language the methods of sampling for inspection from which one may seek guidance for resolving each problem as it may arise. Special methods for dealing with certain classes of products are also available in standard form.[76–79] In addition, there has been afoot a move towards introducing, in standard specifications for specific products and commodities, specially designed clauses for sampling and acceptance criteria, based on statistical principles. This movement, which helps to absolve the inspection and testing authorities of the need for knowledge of mathematical statistics, is relatively recent and has still a long way to go before one may find most standard specifications complete in this respect (*see also* paras 23.28–23.29).

23.26 It may be recalled that the traditional standards also had sampling directives included for the purpose of selecting samples for testing and inspection. But they were seldom based on statistical considerations. One often came across directives to the following effect: "Take one sample from the consignment in such and such a manner and if on test it is found to be satisfactory, the consignment shall be accepted. If this sample fails, take one more sample. If the second sample passes accept the consignment, otherwise reject it." Sometimes a certain percentage of the number of items in the consignment was specified to be taken as a sample. In both these cases statistical principles were ignored and reliance was placed entirely on inadequate evidence which could not guarantee the decisions taken for acceptance or rejection to be within any known margin of safety or, let us say, within reasonable limits of uncertainty. For a long time it had been felt that any departure from the nominal amount of testing and from the *ad hoc* procedures normally specified for sampling in standards would involve undue expenditure. Statistical sampling was considered to be extremely uneconomical, until it was realized, particularly during the Second World War, that without the assistance of statistical methods of sampling inspection no reasonable guarantee of conformity to specifications could be expected. The extra effort and time involved were considered well worthwhile in the interest of safety and reliability of stores on which the lives of the fighting forces depended.

Industry today is also beginning to rely more and more on statistical methods, because it is being gradually realized that, in the long run, unless the utility and suitability of stores is ensured, no long-range profitability could be expected to result. In addition, consumer pressure is building up to force industry to guarantee the quality of its products and their serviceability, which can only be done through statistical approach to sampling inspection.

23.27 Considerable literature is now available which deals with the basic statistical principles involved and presents the relevant tables for statistical sampling.[80–82] Perhaps the best introductory treatment, suitable for study by a non-statistician standards engineer, would be the Indian Standard Manual IS:1548–1969.[68] But without going into the detailed theory, it may be of interest to mention here the different types of sampling plans that are available and to point out their relative merits. The so-called attribute plans which deal with the simple case of determining defective items or the number of defects in the sample include the following:

(1) *Single Sampling Plan (SSP)* One type of acceptance plan, which is simple in procedure, is the single sampling plan in which the evidence supplied by one sample of a specified number of items taken from each lot determines its acceptance or rejection. Thus, a single sampling plan is normally stipulated by specifying a sample size (n) and an acceptance number (a). If the number of defectives found in the sample is less or equal to a, the lot is accepted, otherwise rejected. Hence, rejection number is given by $a + 1$ (also denoted by r). This plan, though simple to operate, is more expensive than other plans because of the relatively larger size of samples it requires to be tested.

(2) *Double Sampling Plan (DSP)* In this type of sampling plan, a first sample is taken from every lot and the evidence is used to accept the lost, to reject the lot or to reserve decision until further information from a second sample is obtained. Thus, while one sample from each lot is taken and never more than two samples from any lot, it would be seen that several lots would be dealt with in the same manner as in the case of SSP but with a much smaller size of sample. Only the lots for which decision is reserved would be required to be sampled twice. The aggregate size of the samples per lot on the average will always be smaller than the sample required for SSP. It is obvious, therefore, that though somewhat more elaborate, the DSP would be more economical to operate than SSP.

(3) *Multiple Sampling Plan (MSP)* This plan is the extension of the

DSP, in which a certain specified number of samples are taken in succession and tested before a lot is declared as accepted or rejected. The MSP is naturally more economical with regard to the overall number of samples per lot to be handled on the average, but is obviously somewhat more elaborate to operate.

(4) *Sequential Sampling Plan (QSP)* While in MSP, a decision is attempted after each sample is inspected and in any case a decision is arrived at after inspecting the last allowable sample, in QSP a decision is attempted on the cumulative evidence gathered at each stage of inspecting the successive series of samples, each consisting of one item alone. This type of inspection is continued till a categoric decision to accept or reject the lot becomes possible. There is, therefore, no question of predetermined sample size or number of samples for QSP. This plan leads, on an average, to the least amount of inspection among all known plans. Obviously, it is the most economical plan as well as the most elaborate requiring a decision to be made after inspecting each successive item.

23.28 In cases where the quality characteristics of a product are defined quantitatively in terms of measurable properties, the same attribute type of plans can be used by classifying the items as acceptable and unacceptable on the basis of measured values of each characteristic. But more sensitive plans are also available – the so-called variables plans which again comprise the single, double, multiple and sequential varieties. Of these only the single sampling plan is normally used, others being found rather cumbersome for direct application. Here again a sample of n items is tested according to the specified method of test for determining each of its characteristics. The values so determined are subjected to certain simple calculations to obtain certain parameters with the help of statistical tables. These parameters are compared with the specification limits of each of the characteristics and if found to bear the expected relationships with the latter, the lot is accepted, otherwise rejected. For the details of calculations involved and the requisite tables, reference is invited to the relevant literature,[63-75] particularly to IS:2500 (Part II)–1965.[70]

23.29 Of the plans described above, a particular one may be found to be preferable under a given set of conditions. But when it comes to writing standard specifications for a specific product, it becomes necessary to take a decision once and for all, which would be applicable under most varying circumstances prevailing in respect of production, distribution and utilization of the particular product. Economy and convenience of operation have to be balanced against efficiency; at the same time, some knowledge of the variability of the product has to be available to help decide on the type of sampling plan to be adopted and the sampling clauses to be designed accordingly for incorporation in the standard. Though single sampling plans are often adopted for their simplicity, double sampling plans are sometimes preferred for their economy. Multiple and sequential sampling plans are seldom chosen, chiefly because of their complexity. Furthermore, the consumer's viewpoint has also to be taken into account. He may be satisfied with the specification limits for the various characteristics, but the sampling inspection procedures cannot give him any guarantee that each and every item of the product in a lot would meet these requirements. The 100 percent inspection which can guarantee this can only be resorted to in very special cases. In a majority of cases, the consumer has to be prepared to take a certain amount of risk that a small percentage of items in a consignment would be substandard or defective, or, alternatively, a small percentage of the total number of lots accepted may contain more than the allowable number of defective items. In statistical language this is called consumer's risk. Similarly, the producer also has to take the so-called producer's risk of having a small percentage of good lots being rejected. The practice has grown that in designing specification clauses for sampling, consumer's risk is taken at 10 percent and producer's risk at 5 percent. This would hardly appear justified – there is no reason why both risks cannot be equalized and perhaps fixed at 5 percent each.

23.30 In the light of these considerations, the standards engineer, should be able to decide, with the help of the statistician, the type of sampling plan most suited to the product in hand, and design the sampling and acceptance clauses of the specifications accordingly. As mentioned earlier (*see* para 23.24), more and more National Standards Bodies are beginning to introduce such statistically designed procedures in national standards[83-88] as would give an adequate assurance of quality to the purchaser at an economically bearable cost of sampling and inspection. In this process, many conflicting claims would have to be considered and compromises arrived at. So far, however, no National Standards Body has laid down a standard procedure and the associated norms for guiding the design of such sampling clauses under different circumstances. As experience accumulates this lacuna too will no doubt be filled in due course. The standards engineer would then be relatively independent of the statistician, for, with a modicum of statistical background, he would be able to design his own sampling clauses.

PRECISION OF TEST METHODS

23.31 When replicate determinations of a given property are made on a material or product by a given test method, the observations are found to vary. Sometimes the differences may be due to incompleteness of instructions in the test method and sometimes to errors in measurement or recording. But even if these sources of discrepancies are removed or minimized, which should be done in any case, the repeat results would still not be found to be exactly the same. The extent to which replicate results may differ, or in other words, the degree of their variability, will depend upon the effectiveness of the test method and uniformity of the conditions under which the test is repeated. While prescribing a test method, therefore, it is useful if this variation between the replicate determinations, or the precision of the test method, is known in quantitative terms so as to serve as a guide to the users of the method for judging the acceptability of the test results. For example, if an operator while determining the oxidation stability of aviation gasoline finds the residue to be 30 mg./100 ml., he may like to check whether this value conforms with the value of 28 as obtained by him previously or with 24 as obtained by some other worker in another laboratory. In the absence of any information about the precision of the test method itself, he would not be able to take a correct decision.

23.32 The precision of a test method assumes different values according as the replicate determinations are made by the same operator under same conditions or by different operators under different conditions. The numerical estimate of the variations under former conditions is known as "repeatability" and under latter "reproducibility." The two terms are defined as follows:

> *Repeatability* is the quantitative measure of the closeness of agreement among the successive results obtained by the same method on identical test material and under the same conditions (same operator, same sample, same apparatus, same laboratory and almost the same time). It is defined as that difference between two such single and independent test results as would be exceeded in the long run in only one case in twenty in the normal and correct operation of the test method.

> *Reproducibility* is the quantitative measure of the closeness of agreement among the test results obtained by the same method on identical test material but under different conditions (different operators, same sample, different apparatus, different laboratories and/or different times). It is defined as that difference between two such single and independent results as would be exceeded in the long run in only one case in twenty in the normal and correct operation of the test method.

23.33 In practice repeatability and reproducibility are quantitatively determined by gathering adequate experimental data from replicate tests run on the same sample in several laboratories by several operators. The procedural details of the methods to be followed for this purpose are gradually being published in standard form.[89] The value of repeatibility is obtained by calculating the standard deviations of the relevant test results obtained within each laboratory and multiplying it by a factor 2.77. Similarly, reproducibility is obtained by multiplying the standard deviation of results from various laboratories by the same factor. Obviously, the less either of these values is the better would be the method. Furthermore, for any one test method, the reproducibility value would generally be larger than the repeatability value. The knowledge of both repeatability and reproducibility is necessary for the comparison of two or more test methods for selection. For, if a number of test methods is available for the same purpose, the one with the least repeatability and reproducibility values should be preferred. Since the reproducibility is of greater importance, the choice may first be made according to the best reproducibility value; but if the reproducibilities of two methods happen to be approximately the same, the choice may be made according to the best repeatability value. However, if the test methods have more or less the same reproducibility and repeatability, the one which is most economical in operation (or least time-consuming) should be preferred. Furthermore, a quantitative knowledge of reproducibility would assist in fixing realistic tolerances for specification values,[90] that is, in such cases where tolerances are dependent on the precision of measurement rather than on the restraints imposed by the utility of the product. Thus, it is recommended that for two-sided tolerances, the specification range may be equal to at least three times the reproducibility.

STATISTICAL TESTS OF SIGNIFICANCE

23.34 Statistical tests of significance are important tools in industrial experimentation and decision-making. They are useful in finding out whether, in case of one sample, the estimates obtained from it differ significantly from certain specified values or whether, in case of two samples, the estimates obtained from them differ significantly from each other. They are also helpful in determining whether a given sample could have arisen from a specified lot. Thus, it may be desirable to find out whether a new germicide is more effective in treating a certain type of infection than a standard germicide, whether a new method of sealing light bulbs will improve their torque test characteristic, or whether one method of preserving foods is better than another insofar as the retention of vitamins is concerned. In such cases, it would be necessary to examine whether the mean values as obtained are the same or different. There may also be cases where it may be worthwhile to find out whether one inspector is more consistent than another or whether a new source of raw material has resulted in a change in properties or consistency of output, or whether the temperature of the bath in which the cocoons are cooked affects the uniformity of the quality of the silk. In these cases it will be necessary to determine whether the variances obtained are the same or not.

23.35 The procedure commonly adopted is to first set up a null-hypothesis regarding the equivalence (or no difference) between the values of the parameters of the lots from which the samples have been drawn. In certain cases, the null-hypothesis may simply refer to the equivalence of the lot parameter to some specified standard values. The appropriate test of statistical significance is then applied on the basis of the statistics computed from the sample and determining whether the observed difference between the statistics could have arisen just due to chance. If the probability of obtaining the observed difference due to chance alone is quite small, it is concluded that the observed difference is statistically significant. The null-hypothesis is then rejected and the values of the lot parameters are taken to be significantly different from each other. In industrial applications, the levels of probability at which the significance is judged are normally taken to be 5 percent or 1 percent. If the observed probability is less than 5 percent but more than 1 percent, it is taken to be significant whereas if it is less than 1 percent, it is taken to be highly significant. Both 5 percent and 1 percent levels of significance are the conventional levels. The particular level to be adopted in practice depends upon the nature of the problem and the seriousness of the decision.

23.36 The three well-known tests of significance suited for most purposes include the t-test, the F-test and the chi-squared test, which have been described in fair detail in most reference books and textbooks on statistics.[91-95] In the case of each test, certain underlying assumptions have to be made; for example, that the observations follow the normal Gaussian distribution and that they are random and mutually independent. If there is any doubt about the validity of the requisite assumptions, it is advisable to seek the guidance of a competent statistician and not to use the tests indiscriminately. A beginning has been made towards incorporating in national standards the basic principles and detailed methods of the various tests together with illustrative examples,[96] thus making them more readily available for use by standards engineers to help solve the practical day-to-day problems.

DESIGN OF EXPERIMENT

23.37 In the constant process of introducing new products and new methods of production and improving the existing ones, the first essential need is to evaluate the superiority of every innovation over what may be presently available. There is also the question of the variables entering in or influencing a process or product. Most problems of this type are relevant to standardization effort at all levels, but at the company level such situations arise perhaps more often than elsewhere. The experiments to be conducted for investigating these questions may be designed in many ways, but the statistical design of such experiments would yield the results most effectively and economically. Methods for statistical design of experiments have already been quite well developed[91-95] and extensively utilized in industry as well as agriculture. However, they have not yet entered the field of standardization in any big way, nor has any attempt been made to present them in a simplified standard form for ready use by the standards engineer.

CONCLUSION

23.38 From several examples given in this and other chapters, it will be seen that mathematical methods have begun to make significant contributions in many ways to the advancement of standardization. With the help of statistical techniques, standards are being made more and more realistic, leaving less and less to chance and guesswork. The potential of standards for contribution to economical production and utilization of goods and services at all levels is being considerably enhanced by the mathematical approach to problems of decision-making, such as the definition of a series of sizes or the modular coordination of components of structures and systems. Many possibilities of mathematical methods have still to be explored and realized. Some of them have been indicated during the course of discussions. Certain new lines of thought are beginning to be pursued, though the ultimate utility of some has still to be determined: for example, Hatanaka's attempt at developing a method for statistical estimation of what he calls the "standardization condition" of components in complex assemblies,[97] his object being to identify the components or classes of components which should receive high priority for further standardization and rationalization. One obvious new development would be that computers would come to be much more extensively used in resolving the really complex standardization problems. Considering the very large number of standards in the world at all levels and even larger number and variety of requirements and other details contained in them, computers have a tremendous potential for service to the standards engineer for information storage and retrieval. Efforts are already afoot at ISO and UNESCO level to set up a worldwide computerized information service on standards.[98] But the class of problems for which computers would be indispensable would be those involving a large number of variables and a variety of possible permutations and combinations for which computational speed is most vital, such as finding the most rational and economical solution of the problem of standardization of shoe and garment sizes on the basis of anthropomorphic measurements of ethnical groups or national populations, the mathematical problem of proving Renard's series as offering the most economical solution for standardizing series of sizes and the like.

REFERENCES TO CHAPTER 23

1. Gnedenko, B. V.: Standardization and mathematics. *Standarty i Kochestvo* (Moscow), 1967, no 6, pp. 88-90.
2. Kvasnitsky, V. N. and Levintov, A. G.: Some trends in the development of scientific foundations for standardization. *Standarty i Kochestvo* (Moscow), 1970, vol 34, no 5, pp. 39-41 (*in Russian*).
3. Gnedenko, B. W.: Mathematical methods in standardization. Normalizacja, 1969, vol 37, no 4, pp. 181-184 (*in Polish*).
4. *Application of mathematics to standardization.* Report of Symposium (collection of 14 papers), 1969, Warsaw, Research Centre for Standardization.
5. Weir, I. R.: A standard aid for designers – preferred numbers. *The Magazine of Standards,* 1962, vol 33, pp. 73-78.
6. Kienzle, Otto: *Normungszahlen* (Preferred numbers). Springer-Verlag, Berlin, 1950, p. 339 (*in German*).
7. Oliver, Bernard M.: The preferred numbers nobody prefers. *Standards Engineering,* 1969, vol 21, no 5, pp. 9-10.
8. *ISO/R 3–1953 Preferred numbers. Series of preferred numbers.* International Organization for Standardization, Geneva, p. 7.
9. *NF X 01–001–1957 Nombres normaux* (Preferred numbers). Association Française de Normalisation, Paris, p. 12 (*in French*).
10. *DIN 323 Sheet 1–1952 Preferred numbers. Basic numbers. Exact numbers. Rounded numbers.* Deutscher Normenausschuss, Berlin, p. 2.
11. *IS:1076–1967 Preferred numbers (first revision).* Indian Standards Institution, New Delhi, p. 29 (earlier edition: 1957).
12. *BS 2045:1965 Preferred numbers.* British Standards Institution, London, p. 20.
13. *Z 17.1–1958 Preferred numbers.* American Standards Association (now American National Standards Institute), New York, p. 22.
14. *ISO/R 17–1955 Guide to the use of preferred numbers and of series of preferred numbers* (*see also* Amendment ISO/R 17/A1–1966). International Organization for Standardization, Geneva, p. 11.

15. *ISO/R 497–1966 Guide to the choice of series of preferred numbers and of series containing more rounded values of preferred numbers.* International Organization for Standardization, Geneva, p. 11.
16. *DIN 323 Sheet 2–1959 Preferred numbers. Decimal geometric series. Explanations, instructions for use, calculating the preferred numbers.* Deutscher Normenausschuss, Berlin, p. 11.
17. Berg, Siegfried: *Angewandte Normzahl* (Preferred numbers in use). Beuth-Vertrieb GMBH, Berlin, 1949, p. 192 (*in German*).
18. Berg, Siegfried: Einführung der Normzahlen in die Praxis (Introduction of preferred numbers in practice). *DIN-Mitteilungen,* 1962, vol 41, pp. 65-69 (*in German*).
19. Erdmann, Hans: Von der "Kunst" des Rechnens mit Vorzugszahlen (Technique of calculations with preferred numbers). *DIN-Mitteilungen,* 1962, vol 41, pp. 197-200 (*in German*).
20. Meriel-Bussy, H.: Trois documents sur la normalisation des nombres (Three documents on standardization of numbers). *Courrier de la normalisation,* 1968, no 201, pp. 222-224 (*in French*).
21. Friedewald, Hans-Joachim: Wesen und Anwendung der Normzahlen (Practice of using preferred numbers). *DIN-Mitteilungen,* 1967, vol 46, pp. 164-169 (*in German*).
22. Gai, A. F.: Discussion on the mathematical analysis of the system of preferred numbers. *Standarty i Kochestvo* (Moscow), 1970, vol 34, no 2, pp. 55-58 (*in Russian*).
23. Zaremba, W. A.: A new approach to size selection. *Mechanical Engineering,* 1968, vol 90, no 2, pp. 23-27.
24. Tkachenko, V. V.: *Method and practice of standardization.* Komitet Standartov Gosudarstvenniy Soveta Ministrov SSSR, Moscow, 1967, p. 612 (*in Russian*).
25. Corbusier, Le: *The modular.* Faber and Faber, London, 1951, p. 243.
26. Corbusier, Le: *Modular 2.* Faber and Faber, London, 1958, p. 336.
27. *Modular coordination in housing.* Report No TAO/GLOBAL/4, 1966, United Nations, New York, 1966, p. 44.
28. *Modular coordination.* Second report of EPA project 174. Organization for European Economic Cooperation, Paris, 1961, p. 224 (*see also* first report published in 1956).
29. *Basic principles of modular coordination.* Housing and Home Finance Agency, Washington, 1953, p. 29.
30. *NF PO1–001–1942 Modulation.* Association Française de Normalisation, Paris, p. 1 (*in French*).
31. *DIN 18000–1970 Dimensional coordination in building construction.* Deutscher Normenausschuss, Berlin, p. 8 (*in German*).
32. *IS:1233–1969 Recommendations for modular coordination of dimensions in the building industry.* Indian Standards Institution, New Delhi, p. 12.
33. *BS 4176:1967 Specification for floor to floor heights.* British Standards Institution, London, p. 4.
34. *BS 4606:1970 Recommendations for the coordination of dimensions in buildings.* British Standards Institution, London, p. 9.
35. *USAS A62.5–1968 Basis for the horizontal dimensioning of coordinated building components and systems.* United States of America Standards Institute (now American National Standards Institute), New York, p. 8.
36. *ISO/R 1006–1969 Modular coordination – Basic module.* International Organization for Standardization, Geneva, p. 7.
37. *ISO/R 1040/1–1969 Modular coordination – horizontal multimodules.* International Organization for Standardization, Geneva, p. 4.

38. *ASA B48.1–1933 (reaffirmed 1947) American standard practice for inch-millimeter conversion for industrial use.* American Standards Association (now American National Standards Institute), New York, p. 10.
39. Nickols, L. W.: Interconversion of inch and metric sizes on engineering drawings. *Engineer,* 25 Mar 1955, vol 199, no 5174, pp. 409-411. Reproduced in *Overseas Edition of Machinery Lloyd* 23 Apr 1955 and *ISI Bulletin,* 1955, vol 7, pp. 183-187.
40. *BS 2856:1967 Precise conversion of inch and metric sizes on engineering drawings.* British Standards Institution, London, p. 16.
41. *ISO/R 370–1964 Conversion of toleranced dimensions for inches into millimetres and vice versa.* International Organization for Standardization, Geneva, p. 11.
42. *AS Z30–1966 Interconversion of inch and metric dimensions.* Standards Association of Australia, Sydney, p. 51.
43. *IS:787–1956 Guide for interconversion of values from one system of units to another.* Indian Standards Institution, New Delhi, p. 19.
44. *ASA Z–25.1–1940 (reaffirmed 1961) Rules for rounding off numerical values.* American Standards Association (now American National Standards Institute), New York, p. 7.
45. *BS 1957:1953 Presentation of numerical values (fineness of expression; rounding off numbers).* British Standards Institution, London, p. 15.
46. *IS:2–1960 Rules for rounding off numerical values.* Indian Standards Institution, New Delhi, p. 12.
47. LaQue, F. L.: Standardization and quality control. *BSI News,* Oct 1971, pp. 17-18.
48. Sen, S. K.: Standardization methodology for developing nations. *ISI Bulletin,* 1971, vol 23, pp. 305-306.
49. Feigenbaum, A. V.: *Total quality control – engineering and management.* McGraw-Hill, New York, 1961, p. 627.
50. Middlecote, A. A.: Total quality control. *South African Standards Bulletin,* 1965–66, vol 19, pp. 257-259.
51. Shewhart, W. A.: *Economic control of quality of manufactured product.* Van Nostrand, New York, 1931, p. 501.
52. Cowden, Dudley J.: *Statistical methods in quality control.* Prentice-Hall, Englewood Cliffs, 1957, p. 727.
53. Hensen, Bertrand L.: *Quality control: theory and applications.* Prentice-Hall, Englewood Cliffs, 1963, p. 498.
54. Grant, E. L.: *Statistical quality control.* McGraw-Hill, New York, 1964, p. 610.
55. Juran, J. M.: *Quality planning and analysis.* McGraw-Hill, New York, 1970, p. 684.
56. *BS 600R:1942 Quality control charts.* British Standards Institution, London, p. 89 (earlier edition; 1935).
57. *ASA Z1.1–1958 and ASA Z1.2–1958 Guide for quality control, and control chart method of analysing data.* American Standards Association (now American National Standards Institute), New York, p. 28 (earlier editions: 1941).
58. *ASA Z1.3–1958 Control chart method of controlling quality during production.* American Standards Association (now American National Standards Institute), New York, p. 35 (earlier edition: 1941).
59. *AS (E)Z.501 to 503–1943 Guide for quality control, control chart method of analysing data, and control chart method for controlling quality during production.* Standards Association of Australia, Sydney, p. 73.

60. *BS 1313:1947 Fraction-defective charts for quality control.* British Standards Institution, London, p. 40.
61. *BS 2564:1955 Control chart technique when manufacturing to a specification.* British Standards Institution, London, p. 77.
62. *IS: 397–1952 Method for statistical quality control during production by the use of control chart (tentative).* Indian Standards Institution, New Delhi, p. 34 (revision due to be issued in 1972).
63. Singh, B. N.: Statistical methods in standardization. *ISO Souvenir.* Indian Standards Institution, New Delhi, 1964, pp. 90-95.
64. *ASTM Designation: E 105–58 Recommended practice for probability sampling of materials.* ASTM Standards, part 30, 1968, pp. 233-237.
65. *JIS Z 9031–1956 Random sampling methods.* Japanese Standards Association, Tokyo, p. 29.
66. *BS 2635:1955 Drafting specifications based on limiting the number of defectives permitted in small samples.* British Standards Institution, London, p. 11.
67. *PD 6452–1970 Guide to inspection procedures.* British Standards Institution, London, p. 11.
68. *IS:1548–1969 Manual on basic principles of lot sampling (first revision).* Indian Standards Institution, New Delhi, 1971, p. 48.
69. *IS:2500(Part I)–1963 Sampling inspection tables – Inspection by attributes and by count of defects.* Indian Standards Institution, New Delhi, p. 22.
70. *IS:2500(Part II)–1965 Sampling inspection tables – Part II Inspection by variables for percent defective.* Indian Standards Institution, New Delhi, 1966, p. 39.
71. *IS:4905–1968 Methods for random sampling.* Indian Standards Institution, New Delhi, p. 33.
72. *IS:5002–1969 Methods for determination of sample size to estimate the average quality of a lot or process.* Indian Standards Institution, New Delhi, p. 20.
73. *MIL–STD–105D–1963 Sampling procedures and tables for inspection by attributes.* US Department of Defence, Washington, p. 64.
74. *MIL–STD–414–1957 Sampling procedures and tables for inspection by variables for percent defective.* US Department of Defence, Washington, p. 110.
75. Singh, B. N.: *Industrial applications of statistical methods* (Paper TS.11-15) Proceedings of the International Conference on Quality Control, Tokyo, 1969, pp. 677-680.
76. *ASTM Designation: E 300–69 Recommended Practice for sampling industrial chemicals.* ASTM Standards, part 20, 1970, pp. 1191-1220.
77. *ASTM Designation: D 300-69 Recommended Practice for sampling of plastics.* ASTM Standards, part 27, 1968, pp. *618–632.*
78. See the following Indian Standards on Methods of Sampling of specific products issued by the Indian Standards Institution, New Delhi:
    IS:436(Part I)–1964: Coal, p. 23.
    IS:436 (Part II)–1965: Coke, p. 12.
    IS:1289–1960: Mineral gypsum, p. 15.
    IS:1405–1966: Iron ore, p. 14.
    IS:1449–1961: Manganese ore, p. 14.
    IS:1472(Part I)–1959: Ferro-alloys, Part I, p. 6.
    IS:1472(Part II)–1962: Ferro-alloys, Part II, p. 9.
    IS:1811–1961: Foundry sands, p. 13.
    IS:1817–1961: Non-ferrous metals for chemical analysis, p. 10.
    IS:1999–1962: Bauxite, p. 15.
    IS:2051–1962: Leather footwear, p. 7.

IS:2109–1962: Dolomite, limestone and other allied materials, p. 16.
IS:2213–1962: Thermosetting moulding materials, p. 7.
IS:2245–1962: Quartzite, p. 16.
IS:2246–1963: Fluorspar (fluorite), p. 14.
IS:2614–1969: Fasteners (first revision), p. 13.
IS:2817–1965: Coated abrasives, p. 8.
IS:3191–1968: Cryolite and aluminium trifluoride, p. 11.
IS:3535–1966: Hydraulic cements, p. 12.
IS:3704–1966: Light metals and their alloy products, p. 11.
IS:4156–1967: Barytes, p. 15.
IS:4166–1967: Ilmenite and rutile, p. 11.
IS:4711–1968: Steel pipes, tubes and fittings, p. 7.

79. Sampling plans in ASTM specifications – a report of a preliminary survey conducted by task group 4 of ASTM Committee E 11 on quality control of materials. *ASTM Bulletin,* July 1957, no 223, pp. 47-51.
80. Pearson, E. S. and Hartley, H. O.: *Biometrika tables for statisticians.* Cambridge University Press, London, vol 1, 2nd ed, 1958, p. 240.
81. Rao, C. R., Mitra, S. K. and Matthai, A.: *Formulae and tables for statistical work.* Statistical Publishing Society, Calcutta, 1966, p. 234.
82. Dodge, H. F. and Romig, H. G.: *Sampling inspection tables.* John Wiley, New York, 2nd ed, 1959, p. 224.
83. *AS Z.17–1962 Fibreboard containers for general purposes,* p. 11, and *AS 0 86–1964 Marine plywood,* p. 12. Standards Association of Australia, Sydney.
84. *C.S. 5:1967 Specification for double edged carbon-steel (untreated) safety razor blades,* p. 14.
C.S. 10:1968 Specification for frozen prawns (shrimp), p. 16. Bureau of Ceylon Standards, Colombo.
85. *DIN 267 Sheet 5–1968 Bolts, screws, nuts and similar threaded and formed parts – technical conditions of delivery, testing and acceptance.* Deutscher Normenausschuss, Berlin, p. 5.
86. See the following Indian Standard specifications, issued by the Indian Standards Institution, New Delhi:
IS:490–1967: Vaccine phials (first revision), p. 10.
IS:1300–1966: Phenolic moulding materials (second revision), p. 10.
IS: 3424–1966: Stainless steel table utensils, p. 9.
IS:3701–1966: Rubber protective sheaths (condoms), p. 11.
IS:4229–1967: Nylon sewing threads for aeronautical purposes, p. 8.
IS:4888–1968: Paper cones for winding yarn, p. 6.
IS:4992–1968: Door handles for mortice locks (vertical type), p. 8.
87. *S.I.143–1964 Canned fruit and vegetables (preserved by heat treatment),* p. 6, and
S.I.727–1969 Rubber safety boots with protective toe-caps, p. 7. Standards Institution of Israel, Tel-Aviv.
88. *BS 771:1959 Specification for phenolic moulding materials,* p. 10.
BS 3704: 1964 Specification for rubber condoms, p. 10.
BS 4590: Part 1:1970 Specification for litre non-returnable soft drink glass bottles, p. 16. British Standards Institution, London.
89. *IS:5420(Part I)–1969 Guide for precision of test methods – Part I Principles and applications.* Indian Standards Institution, New Delhi, p. 4. See also *IS:5420(Part II) Inter-laboratory testing (under preparation).*
90. Singh, B. N.: Use of statistical methods in writing specifications. *Bulletin of Quality Control Association,* 1958, vol 5, pp. 20-24.

91. Cochran, William and Cox, Gertrude M.: *Experimental designs.* John Wiley, New York, 1955, p. 459.
92. Davies, Owen L. *The design and analysis of industrial experiments.* Oliver and Boyd, London, 1954, p. 637.
93. Snedecor, George W.: *Statistical methods.* Iowa State College Press, Ames, 1956, p. 534.
94. Dixon, W. J. and Massey, Frank J.: *Introduction to statistical analysis.* McGraw-Hill, New York, 1957, p. 488.
95. Brownlee, K. A. *Industrial experimentation.* Her Majesty's Stationery Office, London, 1949. *Reprinted* 1957, p. 194.
96. *IS:6200–1971 Statistical tests of significance.* Indian Standards Institution, New Delhi.
97. Hatanaka, Setsuo: Method for statistical estimation of standardized condition. *JSA Technical Report No 3 – Researches and studies applied to standardization and quality control.* Japanese Standards Association, Tokyo, July 1968, pp. 21-24.
98. Computers vital for future development of international standardization. *ISO Bulletin,* Oct 1970, p. 3.
99. Sen, S. K.: Mathematics of interchangeable conversion. *ISI Bulletin,* 1968, vol 20, pp. 258-260.

Chapter 24

Interesting Situations and Solutions

24.0 Standardization, while pervading all walks of life, is quite often called upon to deal with extraordinary situations which may arise from time to time. In resolving the unusual problems arising from those situations, experience and ingenuity are valuable assests, but an intimate knowledge of the actual technology involved is also helpful. As indicated in earlier chapters, the requisite technical knowledge is generally freely available among the representative members of committees drawn from concerned interests and specialized organizations. But it is seldom that the committee members are well enough versed in the basic philosophy of standardization or deeply enough experienced in handling difficult situations arising out of limitations of technological advancement, or a clash of economic interests, or from some other unprecedented serious factor. It must also be remembered that they represent particular interests, the safeguarding of which is their primary duty. Under these circumstances a well-balanced and objective approach by the secretariat or directorate staff is often most helpful, but it is essential that the staff concerned makes a special effort to first acquire the necessary technological background of the subject in hand. Only then can its endeavours be expected to bear fruitful results. In exceptional cases where the size of the task involved justifies it, specialist staff experienced in standardization may have to be recruited.

OVER STANDARDIZATION

24.1 In addition, a certain boldness of approach is essential in handling any situation of this kind. There is, of course, always the danger that one may be tempted to introduce standards in a field or in a manner that may do more harm than good. In this context, the mode and manner of standardization is also very important. Any tendency to over-standardize should particularly be regarded as more dangerous than an attempt at under-standardization. The latter can, at worst, leave the situation unaltered, that is more or less in a condition as might have existed before standardization. But in the case of over-standardization, there is the possibility of upsetting the existing practices in a manner in which standards may not only fail to lead to any benefit, but may even prove to be uneconomical or otherwise harmful.[1]

24.2 Any of the following situations may be considered to be a case of over-standardization:

(1) Standardization at an inappropriate level; for example, standardization of a detailed design of a machine or equipment at the national instead of company level, or adoption of a national standard at company level without any judicious selection of its contents or the elaboration required to fulfil specifically the needs of the company.

(2) Inclusion of too many characteristics of a material or a product, some of which may be inter-dependent and others unessential for determining its effectiveness and serviceability.

(3) Stipulation of tolerances which are tighter than those considered essential from the point of view of usefulness or interchangeability, or those that are finer than the errors of measurement.

(4) Inclusion of avoidable requirements of subjective character or qualitative nature for which neither measurements (direct or indirect) nor comparative assessment is feasible.

(5) Specifying test methods which are unreasonably expensive either in terms of operational cost or initial investment.

(6) Specifying test methods for which the requisite facilities or equipment may not be available in the country and for which overseas testing may have to be arranged at undue expense.

Many instances of other types may arise which may constitute over-standardization, including cases of untimely or premature standardization as well as those where any action may be considered unnecessary or uncalled for.

NOVEL PROBLEMS

24.3 In examining the various propositions and weighing their pros and cons, one may come across such challenging situations as may be free of most of the objections indicated above and yet defy solution for some reason or another. Every organization, national or otherwise, would come across such situations and numerous interesting examples could perhaps be collected if a general survey were to be conducted. But from the author's own limited experience with national and international standardization and his limited knowledge of some of the developing countries, a few instances of interest are presented below, which might prove helpful in meeting some of the novel situations that might arise. The instances chosen are such as may be of interest to most countries but particularly to the developing economies.

24.4 *International Demand for National Standards* National standards of advanced countries are in great demand for consultation and study in most countries outside those of their origin. This demand arises largely from the fact that international exchange of goods and services is still carried out, to a large extent, on their basis. This is partly because the number of international standards and recommendations available is still rather limited and partly because it is the national standards whose authority largely prevails in actual trade transactions. But in the developing countries, the national standards of advanced countries are sought after for other reasons as well; for example, they provide guidance in the preparation of local standards and serve as sources of useful technical information in establishing new ventures. Most National Standards Bodies of advanced countries welcome this demand and meet it gladly partly by sale and partly on a cost-free basis. It is, however, felt that the usefulness of these standards to foreign countries remains rather limited, because of the lack of adequate knowledge of the language (in which the standards of advanced countries are written) that may prevail in those countries.

24.5 Thus it frequently becomes necessary for each interested party to make its own translation of the required overseas standards or to have the translation made by someone more knowledgeable in the two languages involved – the language of the overseas standard and that of the interested party. It is obvious that such translations, carried out by every individual interested party, would lead to a considerable replication of effort on a worldwide scale, some of which at least could be avoided through exchange of information. Moreover, this practice may also introduce certain inaccuracies that may hinder a proper understanding of the intent of the original standard. This has led many NSBs of advanced countries to realize the need for making available official translations of some of their more popular standards in the commonly used languages of the world, such as English and French. It is thus that Japan and Germany have each issued translations of over 2,000 of their standards in English, which is respectively about 31 and 18 percent of the total number issued by each country. Russia has published some 1,200 English translations amounting to about 9 percent of their own total. France and Netherlands have also initiated the move and translated a few of their standards in English.

24.6 But the demand for such translations is limited and the cost involved heavy. Therefore, this solution is by no means inexpensive. Yet, in some quarters, this move has been considered suspect of being designed mainly for the promotion of exports of goods and services. While there can be no doubt that those interested in purchasing goods abroad, according to the standards of the country of origin, would derive considerable assistance from the translated standards, it must be admitted that basically the move constitutes a service for making readily available officially recognized, accurate and uniform translations of standards. Thus, chances of misunderstanding due to varying interpretations are minimized and the time, trouble and expenditure involved in individual translations avoided.

24.7 *Industry-wide Coverage* Occasions arise, particularly in developing countries, when standards have to be prepared on a high-priority basis for covering all aspects and phases of a basic industry, the development of which is planned to be accelerated. Such was the case with the structural steel industry in India soon after the country became independent. The special project prepared and pursued to meet the demand for standards in this field has already been briefly referred to in Chapter 21 (*see* paras 21.9 21.13), in the context of the impact it had on national economy.[2] Timely preparation of standards contributed considerably to the orderly development of the steel industry which, in turn, promoted the advancement of all other industries dependent on steel.

24.8 Another occasion of a somewhat similar character, which arose in India, was the demand for undertaking a comprehensive programme of standardization to cover the whole front of multi-purpose river-valley projects.[3] The various aspects of the programme included the formulation of standard codes of practice for preliminary investigations and preplanning; for design of various types of structures such as dams, power houses and canals; for construction and measurement practices; for inspection procedures and for instrumentation; as also the preparation of standard specifications for materials of construction, instruments, plant and equipment. This was quite a major undertaking for any NSB, but the rich experience that had accumulated in the various departments and agencies within the central government and state governments was extremely useful in effectively tackling the numerous problems involved. A good many practices had already been codified and specifications standardized by various agencies. One of the tasks, therefore, was to coordinate them all in a coherent manner and present them as a part of the systematic structure of national standards, which could be uniformly adhered to by all concerned so that overall national economy could be ensured. In pursuance of this programme, which started about 1964–65, some 70 Indian Standards of special interest to multi-purpose projects have already been issued. This is only about 20 percent of the total number expected to cover all aspects of the programme, and it may take another 5 to 10 years to complete the job. To this must be added the large number of general purpose civil engineering and building material standards, which is already available. If the latter were also included, it could safely be stated that some 50 to 60 percent of the standards requirements in this field has been met.

24.9 *Comparison of National Standards* In certain fields of industry, international exchange of machinery and equipment is so extensive that when it comes to maintenance, the requisite materials and components have to be frequently imported and this is most inconvenient. In certain cases, such as ball and roller bearings, screwed fasteners and the like, adequate enough internationally accepted standards prevail for products of one country to be readily substituted by those of another country. But in the case of alloy and other general purpose steels, international standards have not yet come to prevail to the same extent, and the national standards for composition and quality characteristics differ quite considerably from one another. The position is further complicated by the use of distinctive systems of designations of the different types of steels produced in various countries. Thus, a frequently occurring question as to which British steel could be substituted for such and such American steel had always been difficult to answer, until recently when the National Standards Bodies of certain countries addressed themselves to the task and formulated standards for facilitating comparison of steels of various origins.

24.10 The British Standards Institution was perhaps the first to tackle this job and issue, in 1959, the first standard on comparison of carbon steels of British, American and German origins.[4] Three years later, this British standard was followed by another, this time on comparison of alloy steels.[5] Comparison in these standards was limited to chemical composition. In India, a need arose to make such comparison of steels in respect of physical properties also and to extend it to additional steels of Indian, Japanese and Russian origins as well as to varieties standardized by certain organizations in the USA and Germany other than those covered by the British Standards. This demand was met in 1965 by the issue of a comprehensive Indian standard with an index and commentary.[6-8] The countrywise coverage was further extended in 1968 by the issue of three French national standards, bringing within the orbit of comparison the French, Belgian and Italian steels.[9-11]

24.11 These groups of standards represent the result of exhaustive and intensive efforts, spread over several years of hard work, and cover a great deal of ground. While they do serve the purpose of the countries covered by them, it must be admitted that they do not by any means represent a solution to the worldwide problem of international standards for functionally interchangeable steels. Such a solution can only emerge out of ISO deliberations, and ISO has indeed taken up this problem and published a few recommendations,[12-13] but the work is by no means complete. There are real hurdles in the way of securing worldwide agreements on all the varieties of general interest. Apart from the usual resistance of certain countries to any change in traditional practices, there is the difficulty that the various alloying elements are not available in equal enough abundance in all countries. In any case, as far as ISO Recommendations go, they should invariably guide the formulation of steel standards in all countries, specially the developing ones. But additional varieties will also be required, which will have to be standardized in the light of locally available alloying elements.

24.12 For any developing country introducing its own steel specifications, it would be well, first of all, to consult the ISO Recommendations, the national standards on comparisons referred to above, as well as another excellent compilation of steel properties that has been published in Germany in two languages – German and English – entitled Key to Steel.[14] This last mentioned publication covers the most important categories of steels of several origins, such as Austria, Brazil, Czechoslovakia, France, East and West Germany, Hungary, Italy, Japan, Netherlands, Norway, Poland, Rumania, Spain, Sweden, UK, USA, USSR and Yugoslavia. It will be found that a good many of the steels standardized in different countries differ only in minor respects from one another and any further addition in any country to the already bewildering variety in the world would hardly be justified.

24.13 *Model Legislations* It is not usual for National Standards Bodies to be called upon to prepare standards comprising model legislations or model byelaws for local governments such as municipalities and district boards. But there has been an important exception in the case of New Zealand, where the Standards Association of New Zealand has done pioneering work in this field. It was at the request of the Municipal Association of New Zealand that a unified set of standing orders was prepared in 1954 for governing the proceedings at meetings of city and borough councils.[15] "In addition to providing a uniform code of practice, the standing orders facilitate the interpretation of legal and technical provisions relating to procedures. Although they are primarily intended for use by cities and boroughs they may be readily adopted for town boards, counties, and other local authorities." And, indeed, they were so officially adopted a few years later when, in 1959, another set of standard standing orders was issued for use by counties.[16]

24.14 In addition, New Zealand has gone much further by introducing standards for guiding the practices for running local government. Over the period 1952–60, a large variety of model general byelaws was issued in standard form,[17] which covers diverse aspects of municipal administration, dealing with the removal of refuse, public nuisance, itinerant traders and so on. These byelaws deal not only with matters of general and administrative interest but also with some of the technical requirements of importance. In addition, New Zealand has gone further to produce certain model byelaws of unusual character, which comprise technological and other details requiring government regulation of certain institutions and operations. These include, among others, the model byelaws dealing with hospitals,[18] land drainage[19] and precautions against panic in theatres.[20] An interesting case of internal regulation is the ISI Guide for the Organization of Conventions, which has proved highly valuable in securing smooth operation of large assemblies over the past years.[39]

24.15 Somewhat in the same category are the national standard building codes, which lay down requirements for assuring safety against fire, health and electrical hazards as also hazards that may arise out of inadequacy of building materials or methods of design and construction. Though technical in character, building codes form an essential part of urban byelaws and help control building design and construction activity in the interest of public safety and health. Reference has already been made to such standards which have been published in many countries (*see* Chapter 5, paras 5.22–5.24).

24.16 In any new country where traditions are not so deep-rooted as in some of the older countries, the national standards for good government such as those published in New Zealand can be of great service in establishing efficient local administration. The authorities actually in charge of local affairs would generally appreciate the assistance that can be derived from a well thought-out set of standing orders and model general byelaws such as those of New Zealand, which have been compiled by experienced people after due study and deliberation. Most developing countries stand to benefit considerably from the pioneering work done in New Zealand. Not that the developed countries cannot similarly benefit; the only difficulty in their case would be the established traditions that would stand in their way of securing a widespread acceptance of such model byelaws, which are ordinarily considered a close preserve and a special prerogative of the politician.

24.17 *Interim and Ultimate Solution* By its very nature, standardization is a deliberate and quite often a time consuming process. Thus it becomes advisable sometimes to carry out the implementation of a standard in gradual steps and introduce interim measures while preparing for the adoption of an ultimate solution. In planning a suitable course of action, many factors have to be considered, the chief among them being the ensurance of overall economy, the convenience of production and consumption and the possible speed of growth of consumer demand. But the very recognition of an interim solution sometimes leads to the postponement of the adoption of the ultimate solution. Two examples may be of interest in this context – one in the international field and the other in the national.

24.18 For a long time the International Electrotechnical Commission had been struggling to secure agreements on one unitary series of dimensions of electric motors to replace the two distinct inch and metric-based series, which had been recognized as the only available solution to the varying practices prevailing in the world at the time.[21] Thus the object of achieving interchangeability of motors manufactured in different countries remained far from being achieved. In 1956, India offered a possible solution to this difficult problem, which was duly accepted and a single series of motor dimensions was incorporated for the first time in 1959 in the IEC Recommendation on the subject.[22] This series represented a compromise between the inch and metric-using countries, arrived at by merging the two series in such a manner as to meet the point of view of both the camps without upsetting too much the practices of either. However, in order to give all countries a fair chance to change over gradually, it was agreed that the two existing series may continue to be recognized for an interim period. But, unfortunately, the length of the interim period was not discussed or specified. The result is that both the inch and metric series of motor dimensions still continue to prevail in most of the older countries. Only a few newer countries, like India,[23] have been able to adopt the unitary IEC series, which should find favour with all the newly developing countries that are now planning to go in for electrical machinery manufacture. In the not too distant future, when the metric system becomes more universal, the international interchangeability of electric machines may yet be achieved.

24.19 Another interesting example of interim-cum-ultimate solution is that of the adoption of screw thread standards in India. When this question first came up in the early fifties, it was anticipated in many quarters that India would shortly adopt the metric system but a firm decision had not yet been taken. The screwed fasteners used in India at the time were mostly inch-based, conforming to the relevant British standards, but it was hardly possible to adopt them as Indian standards. Nor was it advisable to go the whole hog and adopt the metric screw threads. In fact, no ISO metric series of screw threads were available at the time. It was, accordingly decided to adopt an interim solution and endorse the British standard for a unified screw thread system in metric equivalents,[24, 25] which had just been published after an agreement had been arrived at between the so-called ABC group of countries (America, Britain and Canada). As soon as the official decision was taken in 1956 to adopt the metric system in India and even before the earlier decision regarding the adoption of unified screw threads could be fully implemented in practice, there arose the question whether the interim solution in favour of unified screw threads should at all be maintained. After considerable debate and deliberation, it was decided, in 1958, in the face of strong opposition from some quarters, to change the course in mid-stream and go straight to metric standards.[26]

24.20 The decision made it possible to avoid the extra effort and cost of an interim changeover and to concentrate all energies on adopting the ultimate solution. This is not to say that the process has been as straightforward as anticipated. There still remain even today many sectors of industry which continue to demand screwed fasteners conforming to the old inch-based standards, some for the purpose of maintenance and replacements, but quite a few because the old inch-based designs have not yet been changed to the metric system. Nevertheless, some 40 to 50 percent of the active capacity of the screw thread industry in India today is devoted to producing metric fasteners and it is expected that another 5 years or so will see a complete changeover. Thus, it would appear that in this case the by-passing of the interim solution has been helpful and that its retention might have further delayed the implementation of the ultimate solution.

24.21 *Resistance to Change in Trade Practices* The resistance put up by trade to a desirable change in its prevailing practices is sometimes so great that even a reasonable alteration required to eliminate the wrongful usage of certain trade terms is quite successfully resisted. This is eminently illustrated by the experience of the Indian Standards Institution in connection with its project for preparing standard specifications for alcoholic beverages. During the pre-independence days, Scotch whisky could freely be imported from Scotland. The corresponding indigenous beverage was known as *desi* whisky (*desi* meaning indigenous) or simply whisky. This was mostly manufactured from indigenously distilled alcohol with the help of imported flavours and colouring matters, and very seldom from a distillation of fermented malted cereals followed by an appropriate ageing process. After independence, considerable restrictions came to be imposed on the import of most consumer goods including Scotch. This stimulated the production of a normal Scotch type of whisky within the country, but not being of Scottish origin it could hardly be labeled as Scotch or distinguished reasonably from the brew currently made from alcohol and additives.

24.22 When it came to standardization, the question of terminology came up with full force. Producers of the Scotch type of whisky claimed that the additive blend could not be called "whisky" at all and that its nature should be clearly brought out by an appropriate nomenclature that may be adopted to designate it. Several names were suggested including "synthetic type whisky" The producers of this latter type claimed that their product had been selling in India for almost a century under the label of whisky and their clientele were likely to be misled if any change were introduced in its nomenclature. The argument went on for several years until a happy compromise was found for labeling "Scotch type" of whisky as "malted whisky" and the other as simply "whisky".[27] In addition, with a view to making the story complete, a third term – "blended whisky" – was adopted and defined as being a mixture of at least 10 percent of "malt whisky" with 'whisky" or neutral spirit.

24.23 *Circumventing Trade Secrets* In certain fields of industry, the work of standardization is handicapped by an insistence of producers on the preservation of certain trade secrets. Under certain circumstances, the guarding of trade secrets may be considered quite justified, especially when this would tend to prevent the competing producers from making inroads into one's own trading areas. Standardization as such is not interested in probing into the confidential nature of production processes or the secret compositions of products in such a manner as to ignore the interests of the producer. But it does and should insist on safeguarding the interests of the consumer. This latter concern of standards could be satisfied much more effectively by specifying performance characteristics of a product and indicating such qualities as would influence its reliability. The questions of processes and compositions become secondary and even unnecessary from the consumer's viewpoint, so long as he is assured of performance and reliability. However, in certain industries, it is not always possible to specify methods of test and lay down the limiting values of characteristics, which would assure a given level of performance and reliability under most service conditions. Such a situation may arise out of the fact that the technology of testing certain products for their durability such as paints, is not yet far enough advanced. In other cases, such as lubricants, the difficulty may further be aggravated by the existence of many secret compositions and brands produced by several competitors, each maintaining that only a specific brand of his own make will serve a given purpose. In the case of lubricants, another complication is introduced by the machinery and equipment manufacturers tending to recommend specific brands of certain lubricants for use with products of their manufacture, which is perhaps not so much due to their being in collusion with lubricant manufacturers as to their need to safeguard themselves.

24.24 The users of machinery and equipment, particularly of the highly specialized types, find that in such a situation they have no option but to use what is recommended. Fortunately, the quantity of the lubricants required and their costs are insignificant compared to the investment involved in the machinery and to the value of its output. But, in any large industrial complex, the total lubricant costs do aggregate to an appreciable amount, so much so that competitive purchasing becomes capable of bringing about considerable savings. Such purchasing, however, requires the existence of standard specifications for lubricants, national, international or otherwise. But the science of lubrication and lubricants has not advanced far enough to enable an objective assessment to be made of lubricating oils and greases in relation to given services. It is so far largely a matter of experience, which is no doubt guided by certain chemical and physical tests used chiefly for identification purposes. A most interesting beginning appears, however, to have been made in regard to competitive buying of lubricants in Germany, where a machinery manufacturer, after intensive research and study, has managed to arrive at a company-level standard covering its own requirements of lubricating oils and greases; cooling, cutting, heat-treatment, batching oils, etc.[28] Besides reducing the number of varieties of products to be purchased from 179 to 25, this standard has made competitive buying possible from among the 15 possible suppliers for the first time, leading to a saving of almost 50 percent in the annual bill. Furthermore, the overall improvement in operations has been such that the frequency of machine breakdown through the use of wrong lubricant has been reduced from 29 to 2 per annum.

24.25 This is not to say that the development offers a final solution to the problem of providing objective specifications for lubricants, but it may be considered a definite step forward. Actually, the company standard referred to above simply groups together machines and situations in such a manner that each group would require the same lubricant. It also lists, against each group, the various brands of the 15 different manufacturers, which would meet the need of each of the groups. To each of the 32 distinct groups so listed, has been allotted a definite symbol. This symbol is applied to the containers of lubricants on the one hand and at the point of application on the machines on the other, thus minimizing the possibility of wrong usage. Competitive buying is made possible, as the purchase department can invite tenders by brand names from the several manufacturers of the equivalent lubricants listed in the standard against each group. Based on this company standard, an interesting national standard for marking lubricants and lubricating points has now been issued by the German National Standards Body,[29] which may yet be the precursor of further standardization in this field leading to the ultimate solution of the problem of lubricants standards of the specifications type.

24.26 *Utilization of Wastes* Economic utilization of large-scale industrial and agricultural wastes is a serious problem for most countries, but in the developing countries, the problem is even more complex. This is because their infant economies can hardly afford to write off the wastes and yet their industrial structure may not be fully enough developed to make the best use of them. Thus large quantitites of sugarcane bagasse may simply be burnt as fuel, or molasses for sugar manufacture be allowed to rot and pollute the atmosphere. Piles of blast furnace slag are often observed to grow into small hills near steel plants, largely because its take-off depends on the development of slag-consuming industries like cement. Now, this problem of slag utilization arose in India when the absence of a standard for slag cement was found to be in the way of establishing a new industry for its manufacture. The formulation of such a standard, however, required the chemical and physical examination of the available slags and the determination of the qualities of the cement which would result from their use as a raw material on a large scale. Adoption of overseas standards in a case like this was considered quite inadvisable, though some guidance could nevertheless be obtained from them. After extensive investigations spreading over several years into the chemical and physical nature of available slags and a close study of the published literature, it became finally possible to establish a standard on the subject in 1953.[30] This helped to initiate production and utilize slag cement on a large scale. It also stimulated a considerable amount of research towards the improvement of the nature and knowledge of the composition of the slag itself and the cement made therefrom.[31-33] Thus, as the knowledge grew, the standard was progressively amended and improved through successive revisions in 1962 and 1967.[30]

24.27 If a survey were to be made, it is quite likely that hundreds of instances would be found where utilization of wastes has been brought about through standardization. But one more case from Indian experience may be of some interest, in which a by-product of a forest produce, which normally went to waste, was converted into a useful raw material for two of the most important industries of the country, namely jute manufacture and cotton textiles. Tamarind, a tropical tree, is valued mainly for its pods from which a sweet-sour pulp is derived for use mainly in the kitchen. The seeds from the pods found little use, if any, until during the Second World War when supplies of starch became extremely restricted and the powder of the seed kernels came in handy as a substitute for sizing starch required for the textile industry. Though it did serve the purpose then, it was not found to be a very good substitute. After the war, this practice would have disappeared, had the supplies of starch become abundant again. But, as it happened, the food cereals could hardly be diverted for starch manufacture in large enough quantities because of the food shortage in the country. Intensive research on tamarind kernel powder was, therefore, initiated with a view to improving the product and making it a worthy substitute for maize starch. However, in order to establish a viable industry, a national standard was found to be essential, and this was issued, first in 1951 on a tentative basis, then revised in 1956 and further amended in 1969.[34] This standard was for the use of the powder in cotton textiles, but in due course experimentation proved the usefulness of the product for the jute industry as well and a similar standard was issued for this purpose in 1962.[35]

24.28 These two standards represent a typical example of how standardization may help to establish an entirely new industry based on waste material. But in this case, for the standard to have been able to do this, it had to be more than just a specification. For example, in both the tamarind kernel powder standards, appendices were included containing helpful hints to assist the producer as well as the user of the product. To the producer, a code of practice for the manufacture of powder was offered to help him produce standard quality material. To guide the textile technologist in the proper formulation and application of the size to his yarn and fabric, methods of preparation of size mixes and hints for avoiding pitfalls in the process were included. The inclusion of all this information has proved to be the key to the successful establishment of a viable new industry, which has a good future. It has indeed been estimated that an adequate quantity of tamarind kernel is available in the country to produce enough powder to meet all the demand of the cotton as well as the jute industries.

24.29 *Adoption of Overseas Standards.* In the absence of an internationally agreed standard on a given item, it has often been found feasible to adopt a standard from an industrially advanced country with which there may have existed close trade relations or other historical, cultural or linguistic links. In adopting such a standard, it is quite important to ensure that the specific needs of one's own country are fully provided for in the standard in question. In its pre-independence days, India used to import most of its electrical fans from England, as it manufactured only a few indigenously. All fans were expected to comply with the relevant British Standards (BS 367 and BS 380). But, when it came to patterning the Indian Standards after the British Standards, it was found that the latter were incomplete, inasmuch as they did not specify either the air delivery or the service value of the different types and sizes of fans, though the requisite methods of test for determining these qualities had been given. It would be appreciated that from the consumer's point of view, the specification of performance limits for both these characteristics were as important as the test methods. Without any knowledge of how much air a given fan would deliver per minute and how much electricity it would consume per unit of air delivery, an individual consumer would be quite unable to make a wise choice. If, however, he happened to be a large-scale buyer, he would be able to obtain from various manufacturers the guaranteed values of these characteristics and base his judgement on a comparative assessment of the values and the prices quoted. But, even in this case, no guidance would be available from the standard as to what level of values may be expected to be considered adequate or satisfactory.

24.30 The consumer demand for completely consumer-oriented fans standards was, therefore, pressed by the National Standards Body, but the indigenous manufacturers of the fans put up quite a resistance and took their own time to agree to it. Ultimately, the first Indian Standard on the subject was issued in 1951 and other standards also followed in slow succession.[36] Eventually, the Indian Standards Institution, as the secretariat of the IEC committee on the subject, was successful in processing a set of international specifications on electric fans which also included the values of these performance characteristics of interest to the consumer.[37] Unfortunately, the position changed once again, later on, and within 5 years of their publication, it was decided to delete the performance values from the fans specifications of IEC. The reason, which apparently led to this reversal of position, was that a greater emphasis had begun to be placed on the safety requirements of all types of domestic electric equipment and manufacturers did not perhaps wish, in addition, to have to continue to meet the minimum performance requirements as agreed to earlier. Consumers would undoubtedly welcome the new IEC recommendations on safety of domestic fans,[38] but they would, nevertheless, like to see the performance values also continue to be retained not only in fans specifications, but also in specifications of most other items of interest to them.

CONCLUSION

24.31 The few instances of the interesting situations briefly described above would illustrate how challenging and intriguing the task of standardization can possibly be and how satisfying when acceptable solutions are found and implemented. To accomplish this requires quite a good deal of ingenuity and tact, but, most of all, patient application and dogged persistence. Except perhaps in the case of company standardization, very few case histories of standardization at other levels are discussed in published literature, which would highlight the challenging character of the task of the standards engineer. Even in the case of company level case histories, discussions usually centre on the economic gains accruing from a given standardization effort. It would be most useful if active standards engineers at all levels – particularly the national and the international – would take time to discuss more freely their unusual experiences and commit them to writing for the information and benefit of others. Even the presentation of unsolved and difficult problems would be of great interest and may result in the emergence of useful suggestions for solution from unexpected quarters.

REFERENCES TO CHAPTER 24

1. Kinzel, Augustus B.: Specifications – their evolution, uses and abuses. *ISI Bulletin,* 1963, vol 15, pp. 61-66 (*see also* reference 4 of Chapter 21).
2. See references 10, 11 and 26 to 33 of Chapter 21.
3. Handa, C. L. and Krishnamachar, B. S.: Economy in river-valley projects through standardization. *ISI Bulletin,* 1967, vol 19, pp. 178-182.
4. *BS 3179 Part I: 1967 Comparison of British and overseas standards for steels – chemical composition of wrought carbon steels (first revision).* British Standards Institution, London, p. 34 (*earlier edition*: 1959).
5. *BS 3179: Part 2: 1962 Comparison of British and overseas standards for steels – chemical composition of wrought alloy steels.* British Standards Institution, London, p. 56.
6. *IS:1870–1965 Comparison of Indian and overseas standards for wrought steels for general engineering purposes.* Indian Standards Institution, New Delhi, p. 131.
7. *Index to IS:1870–1965 Comparison of Indian and overseas standards for wrought steels for general engineering purposes.* Indian Standards Institution, New Delhi, p. 32.
8. *IS:1871–1965 Commentary on Indian standard wrought steels for general engineering purposes (complementary to IS:1570–1961).* Indian Standards Institution, New Delhi, p. 69.
9. *NF A 35 600–1969 Aciers de construction d'usage général – comparaison de nuances normalisées françaises et étrangéres.* Association Française de Normalisation, Paris, p. 6.
10. *NF A 35 601–1968 Aciers non alliés et alliés spéciaux pour traitement thermique – comparaison des nuances normalisées françaises, allemandes et américaines.* Association Française de Normalisation, Paris, p. 8.
11. *NF A 35 602 1969 Aciers inoxydables comparaison des nuances normalisées françaises, allemandes, américaines et anglaises.* Association Française de Normalisation, Paris, p. 7.
12. *ISO/R 1052–1969 Steels for general engineering purposes.* International Organization for Standardization, Geneva, p. 11.
13. *ISO/R 683 Heat-treated steels, alloy steels and free cutting steels:*
    Part 1–1968 *Quenched and tempered unalloyed steels,* p. 12.
    Part 2–1968 *Wrought quenched and tempered steels with 1% chromium and 0.2% molybdenum,* p. 15.
    Part 3–1970 *Wrought quenched and tempered unalloyed steels with controlled sulphur content,* p. 15.
    Part 4–1970 *Wrought quenched and tempered steels with 1% chromium and 0.2% molybdenum and controlled sulphur content,* p. 16.
    Part 5–1970 *Wrought quenched and tempered manganese steels,* p. 15.
    Part 6–1970 *Wrought quenched and tempered steels with 3% chromium and 0.5% molybdenum,* p. 12.
    Part 7–1970 *Wrought quenched and tempered chromium steels,* p. 15.
    International Organization for Standardization, Geneva.

14. *Stahschlüssel* (Key to steel). Verlag Stahlschlüssel Wegst, Marbach (Germany). 8th ed, 1968, p. 295.
15. *N.Z.S.S. 1167:1954 Standing orders for municipalities.* New Zealand Standards Institute (now Standards Association of New Zealand), Wellington, p. 40.
16. *N.Z.S.S. 1501 1959 Standing orders for counties.* New Zealand Standards Institute (now Standards Association of New Zealand), Wellington, p. 42.
17. *N.Z S.S. 791 Model general byelaws:*

Part I:1952	*Public places,* p. 18.
Part II:1952	*Scaffolding and deposit of building materials,* p. 6.
Part III:1952	*Hawkers, pedlars and itinerant traders,* p. 9.
Part IV:1952	*Public libraries,* p. 8.
Part V:1952	*Billiard-rooms,* p. 9.
Part VI:1952	*Cemeteries,* p. 15.
Part VII:1952	*Lawn cemeteries,* p. 4.
Part VIII:1952	*Cremation and the crematorium,* p. 4.
Part IX:1952	*Household refuse purchasers,* p. 5.
Part X:1952	*Removal of refuse,* p. 5.
Part XI:1952	*Nuisances,* p. 6.
Part XII:1952	*Offensive trades,* p. 6.
Part XIII:1952	*Sale of second-hand clothing and hiring out of clothing,* p. 4.
Part XIV:1952	*Public baths and swimming-pools,* p. 7.
Part XV:1952	*Beaches: bathing and control,* p. 6.
Part XVI:1952	*Parks and reserves,* p. 8.
Part XVII:1952	*Water-collection areas,* p. 5.
Part XVIII:1954	*Licensing of plumbers,* p. 7.
Part XIX:1954	*Licensing of drainlayers,* p. 7.
Part XX:1954	*Amusement devices and shooting-galleries,* p. 7.
Part XXI:1954	*Licensed vehicles and stands,* p. 15.
Part XXII.1954	*Water supply,* p. 19.
Part XXIII:1954	*Abattoirs,* p. 11.
Part XXIV:1954	*The keeping of animals, poultry, and bees,* p. 7.
Part XXV:1954	*Signs and hoardings,* p. 13.
Part XXVI:1960	*Licensing and control of boardinghouses and apartment buildings,* p. 19.

New Zealand Standards Institute (now Standards Association of New Zealand), Wellington.
18. *N.Z.S.S 1166:1970 Model byelaw for Hospitals.* New Zealand Standards Institute (now Standards Association of New Zealand), Wellington, p. 18.
19. *N.Z S.S. 1711:1962 Model land drainage byelaw.* New Zealand Standards Institute (now Standards Association of New Zealand), Wellington, p. 11.
20. *N Z.S.S. 789–1953 Byelaw for precautions against fire and panic in theatres, public halls and assembly halls.* New Zealand Standards Institute (now Standards Assocation of New Zealand), Wellington, p. 15.
21. *Report on IEC work on standard dimensions for electric motors.* International Electrotechnical Commission, Geneva, 1956, p. 12.
22. *IEC Pub 72–1971 Dimensions and output ratings for rotating electrical machines – frame numbers 56–400 and flange numbers F 55–F 1080 5th ed.* International Electrotechnical Commission, Geneva, p. 42.
23. *IS:1231–1967 Dimensions of three-phase foot-mounted induction motors (second revision).* p. 12.
IS:2223–1971 Dimensions of flange mounted ac induction motors (first revision), p. 15. Indian Standards Institution, New Delhi.

24. Unification of screw-thread standards. *ISI Bulletin,* 1949, vol 1, pp. 44-46.
25. BS 1580:1953 Unified screw threads (with metric equivalents). British Standards Institution, London, p. 101.
26. Changeover to metric screw threads and fasteners. *ISI Bulletin,* 1963, vol 15, pp. 224-231.
27. *IS:4449–1967 Specification for whiskies.* Indian Standards Institution, New Delhi, p. 6.
28. Riebensahm, Hans E.: Cost savings through in-plant standards on lubrication and lubricants – a case study. *ISI Bulletin,* 1965, vol 17, pp. 445-448.
29. *DIN 51502–1967 Lubricants and related materials – marking of containers for lubricants, lubrication equipment and lubrication points.* Deutscher Normenausschuss, Berlin, p. 6 (*see also* DIN 51567, DIN 51574 and DIN 51588).
30. *IS:455–1953 Specification for portland blastfurnace slag cement (second revision).* Indian Standards Institution, New Delhi, p. 24 (*see also* first revision. 1962, p. 12 and second revision: 1967, p. 11).
31. Chopra, S. K. and Taneja, C. A.: Investigations on Rourkela slag for cement manufacture. *ISI Bulletin,* 1966, vol 18, pp. 365-370.
32. Kumar, R. L.: Blast furnace slags: Standardization for efficient use. *ISI Bulletin,* 1968, vol 20, pp. 171-176.
33. Taneja, C. A.: Blast furnace slags. *ISI Bulletin,* 1969, vol. 21, p. 351.
34. *IS:189–1951 Specification for tamarind kernel powder for use in the cotton textile industry (tentative).* Revised and issued as firm standard in 1956. Amendment no 1, July 1969. Indian Standards Institution, New Delhi, pp. 7 and 9.
35. *IS:511–1962 Specification for tamarind kernel powder for use in the jute textile industry.* Indian Standards Institution, New Delhi, p. 8.
36. *IS:374–1966 Specification for electric ceiling type fans and regulators (second revision),* p. 18 (*earlier editions*: 1951 and 1960).

    IS:555–1967 Specification for electric table type fans and regulators (second revision), p. 22 (*earlier editions*: 1955 and 1960).
    IS:1169–1967 Specification for electric pedestal type fans and regulators (first revision), p. 22 (*earlier edition*: 1957).
    IS:2997–1964 Specification for air circulator type electric fans and regulators, p. 22. Indian Standards Institution, New Delhi.
37. *IEC Pub 175–1965 A.C. electric table type fans and regulators,* p. 41.
    IEC Pub 176–1966 A.C. electric ceiling type fans and regulators, p. 37.
    IEC Pub 174–1966 A.C. electric pedestal type fans and regulators, p. 41. International Electrotechnical Commission, Geneva.
38. *IEC Pub 342–1971 Safety requirements for electric fans and regulators.* International Electrotechnical Commission, Geneva.
39. *Guide for the organization of ISI conventions.* Indian Standards Institution, New Delhi, 1964, p. 34.

Chapter 25

Education and Training

25.0 In any new and fast developing discipline like standardization, which owes its advancement, indeed its very existence, to a handful of self-taught men of broad vision and keen foresight, the problem of education and training is bound to be full of many question marks. Indeed, this problem has assumed serious proportions only during the past twenty years or so, when the demand for standardization began to become much more universal than ever before and when standards began to enter into many new walks of human endeavour. Simultaneously, the discipline itself became more complex and developed new branches of activity. The questions that now have to be answered include: who are to be taught and for what purpose, what should be taught and to what depth, what should be the methods of teaching, who should be the teachers, under what auspices should the teaching be organized, and so on. Detailed answers to these questions are not very easy to find, nor could any attempt be made towards this end within the bounds of a brief chapter like this. But it would be interesting to review what has been done so far and to indicate the possible avenues to be explored.

25.1 In order to appreciate the various efforts at education and training that have been made during the recent years in different countries, it is important to recognize that there are two broad classes of individuals involved to whom the programmes of teaching and/or training need be addressed, namely those who are concerned mainly with the use of standards and those who are concerned primarily with their preparation. The former class includes a vast variety of people, in fact almost every one. But the most important among them are those who have constantly to make professional use of standards, including the technologists and engineers from all branches of technology and engineering sciences, be they concerned with design or production, with planning, execution or assessment, with construction or control, with sales or procurement. Of all the users of standards, this group has to be given the highest priority in helping it to acquire familiarity with the significance of standardization and how and where to seek guidance from standards in the discharge of their normal functions (*see also* Chapter 14, para 14.14). As far as the education of the general public and the common consumer is concerned, the question has been briefly dealt with in Chapter 14 on implementation (*see particularly* paras 14.9 and 14.10).

25.2 The second broad class, which is concerned mainly with the preparation of standards and needs varying kinds and amounts of specialized training, may be divided into two general categories having somewhat different needs; the first comprising personnel engaged in company-level standardization and the second, those concerned with work at other levels, among whom the national-level workers would perhaps outnumber all the others. These two categories are clearly distinguished chiefly because their functions and modes of operation are somewhat different, though the basic principles of standardization involved are the same. Most of the training and educational programmes so far organized are addressed to one or the other of these two categories of workers. Attempts at evolving more generalized and comprehensive courses have also been made in several countries, but there is no indication of the extent to which these various courses may be considered as being in line with one another.

BRIEF REVIEW OF VARIOUS EFFORTS

25.3 It would appear that the first attempt at organizing a training programme of any sort was that by Dr. John Gaillard of the American Standards Association in 1947, when he offered a 5-day seminar for company standards engineers, educators planning a college instruction course on industrial standardization and others interested.[1] This course continued to be the only one in the world on the subject for many years and during the fifties it came to be offered twice a year[2] in open sessions and was additionally made available in private sessions to interested companies on their own premises. Gradually, as standardization activity began to assume more and more importance among the advanced countries as also to embrace a number of developing countries, the demand for standards personnel both at company and higher levels began to increase. To meet this demand, courses and programmes of training and education began to be organized in several other countries besides the USA. These included Australia, Czechoslovakia, France, Germany, India, Iran, Japan, Latin America, Poland, UK and USSR. Some of these courses have specially been designed to train the personnel of the developing countries where the demand for such training is the greatest. But even in developed countries, emphasis is being laid on the training and education of technical and engineering personnel in general, because of the importance that is being attached to ensuring a more widespread implementation of standards at all levels.

AUSTRALIA

25.4 Australia is perhaps the latest to join the group of countries offering training courses to the personnel of National Standards Bodies of developing countries. Thus, a 13-week course was organized in 1971 for the first time. It was financed under some of the mutual aid programmes of the government of Australia, namely, the Colombo Plan and the Special Commonwealth African Assistance Plan.[3] "The main purpose of the course is to give trainees first-hand experience in how standardization is carried out in Australia, both at the theoretical and practical levels. Particular emphasis is being placed on the relationship of standardization to industrial development."

CZECHOSLOVAKIA

25.5 Czechoslovakia appears to have organized a post-graduate university course for citizens of developing countries who have passed the faculty of social sciences at the University of the 17th November in Prague and who wish to specialize in technical standardization at home.[4]

FRANCE

25.6 In France, considerable progress appears to have been made, as early as in 1960, in injecting the knowledge and practice of standardization into the regular educational curricula of technical schools.[5] The training is designed not only to demonstrate how to apply standards, but also to show how to participate in the development of new standards.

"Prizes are regularly awarded in over twenty engineering training colleges . . . for those who achieve distinction in their studies of standards." Students are encouraged to pursue these studies with a view to pinpointing omissions, imperfections and lack of coordination or the need for amendments, to survey the application of standards in actual practice and assess their effect on cost reduction, material and labour savings and national economy.

25.7 Another significant French effort centres around a 21-week course offered for the benefit of developing countries.[6–8] This course has been organized almost regularly at intervals of roughly two years since 1961–62 by the French National Standards Body, AFNOR, in collaboration with the Association pour l'Organization des Stage en France (ASTEF) which was set up under the auspices of the government, but on the initiative of the industry, with the specific purpose of sharing the French know-how with the developing countries. For trainees who lack knowledge of the French language, special arrangements are made to put them through an accelerated 3-month language course before the training starts. In addition, two 12-week training courses have been arranged over the past decade, specially for the nationals of the French-speaking African countries.

GERMANY

25.8 In East Germany, under an official decree of 1966, post-graduate courses are being offered by technical universities and engineering institutes to impart training in standardization to students for qualifying them as standards engineers.[9] Similarly, in certain states such as Westphalia in West Germany, a decree of 1961 has enabled the University of Aachen to conduct courses on standardization, variety reduction and unification.[10]

25.9 At company level, the National Standards Body of Germany, DNA has been very active in imparting training to standards engineers in the theory and practice of standardization at company level through seminar courses and lectures organized periodically, in which experienced engineers drawn from industry take regular and direct part.[10] Up to 1966, some 26 such programmes had been arranged, in which over 1,000 trainees had participated, among whom 70 were from outside Germany. DNA is also understood to be compiling textbooks for the introduction of standards instruction early in student life in schools.

INDIA

25.10 Training in the field of standardization in India had its origin in 1958 when the Indian Standards Institution organized, for the first time, a 2-year comprehensive course of training for its own newly recruited staff of engineers who had freshly graduated from universities and engineering institutions and who had had no direct contact with industry, least of all with standardization. This course included lectures, discussions, visits to laboratories and industries, and on-the-job training within the Institution. With the success of this course and the expansion of demand for standards engineers generally within and outside the country, it became necessary to expand the training activities of ISI until now they envelop, besides standardization, other closely allied fields,[7-8] such as statistical quality control and industrial testing. The training programmes on standardization have also become much more diversified and specialized under various heads, as follows:

National Standardization

(1) A 2-year comprehensive course designed specially for ISI probation officers.

(2) A 15-16 week intensive course intended for national-level workers of developing countries, financed through the Colombo Plan and other international assistance plans, and now offered regularly during winter and spring every year.[11]

(3) A 3 month course of similar type to (2) above, offered to be organized locally in a developing country on request.

Company Standardization[12, 13]

(4) A 1 to 2-day management conference programme addressed to top level management.

(5) A sandwich survey training programme, designed for company standards engineers, in which the first 4 to 5 days are devoted to collective briefing, the second period of 4 to 6 weeks to individual in-plant survey and assessment of needs within the plant to which the trainee belongs, and the third period of 2 to 3 days to a collective group review of individual reports and exchange of experiences.

(6) A 10-day intensive course specially designed to create a cadre of company standards engineers out of the personnel already engaged in industry.

So far nearly 2,000 individuals have taken advantage of these training programmes, of which 50 were from overseas developing countries.

25.11 As a result of these training and other promotional activities, a great deal of interest has developed in India for the advancement of company standardization activity. Since the first ISI training course in this field was given in 1963, several seminar discussions have taken place in the country, among which the one on "Training in standardization technology," held in 1966, was perhaps the most important and widely attended.[14] More recently, in another nation-wide seminar, the importance of industrial standards in technical education was examined and the need was stressed to inject information on standards and standardization into the engineering and technical curricula of universities and other institutions of learning, with a view to securing a better alignment of education with the industrial needs of the country.[15-17] As a result of these and other discussions, some 16 orientation programmes on standards for technical teachers have been held in 10 states of the Union, in which well over 1,000 teachers, drawn from nearly 200 technical institutions, have participated.[18]

IRAN

25.12 The Institute of Standards and Industrial Research of Iran has been offering, since 1968, to officials of National Standards Bodies of developing countries, facilities for training in standardization, in particular relation to export inspection of agricultural produce. Under this programme, the trainees are expected to bear the cost of international travel, but Iran meets the subsistence and other local expenses. It is understood that some 30 trainees have taken advantage of these facilities.

JAPAN

25.13 Soon after the end of the Second World War, when Japan began to rebuild its industries, an unprecedented interest developed in that country among entrepreneurs, industrialists and engineers for the adoption of statistical quality control measures in production and taking advantage of the wartime experience in the USA, it outdid America. Hundreds of courses and seminars were organized for imparting the knowledge of quality control techniques. It has been estimated that nearly 80,000 participants have so far participated in such programmes which, while emphasizing quality control principles and practices, also touched upon standardization, of which quality control is an integral part. It was only in 1968 that Japan began to organize courses laying particular emphasis on standardization as such, with quality control as an important constituent. These latter programmes, which are specifically designed to meet the requirements of trainees from developing countries, are conducted under government auspices through the Overseas Technical Cooperation Agency, with the active assistance of the Standards Division of the Industrial Science and Technology Ministry of International Trade and Industry and the Japanese Standards Association.[19] In these courses, something like two-thirds of the time continues to be devoted to quality control topics but principles and organization of standardization are also discussed in fair detail both from the national and the company-level points of view.

LATIN AMERICA

25.14 Under the leadership of COPANT (Pan American Standards Commission) courses of training for standards engineers, to work at the national level, have been continuously organized since 1962. Available to participants from all the Latin American countries, this is perhaps the only training programme in this field which is a result of cooperative effort among several countries. Starting from 1962 and up to 1970, over 100 trainees have participated in these programmes which were held mostly in Buenos Aires and ranged in duration from 2 weeks to 3 months.[20]

POLAND

25.15 Since the early fifties the Polish Standards Committee has been organizing courses on standardization for engineers and technicians in various branches of industry.[21] These courses are held several times a year and, judging from the number of participants in each, they would appear to be quite popular. In addition to this training, which may be assumed to be at the company standardization level, the Polish Standards Committee organizes specialized courses for specific branches of industry. Approximately 5,000 trainees have taken advantage of more than 120 courses arranged already. Standardization has also been introduced in something like 40 educational programmes of higher professional and technical schools. For this latter purpose, special courses are offered for training of professional teachers and lecturers. It is also understood that a one-year post-graduate course is proposed to be organized in one of the technical universities with a view to assisting the advancement of the discipline of standardization.

UNITED KINGDOM

25.16 There has been some discussion in the UK on the necessity for giving training on standards matters to students in technical colleges,[22, 23] but no important step appears to have been taken to introduce standardization as a special subject of study in the technical curricula. However, references to British standards continue to be made and their use indicated when teaching some technical subjects in civil, mechanical and other branches of engineering. Application of workship technique to standards instruction has been proposed,[24] but the most significant development has been the introduction from 1967 of a residential 3-5 days course for standards engineers, organized by the Standards Associates Section of the British Standards Institution in cooperation with the Department of Management Sciences of the Institute of Science and Technology, University of Manchester.[25, 26] This course has now become a regular annual feature. For company level workers, it has proved a useful opportunity to exchange views among themselves and with those who operate at the national level. In addition, the British Standards Institution offers tailor-made courses for individuals from overseas to suit the requirements of each particular candidate.[8] These courses may extend from 6 to 12 weeks and include introductory talks, attachment to the relevant technical divisions and committees of BSI and visits to outside laboratories and establishments of interst. The candidates are sponsored either by the British Council or by the Ministry of Overseas Development under the technical assistance training programme.

UNITED STATES OF AMERICA

25.17 Reference has already been made to the pioneering effort of Dr. Gaillard in the United States (*see* para 25.3). Apart from this short-term seminar type of course given by Dr. Gaillard over many years, the subject of standardization has also found an important place in the business management course conducted by Professor Leo B. Moore in the Massachusetts Institute of Technology. In addition, Milek[27] is recorded to have successfully taught a one-year course in standardization at the Los Angeles Trade Technical College, the objectives of which, among other things, were to "attract students into the field of standards engineering; establish a basic teaching programme to meet the requirements of the standards engineering profession; and to offer guidance to individuals seeking vocational guidance in the field of standards engineering." Other suggestions have also been made for evolving a practical diploma course intended for fulfilling the basic educational requirements of a standards engineer.[28]

USSR

25.18 In the USSR, educational and training activity in standardization appears to have developed on quite an intensive as well as extensive scale. It permeates down to the secondary school level on the one hand and continues up to the topmost level of leading engineers and highly qualified specialists,[29-31] working at the national and international levels of standardization. Trainees of the top echelon are expected to write dissertations and defend them in the same manner as in the case of doctorate theses examinations. Extensive correspondence courses are also organized for those who may not be able to take time off from their employment. According to one account,[30] some 23,000 persons had undergone training of various types between the years 1965 and 1970. For those working in developing countries, specially selected experienced engineers are given a 3-year training in the Academy for Foreign Trade and a 2-year training in the Institute of Oriental Languages of the Moscow State University.

25.19 The emphasis appears to be on giving education in basic principles and application of standards at all levels and in all branches of technology, both for the building up of specialized cadres of standards engineers as also for the information and guidance of engineers, technologists and skilled tradesmen in the application of standards in their normal specialized fields of activity. So much importance is being attached to educational activity that a Personnel Training Section has been organized under the Scientific and Technical Council of the State Standards Committee of the USSR. Under its recommendations, the Minister of Higher and Secondary Specialized Education has issued an order charging the respective boards of universities with the task of including a syllabus on the principles of standardization and quality inspection in the curricula of engineering and economic higher schools and secondary technical schools.[32] In pursuance of this order, intensive activity has been initiated for framing the detailed curricula for various levels and classes of institutions and for writing specialized textbooks for teaching the subject.

INTERNATIONAL EFFORT

25.20 There has been a great deal of discussion on the subject of training and education on standardization in various forums of the United Nations as also within ISO. Some of the seminars and workshops held under UN auspices have helped disseminate knowledge about standardization among the personnel of the developing countries. These seminars and workshops have also come up with some definite suggestions for international action to help meet the growing demand for trained standards engineers in these countries.

UN INTER-REGIONAL SEMINARS, HELSINGÖR

25.21 This well-attended seminar on the promotion of industrial standardization in developing countries, in which participants from 27 countries of Africa, Asia, Europe and Latin America took part, was organized by the United Nations in 1965[33] in cooperation with the Royal Government of Denmark. It was addressed by as many as 21 specialists from both the developed and developing countries. Having discussed most of the important aspects of standardization work – objectives, levels, organization, preparation and implementation of standards, consumer needs, certification and so on – the seminar took note of the various efforts that had been made in different countries for the training of standards engineers at the national and company levels.[7] It was also pointed out that the United Nations was ready to give high priority to fellowship and scholarship awards for candidates from developing countries to attend such courses as were available in different countries. "The participants nevertheless thought that the establishment of one or more international centres for training in standardization was the need of the hour."

TRAINING WORKSHOP, ADDIS ABABA

25.22 Confined to the African continent, this workshop, organized by UNIDO in 1970, was attended by participants from 10 developing countries and addressed by 6 specialists.[34] Having dealt with the broad spectrum of subjects related to standardization, the workshop participants were familiarized with the facilities which existed in several countries for the training of standards personnel.

TRAINING WORKSHOP, SANTIAGO

25.23 This workshop, held in 1971, was also sponsored by UNIDO and was addressed mainly to the needs of countries of Latin America.[35] Here, again, the question of training was dealt with in some detail, the basis of discussion being provided by a comprehensive paper contributed by India.[8] The participants found the topic important enough to adopt a detailed resolution for further action to be taken by the National Standards Bodies, governments of the countries and by UNIDO, which recommended as follows:

1. Steps be taken with governments for the inclusion of courses on

standardization at the medium, higher and post-graduate education levels.
2. National standards institutes collaborate in the preparation, prescription and expansion of these programmes and give them a practical shape.
3. Research bearing on standardization be favoured in the training of professionals.
4. UNIDO study the possible coordination of training programmes of various organizations for the training of experts in standardization and assist the National Standards Institutes in the development of their training programmes.
5. UNIDO in collaboration with CEPAL, support these requests underlining the economic importance of standardization and its direct effects in the application of systems of quality control in production.

UNIDO SYMPOSIUM ON INDUSTRIAL DEVELOPMENT, ATHENS

25.24 In the otherwise overcrowded programme of this symposium held in 1967, the subject of standardization and training of personnel for it found a place on the agenda. It was recorded that the United Nations Development Programme may be approached to help set up the international training centres of the type suggested by the Helsingör seminar.[36]

ASIAN STANDARDS ADVISORY COMMITTEE

25.25 This consultative committee of the UN Economic Commission for Asia and the Far East (ECAFE), which has already been referred to in Chapter 12 (*see* paras 12.9 and 12.10), in its very first session in 1968, arrived at an estimate that more than 200 persons would be required for manning the upcoming NSBs of the ECAFE region countries within the next 3 years and that there was an urgent need for augmenting the existing training facilities in India, Iran and Japan, for which United Nations assistance might be requested.[37] Later, during the same year, a Consultative Group appointed by the Asian Standards Advisory Committee considered a proposal for organizing under UN auspices a peripatetic or roving type of course of training for national standardization, in which the trainees over a period of about 4 months would travel from Iran to India to Japan receiving in each country lecture-cum-workshop training on the subjects in which each country had specialized, terminating with a brief period of review and report writing at the ECAFE headquarters in Bangkok.[38] While recognizing that each of the three countries had something special to offer in regard to the nature of training facilities, it was not possible to secure an agreement on the roving course scheme either within the Consultative Group or in a later meeting of ASAC itself.[39]

25.26 The second session of ASAC did, however, agree to recommend that United Nations fellowships be made available more liberally to developing countries to enable them to take better advantage of the training facilities already available in India, Iran and Japan, as well as those for on-the-job training which Australia was prepared to organize on the pattern of the UK.[39] An important step taken by ASAC at this session was to recommend the adoption of an agreed form of a curriculum for a 20-week training course in standardization for national level workers, which had been originally prepared on the basis of mutual consultation among the countries of the region.[40]

INTERNATIONAL ORGANIZATION FOR STANDARDIZATION

25.27 The Development Committee of the International Organization for Standardization, ISO/DEVCO, has been engaged, among other things, in working out a typical set of programmes for the training of national level and company level staff workers as also "officials in the economic field," which would, it is presumed, include top level management.[41] The ISO Committee on Scientific and Technical Information, ISO/INFCO, is understood to have undertaken to conduct a worldwide survey on the training of standards engineers and technicians.[42] The objective of this survey appears to be more to prepare an informative compilation than to present a critical or comparative evaluation of the various programmes.

CONCLUSION

25.28 From the very brief review of the present position as given above it would be clear that in most countries, advanced as well as developing, the need for education and training has been widely felt and that several categories of personnel are involved. In the developed countries, attention has largely been concentrated on specialized instruction in company standardization and to an extent at the student level for imparting general familiarity with standards and standardization to future generations of engineers and technologists. In addition, several advanced countries like France, Japan, UK and USSR are also making efforts, each in its own way, to help fill the need of developing countries for trained personnel to work at the national level. In this latter sector, some developing countries, like India, Iran and the Latin American group, have also joined hands. The international effort, on the other hand, has been very largely limited to the one seminar and two workshops held so far by the UN or UNIDO in Europe, Africa and South America. However, some proposals for exploring possible avenues for concerted international action have been discussed within the ECAFE, UNIDO and ISO forums.

25.29 Another conclusion which one may derive from this brief review is that there is a great deal of divergence between the efforts of the various agencies engaged in developing and offering the different courses of instruction. Nor is there any significant evidence of an organized exchange of information and experience on a worldwide scale in the field of training and education. ASAC deliberations within ECAFE are perhaps the lone exception in this respect. This is certainly not a very healthy sign for the future development of a newly emerging discipline which depends for its advancement mainly on mutual consultation and exchange of views aimed at arriving at general consensus. Question may be asked as to how could such a situation have arisen in regard to the most basic issue of education and training in this field. Was it due perhaps to a lack of agreed codification of the basic principles and fundamentals of the discipline? Or was it due to a tendency on the part of certain national interests to promote their own brand of standards in competition with others and use training as a means for the indoctrination of the new generation of workers from the developing countries? It certainly could not have been due to the non-availability of an international forum for mutual discussions, when the ISO forum has not only existed for a long time but has been extremely active on many fronts. ISO had indeed organized during early years a special committee to deal with the problems of the developing countries, namely the ISO/DEVCO. Admittedly, the question of education and training is a much wider problem and touches all other countries as well, and DEVCO could perhaps have dealt with only a part of it. Could the ISO Committee on Scientific Principles of Standardization, ISO/STACO, perhaps tackle the problem? Whatever the real reason or reasons may be for the present state of affairs, it would be more appropriate and to the point to deliberate on what constructive action or actions might now be taken to meet the future needs.

25.30 Under normal conditions, universities and other institutions of learning could be depended upon to develop the necessary training and educational programmes, as also to stimulate advanced research on the subject. But in the present case it would appear that most efforts at introducing standardization in the academic institutions have at best been only partially successful. Nor do universities have available the means of acting in concert internationally, which would appear to be extremely important in this case, at least in the immediate future. It would thus seem necessary that reliance would have to be placed on international agencies such as ISO, UNIDO and UNESCO – ISO for its direct contacts with national standardization agencies which have had years of actual experience in developing training and educational programmes, UNIDO for its interest in promoting the advancement of standardization as an important infrastructure to industry, and UNESCO for its concern with the development of education, science and technology of which standardization is fast becoming an important and integral part.

25.31 If these three international bodies could agree to address themselves to the development of the discipline of standardization and promote the education and training of specialized personnel from all countries of the world, it would be most appropriate that they establish collectively an international institute for the purpose, with financial assistance derived perhaps from the United Nations Development Programme and/or from independent private and public endowments. Some proposals for the establishment of one or more regional or inter-regional centres under UN auspices have indeed been already considered; for example, those made at the Helsingör seminar.[33] But no tangible steps have so far been taken. If any of these proposals at all matures, it would be well worthwhile considering whether there should be one central institute or several centres of training. In the latter case, the effort and resources would naturally be divided and the result may considerably be diluted. A unitary and multilingual centre or institute, working under joint international direction, would perhaps be much more effective than a number of them, organized on regional or linguistic basis. Only then can a truly international direction be given to the evolution of a truly international movement on standardization. When, in due course, the movement spreads wide enough and the facilities of one unit are found to be inadequate, thought could usefully be given to creating additional regional or linguistic sub-centres.

25.32 The objectives of such a unitary international centre or institute as envisaged above should be comprehensively conceived and may, among other things, include the following:

(1) To develop training programmes and related curricula for training standardization workers at all levels of standardization in all sectors of human endeavour.

(2) To train national-level workers, who would mostly be from developing countries, but may also be from others.

(3) To train instructors from all countries for organizing the training of company-level workers.

(4) To suggest and make recommendations for the introduction of standards teaching in the regular curricula of professional engineers and technologists as also of students at the secondary level of general education.

(5) To compile and publish textbooks and reference works on standards for use in connection with the various programmes of instruction.

(6) To promote the development of the discipline of standardization and help expand the frontiers of knowledge.

(7) To undertake and promote research and investigations on specific problems of standardization (some of which have been indicated in earlier chapters).

(8) To issue one or more journals of research in several important languages for publishing and encouraging original contributions and innovations.

(9) To organize periodical conferences and symposia for the exchange of information and new knowledge and experience, with a view to furthering the advancement of the discipline and its systematic codification.

(10) To encourage the creation of new National Standards Bodies and to assist those newly formed in strengthening themselves, through consultancy and other services.

25.33 The creation of such a comprehensive international institute would require the concerted effort of all advanced and developing countries. One of the first tasks to be tackled as a preliminary to the planning of such an institute would perhaps be to survey and assess the results so far achieved by the various programmes and to estimate the worldwide need for education and training at all levels, starting from company level engineers up to the national level chief executives. Only then could appropriate provisions be made in the plan, both qualitatively and quantitatively. If this effort succeeds it is bound to accelerate the pace of development of world standards, smooth the road for international trade and exchange, and ultimately help make the one-world concept an early reality.

REFERENCES TO CHAPTER 25

1. Gaillard offers standardization course. *Industrial Standardization,* 1947, vol 18, p. 108.
2. Gaillard now offers private company seminar in addition to New York sessions. *Magazine of Standards,* 1956, vol 27, p. 90.

3. International training course in industrial standardization. *Standards Association of Australia Monthly Information Sheet,* Aug 1971, pp. 15-17.
4. Filip, Zdeněk: Postgraduate courses on technical standardization at the University of the 17th November in Prague. *Normalizace,* 1970, vol 18, pp. 61-63 (*in Czech*).
5. Marshall, Francis M.: Education in standards. *Standards Engineering,* 1960, vol 12, no 5, pp. 5-7.
6. Raghupathy, M.: Training of standards engineers – ASTEF course in France. *ISI Bulletin,* 1966, vol 18, pp. 355-356.
7. Sen, S. K.: Training of standards engineers. *Complete collection of papers presented at the United Nations international-regional seminar on promotion of industrial standardization in developing countries.* Vol 1 (Ref M 40/1–65) Danish Standards Association, Copenhagen, 1965, pp. 347-365.
8. Krishnamachar, B. S.: Training in standardization. *Paper presented at UNIDO workshop on standardization.* United Nations Industrial Development Organization, Vienna, 1971, pp. 1-21.
9. Meissner, E.: Postgraduate study for standards engineer. *Standardisierung,* 1968, vol 14, pp. 260-262 (*in German*).
10. *Fifty years of German Standards Institution.* Deutscher Normenausschuss, Berlin, 1967, p. 117 (*see* pp. 85 and 91) (*in German*).
11. *Training in standardization for developing countries.* Indian Standards Institution, New Delhi, 1971, p. 6.
12. Gupta, A. K. and Kaul, Jainath: Emergence of standards engineering profession in India. *Souvenir of the First National Conference of the Institute of Standards Engineers,* Bombay, 1969, pp. 82-86.
13. Gupta, A. K.: Company standardization in Indian industry. *The Standards Engineer (India),* 1970, vol 4, pp. 41-46.
14. Training in standardization technology. *Papers for Session S-5 of Tenth Indian Standards Convention (Ernakulam).* Indian Standards Institution, New Delhi, 1967:
    S-5/1 Mukerji, A. K.: Standard technologists – The need of the day.
    S-5/2 Menon, V. K. S.: Training in standardization technology.
    S-5/3 ISI: Training of standards engineers.
    S-5/4 Raghupathy, M.: Training of standards engineers – ASTEF course in France.
    S-5/5 Lodh, M. R.: The need for training in standardization technology.
    S-5/6 Khan, Hamid Ahmed: A national society of standardization technologists or a nation-wide specifications-minded society.
    S-5/7 Raghavachari, S.: Training in standardization technology.
    S-5/8 Subramanian, V. R.: Standards engineer – jack of all trades and master of standardization.
15. Importance of industrial standards in technical education. *ISI Bulletin,* 1969, vol 21, p. 57 and pp. 63-67.
16. National standards in technical education. *ISI Bulletin,* 1971, vol 23, pp. 313-314.
17. Standards in polytechnic curricula. *ISI Bulletin,* 1971, vol 23, p. 528.
18. Standards in technical education – Orientation programme for teachers:
    Madras. *ISI Bulletin,* 1969, vol 21, p. 488.
    Ahmedabad. *ISI Bulletin,* 1970, vol 22, pp. 271-272.
    Chandigarh. *ISI Bulletin,* 1970, vol 22, p. 462.
    Patna. *ISI Bulletin,* 1971, vol 23, pp. 313-314.
    Calcutta. *ISI Bulletin,* 1971, vol 23, pp. 369-370.
    Allahabad. *ISI Bulletin,* 1971, vol 23, p. 420.

Mysore. *ISI Bulletin,* 1971, vol 23, p. 475.
Dibrugarh. *ISI Bulletin,* 1972, vol 24, p. 32.
19. *Information on group training course in industrial standardization and quality control.* Government of Japan, Tokyo, 1971, p. 8.
20. *Pan American Standards Commission* – a COPANT document dated Aug 1971. COPANT, Buenos Aires, p. 22.
21. Maraszkiewicz, Z. and Wiszniewicz, E.: *Standardization in Poland – origin, development, activities.* Polish Standards Committee, Warsaw, 1961, p. 49 (*see* pp. 43-44).
22. Newsome, R. I.: Standards training in the technical colleges – an educationalist's viewpoint. *BSI News,* Jan 1959, pp. 9-11.
23. Standards training in the technical colleges – Readers' Viewpoint. *BSI News,* Apr 1959, pp. 9-11 and 14.
24. Corkill, T. M.: Education for standards – one of the broadest education problems in industry. *BSI News,* Nov 1967, pp. 12-13.
25. First residential course for standards engineers – BSI and Manchester University. *BSI News,* Sep 1967, pp. 21-22.
26. Standards course at Manchester covers wide range of topics. *BSI News,* Sep 1971, pp. 14-15.
27. Milek, John T.: A search for basic standards knowledge. *Standards Engineering,* 1961, vol 13, no 5, pp. 12-14.
28. Glie, Rowen: Education for a standard engineer. *Proceedings of the 17th annual meeting.* Standards Engineers Society, Boston, 1968, pp. 42-51.
29. Ogryskow, W. M.: The training of cadres for standardization and metrology in the USSR. *Standardisierung,* 1968, vol 14, pp. 263-266 (*in German*).
30. Kochin, V. P.: Committee of standards and local standardization authorities in professional training and qualification improvement of specialists in standardization and metrology. *Standarty i Kochestvo,* 1970, no 4, pp. 36-38 (*in Russian*).
31. Zelenko, G. I.: Standardization and personnel. *Standarty i Kochestvo,* 1970, no 8, pp. 6-9 (*in Russian*).
32. Zelenko, G. I.: Standardization personnel training meets new tasks. *Standarty i Kochestvo,* 1971, no 6, pp. 31-32 (*in English*).
33. *The promotion of industrial standardization in developing countries, Helsingör, Denmark, 4 to 25 Oct 1965.* United Nations, New York, 1966, p. 30.
34. *Training workshop for personnel engaged in standardization, Addis Ababa, Ethiopia, 17 to 24 Nov 1970.* United Nations Industrial Development Organization, Vienna, 1971, p. 82.
35. *Training workshop for personnel engaged in standardization for Latin America – Santiago de Chile, 27 Sep to 2 Oct 1971.* United Nations Industrial Development Organization, Vienna (*in Spanish*).
36. *International symposium on industrial development, Athens, 29 Nov to 20 Dec 1967 – Issues for discussion – Standardization.* United Nations Industrial Development Organization, Vienna, 1967, p. 8.
37. *Report of the Asian Standards Advisory Committee (first session) to the Asian Industrial Development Council (third session).* Doc: AIDC(3)/9. Economic Commission for Asia and the Far East, Bangkok, 1968, p. 66 (*see* p. 9).
38. *Report of the Consultative Group of the Asian Standards Advisory Committee.* Doc: AIDC(4)/7. Economic Commission for Asia and the Far East, Bangkok, Dec 1968, p. 34.
39. *Report of the Asian Standards Advisory Committee (second session) to the Asian Industrial Development Council (fifth session).* Doc: AIDC(5)/2. Economic Commission for Asia and the Far East, Bangkok, June 1969, p. 27.

40. See Annexure III of reference 38.
41. *Training in the field of standardization.* Doc: ISO/DEVCO (Secretariat-38) 49E. International Organization for Standardization, Geneva, June 1971, p. 4.
42. ISO to publish new information catalogues. *ISO Bulletin,* Feb 1971, vol 2, no 2, p. 5.
43. Bartzsch, E.: Kolloquium – and Colloquium "Standardization" of the Institute for Efficiency and Standardization of the Technical Highschool, Dresden. *Wissenschaftl. Veröff. TH Dresden,* 1957, vol 6, pp. 830-834 (*in German*).
44. Hering, D.: The contribution of educational theory to standardization training. *Standardisierung,* 1960, vol 6, pp. 1/257-262 (*in German*).
45. Koloc: The teaching of standardization. *Standardisierung,* 1960, vol 6, pp. 1/129-137 (*in German*).
One or two problems of standardization theory. *Standardisierung,* 1959, vol 5, pp. 1/502-507.
46. Ogryzkov, V. M. and Kobylyansky, D. A.: Problems of organization and methods of teaching standardization in higher and secondary technical training schools. *Standarty i Kochestvo,* 1968, no 5, pp. 69-73 (*in Russian*).
47. Meissner, Erwin: System of training and further education in standardization in the German Democratic Republic. *Normalizace,* 1968, vol 16, pp. 284-286 (*in Czech*).
48. Cordasevschi, A. L.: Training of standardization specialist. *Standardizarea,* 1968, vol 20, pp. 564-566 (*in Rumanian*).
49. Shapovalenko, A. M., Khutsiyev, A. I. and Dmitriyev, M. M.: Standardization in teaching procedure and in preparation of yearly and diploma design projects. *Standarty i Kochestvo,* 1970, no 2, pp. 52-54 (*in Russian*).
50. Ogryskow, W. M.: System for training and qualifying for standardization in the USSR. *Standardisierung,* 1967, vol 13, pp. 398-401 (*in German*).
51. Krieg, K. G.: Characteristics of participation in standards departments. *DIN-Mitteilungen,* 1970, vol 49, pp. 214-222 (*in German*).
52. Bartzsch, E. and Hentschel, B.: New paths in standardization training. *Standardisierung,* 1969, vol 7, pp. 243-250 (*in German*).

Chapter 26

The Future

26.0 Considering the comprehensive character that is being assumed by this newly emerging discipline and the rapidly increasing pace of the present-day technological advancement, it is not quite a simple matter to be able to predict with any degree of accuracy the future of standardization. As the worth of standardization becomes more and more appreciated, the demand for standards is bound to grow in volume not only for application in conventional fields of endeavour but also for the newly developing technologies. It is these latter fields that pose the real problem of visualizing the shape that standardization may assume within the short span of another generation. Furthermore, the character of standardization is bound to acquire new dimensions and assume new complexions. Its ever-increasing importance as an infrastructure to all economic and cultural advancement – industrial, agricultural, education and other fields – may win for it a status in human society, which it has never aspired to enjoy so far.

26.1 Most otherwise well-informed people, who have not come in intimate contact with standardization in any of its manifold ramifications, are so unfamiliar with its basic tenets and working details that they entertain most surprisingly naive ideas about it. For one thing they greatly fear that, given free reign, the standards engineer would reduce every detail of existence to such uniformity that life would lose all is flavour and variety. They entirely overlook the fact that by eliminating unnecessary and wasteful variety, standards help to canalize the creative faculties of man, to open new avenues of endeavour and make life more colourful, interesting and creative. It is this transformation in understanding about the nature of standardization by people in general which is most essential, if those engaged in its practice are to find a prestigious place in society. It is only then that standardization will be able to make its full contribution to the betterment of man's life on earth. Judging from the developments during the past two decades, it would appear that movement in that direction has been well initiated and it will not be long before standardization and standards engineers may be called upon to take an important part in the planning and execution of programmes of economic and social development in most countries of the world. This movement of sharing with the standards engineer the responsibility for planning development may even extend to the international sphere, that is if progress continues to be registered in regard to greater cooperation among nations.

26.2 In more specific terms, the tempo of standardization is bound to increase because of the growing demand from three distinct spheres, namely:

(1) rapidly advancing technologies, particularly those of recent origin and those yet to be conceived of;

(2) new fields of human endeavour in which standardization has not yet entered; and

(3) the international movement for standardization, which is presently gaining momentum.

ADVANCING TECHNOLOGIES

26.3 For convenience of discussion, three categories of technologies may be distinguished: those in which standardization has already been active and to the development of which it has made some contribution; the technologies which are of recent origin and in which standardization is yet to take a significant part; and lastly those which may yet come into being as a result of the intensive research activity presently under way, the pace of which is bound to accelerate in the future. The existing fields of technology could well continue to furnish the proving ground for enhancing the utility of standards and their application, so as to facilitate their future extension to the more sophisticated fields, where requirements will be much more rigid and vastly more critical. As far as new technologies which have yet to make their appearance are concerned, little need or can be said, except to express the hope that standardization techniques by that time will have become far enough refined to meet the challenge of the day.

26.4 In many of the existing fields where standardization has been practised for quite some time, there is a great need for changing the attitudes of all those concerned to make them much more conscious of the value of standards than they are at present. For example, in the mechanical engineering field, the rationalization of sizes to ensure optimum economy and maximum interchangeability on a worldwide scale is yet to be achieved. In the building and construction industry, general use of modular coordination and the ex-stock availability of standardized pre-fabricated building components and elements are still a matter of the future. Could it not be expected that one should be able to purchase all the components for one's house ex-stock and put it up practically overnight in accordance with one's own individual design? A good deal of progress has been made in this direction, but individual construction on site is still the order of the day. Electrical engineers are continuing their struggle to achieve standardization of distribution voltages and ratings of machinery and equipment, and interchangeability of domestic plugs and sockets. In some countries consumer needs have not received the attention of the standards engineer they deserve. It is only relatively recently that the consumer is beginning to organize himself. His pressure for pursuing research and development in his own interest on the one hand and his demand for performance standards on the other are yet to bear tangible fruit. The pollution problems which pose a threat to human life on this planet constitute a real challenge to standards men. It is within the realm of standardization to provide the means for controlling industrial and other processes which lead to pollution, to provide means for monitoring the prevailing degree of pollution and to lay down norms and tolerance limits beyond which pollution should not be allowed to exceed.[1]

26.5 The list may be extended indefinitely to indicate how inadequate is the coverage of even those fields of endeavour where standardization has made some positive inroads. On the other hand, there are a number of fields in which standards are just beginning to make their appearance and which present vast areas of almost virgin territory for the standards engineer to explore. Take, for instance, nuclear energy, computers and data processing, laser technology, holography, space and rocket travel satellite communication and biophysics. It is in such fields as these which are in the early stage of development that standardization can make some real impact – starting from the basic terminology of concepts, which may come to represent a universal language, and up to the norms of performance and interchangeability of parts and equipment – functionally as well as dimensionally.[2 3] Parenthetically, the standardization of vocabularies is an important enough field in itself which has a great potential for the future. It is through success in this field that the welter of existing languages can be reduced to a reasonable level, at which the scientist and technologist of tomorrow will be able to handle conveniently the problems of worldwide communication.

NEW FIELDS

26.6 As far as new fields of endeavour are concerned in which standardization has not yet entered, there is a vast scope for valuable contributions to be made. But such fields will become accessible to the standards engineer only after he has effectively demonstrated to all concerned the value of his work in the technological fields in which he is now mainly engaged and has established for himself a much more prestigious position in the society of man than he holds at present. Standards are already entering into the fields of agriculture in a big way in many countries. To some extent they have touched upon the field of education as well, but only as far as technology becomes involved; for example, in the matter of schoolroom equipment or laboratory apparatus. The larger educational issues of standards of teaching and standards of student accomplishment remain quite untouched, except by academicians otherwise unfamiliar with the technology of standardization. There is the question of standardizing intelligence tests, aptitude tests, vocational guidance and so on. The very problem of finding the right man to fill a given job of work is long outstanding, for lack of a reliable solution.

26.7 In the fields of economies and econometry a great deal of experimentation of vital importance is going on. Though it is hardly possible to expect any radical developments in this sphere in the near future, yet the time may well come when certain methods and practices may usefully lend themselves to standardization. One hears so much of standards of living of people, but is there not a need for a quantitative definition of such a standard. The New Zealand National Standards Body has already pioneered in the most difficult field of local governments (*see* Chapter 24, paras 24.12 and 24.13). The question arises whether standardization in respect of government at higher levels could also be usefully attempted. When one talks of standardization in respect of governments, one must recognize that constitutions, laws, decrees, rules

evolution of a strong international standards movement.

26.12 The fact that ISO has started issuing international standards instead of ISO recommendations[5] with effect from 1972 is indicative of future trends, when ISO will begin to exercise greater authority and depend less on a persuasive approach to implementation of international decisions. While it may not be denied that it is a move in the right direction, it is also important to ensure that the international decisions constituting the international standards of ISO do, in fact, represent a real consensus among all the member bodies. It must always be borne in mind that any decision arrived at under pressure of a small group of otherwise influential nations is bound to lead to a weakening of the international fabric. Every effort, therefore, would be justified in strengthening the weaker members of the family and enabling them to shoulder an equitable burden of responsibility. Here is where ISO/DEVCO's role will have to be further enlarged.

26.13 The day does not appear to be far when along with tariff barriers which stand in the way of freer international trade, the technological barriers arising out of varying standards and official regulations will crumble and the flow of trade will be made more smooth through international standardization.

WAYS AND MEANS

26.14 If all that has been visualized above, or an appreciable part of it, is to be accomplished, our methods of approach and procedures will have to undergo significant evolution. Any change of revolutionary character is, of course, out of the question. The foundations for the future development of the discipline should be considered to have been well and truly laid by the pioneers during the past half century or so. What appears now to be called for is a forced pace of evolution – building a strong and enduring superstructure on the already existing foundations and doing it at a pace commensurate with the rapid pace of technological innovations.

26.15 An important test of how far standardization may be able to assist the developing technologies would be to see how far in advance could standards be prepared before practices become frozen in varying moulds. Fortunately, it is becoming appreciated much more widely than ever before that standardization and developmental research should proceed more or less hand in hand, if chaos is to be avoided. There is little that can be done in this regard so far as the current practices of the existing technologies are concerned, except to draw lessions from the prevailing conditions, in which almost every attempt at rationalization of practices on an objective basis runs up against the serious hurdle of the interests already vested, which naturally resist any change requiring fresh investments. If, in future, adequate vigilance is exercised, of which there are ample hopeful signs, the newly developing technologies will be able to avoid the pitfalls from which the existing world economy suffers. It will be ever more necessary to insist on standards to emerge simultaneously as an integral part of the industrial development process. This is more easily said than done, because in competitive economies the urge to be a little ahead of one's competitor is to be reckoned with. Would the standards engineer be able to find a way around such situations, or would the competitive nature of economy have to undergo some adjustments?

26.16 We can also expect future standards to become more and more realistic in view of the emergence of statistical methods and techniques which enable us to adjudge within a reasonable margin of probability the process capabilities, the average outgoing quality of manufacture, the sampling errors, the significance of test results and so on. Not only that, but the statistical methods available for the control of quality enable the output quality to be maintained well within such limits as may be determined by consumer needs, without in any way increasing the cost of production and, in fact, sometimes even reducing it. In the context of developing economies, the value of these methods is being appreciated so much that standardization procedures are being suggested which will primarily be based on statistical data.[10] In certain industrially advanced countries like Japan, such a practice appears already to have taken quite a deep root. In relation to the newly emerging industries, however, statistical tools will have to be brought into action at the early stage of developmental research so as to guide rational and realistic standards to be drawn up.

26.17 Attention has already been drawn to the complex and time-consuming nature of procedures that have come to be generally adopted for the preparation of standards (*see* Chapter 10, paras 10.36–10.47). Some of the palliative measures taken to help accelerate them have also been discussed. However, in order to meet the needs of the future, much more radical solutions may have to be found. More effective and more efficient procedures would become essential for catering to the requirements of standards for freshly emerging technologies. Furthermore, revisions of standards will have to be made much more frequently than at present to take care of the rapid changes in requirements which are bound to take place with the rapid advancement of research. All this means that some radical thinking will have to be done. Would the answer lie in giving greater authority and making available much greater means to the standards making bodies than they possess today? In the interest of common good, would authoritative imposition of standards become more acceptable than it is today? Whatever may be the answer to these questions and whatever other solutions may or may not be found as adequate, one thing is quite certain that the present procedures would not be able to meet the growing demands of the future.

26.18 An interesting suggestion, which could greatly help in accelerating the procedures, has come from Kirkpatrick and Rosenfeld: "Standards and standardization essentially represent agreement following negotiations. The extent and significance of agreements reached depend on the level, character and scope of the negotiations undertaken. Further progress of standardization is inextricably linked with such negotiations. The tenor of our days (and more so of the future) demands recurring factual accomplishments – not theoretical or potential benefits. These benefits accrue only after agreements in specific areas – and far too often agreements are not reached (or delayed to varying extents) due to the lack of capability and talent of those negotiating. Witness the development of skill in labour negotiations – much study, effort and training go into the conference room with the persons involved. Is our area less critical – less important? Would not an increase in agreement produce more (and rapid) standards and would not more skill in negotiating produce more (and quicker) agreements?"[11] Thus, it will be essential for the standards engineer to become, among other things, a much more proficient negotiator than perhaps he is today and begin to learn lessons from the trade unionist – a person with whom he has not so far come in much contact.

26.19 Another question mark for the future is whether some new areas would come to be added to the orbit of legally mandatory standards. Judging from the growth of the consumer movement in some countries and the increasing demand for health, safety and pollution related standards being made mandatory, it would appear that the general trend would perhaps be the gradual restriction of the area of voluntary standards. In the context of the forced pace of industrial and social development being pursued in the newer countries, some compulsion in the use of standards is also being brought about in certain important areas such as that of export quality control. Furthermore, in times of emergency, even among the most developed countries, resort has often to be taken to the application of mandatory standards. Though, it is hoped, that in future such emergencies as those created by wars may not arise or arise rather rarely, yet widespread implementation of standards in the interest of national well-being and economic advancement may have to be secured through legislative means. Mexico is understood to have already made government subsidies and tax exemptions for essential and new industry conditional on the fulfilment of the requirements of national standards. Is it not feasible that under special conditions as may develop in future, the industrially advanced countries may also come to consider the possibility of adopting some such legal measures for the enforcement and implementation of certain types of standards. In the centrally controlled economies, like the USSR and some other East European countries, all national standards are already mandatory. It is only in the context of free economies that this question arises. But even in such environment, circumstances appear to be slowly developing which tend to be gradually reducing the sphere of voluntary standards, which nevertheless, have their place, even in controlled economies. For example, at the company level, while complying strictly with all relevant national standards, different firms in a centrally controlled economy may produce goods to their own designs and specifications.

26.20 Another development, which will constitute a departure from the present, is the possibility of much more general adoption of the techniques of planning than has been the case hitherto. Planning for standardization has proved its worth at the national level in countries like India and the USSR (*see* Chapter 20). There are some signs of its being appreciated in other countries as well. Frontard, for example, refers to a plan for augmenting the staff of the French National Standards Body by 25 percent with a view to increasing its output of standards by 100 percent.[12] In a sense, the USA is also planning, and planning in a big way, to change over to the universal use of standards of measurements on the basis of the metric system.[13]

CONCLUSION

26.21 Any speculation about the future is always risky. Based, as it has to be, on the present-day trends, it cannot take account of the unexpected. The world today is, so to say, at a cross-roads. One way points to more and more disruption and disharmony leading ultimately to disaster; the other, to greater understanding and accord, leading to elimination of conflicts between nations and full exploitation of all resources available to man for the benefit of man. It is everyone's hope that the choice will rightly be made and that future generations of man will live in a world of unimaginable possibilities.

26.22 As regards the future of standardization, it can only be visualized in the latter context. But, in order that some of the expectations may come about, it is prerequisite that a full-fledged discipline of standardization be created, that standardization ceases to be an activity ancillary to engineering, industrial management, economic planning or whatever else. This would need intensive thinking and extensive organizational work. Many more men of high calibre would need to be attracted to adopt standardization as a profession. Those within the profession will need to reorient their thinking. Fortunately, it is a basic principle of standardization to work and think collectively and cooperatively. This feature would have to be exploited to the fullest possible extent by pooling all the available resources – material as well as mental. At the same time, individual thinking and original contributions to the advancement of the discipline should be encouraged to the fullest. Both individual and cooperative efforts are essential. When the discipline emerges from its present status and wins for itself all-round recognition, the time will be ripe to explore and promote the numerous possibilities of the future in the context of the world that is still to come.

REFERENCES TO CHAPTER 26

1. Ghosh, A. N.: Some problems of the seventies. *Science Reporter,* 1970, vol 7, pp. 570-575.
2. Tomorrow's challenges for voluntary standards. *The Magazine of Standards,* 1969, vol 40, pp. 4-5.
3. A glimpse at the brave new world of the eighties. *ISI Bulletin,* Jan 1970, p. 3.
4. Medaris, J. B.: Standards for survival. *The Magazine of Standards,* 1959, vol 30, pp. 374-377.
5. Sturen, Olle: A turning point in ISO history. *ISO Bulletin,* Jan 1972, vol 3, no 1, p. 1.
6. LaQue, Francis L.: *Message from the ISO President on the 25th anniversary of ISO.* International Organization for Standardization, Geneva, 1972, p. 1.
7. Sherfield, Lord: *BSI annual report 1970 to 1971.* British Standards Institution, London, p. 100 (*see* p. 1).
8. *US metric study – International standards (Special Publication 345–1).* National Bureau of Standards, Washington, 1970, p. 145.
9. See references 2 and 3 of Chapter 14.
10. Sen, S. K.: Future of standardization in India. *ISI Bulletin,* 1971, vol 23, pp. 213-216.

11. Kirkpatrick, Daune M. and Rosenfeld, David A.: Tomorrow's standardization – where and how far is standardization going? *Standards Engineering,* June/July 1967, vol 1, no 1, pp. 10-12.
12. Frontard, R.: Normes de papa.– Normes de demain. *Courrier de Normalisation,* Jan-Feb 1970, no 211, pp. 5-10 (*in French*).
13. De Simone, Daniel V.: *A metric America – a decision whose time has come (Special Publication 345).* National Bureau of Standards, Washington, 1971, p. 170.
14. Toffler, Alvin: *Future shock.* Pan Books, London, 1970, p. 517 (*see* pp. 240-246).

Index

– indicates "with reference to"